Farm-level Modelling

Techniques, Applications and Policy

FSC
www.fsc.org
MIX
Paper from
responsible sources
FSC® C013604

To Jenita, Krish, Neeva and Kritika
SHAILESH SHRESTHA

To Jude and Gayle
ANDREW BARNES

To Afrouz and Sam
BOUDA VOSOUGH AHMADI

Farm-level Modelling

Techniques, Applications and Policy

Edited by

Shailesh Shrestha

Agricultural Economist, SRUC (Scotland's Rural College), Edinburgh, UK

Andrew Barnes

Reader in Innovation and Behavioural Change, SRUC, Edinburgh, UK

and

Bouda Vosough Ahmadi

Agricultural Economist, SRUC, Edinburgh, UK

CABI

CABI is a trading name of CAB International

CABI
Nosworthy Way
Wallingford
Oxfordshire OX10 8DE
UK

Tel: +44 (0)1491 832111
Fax: +44 (0)1491 833508
E-mail: info@cabi.org
Website: www.cabi.org

CABI
745 Atlantic Avenue
8th Floor
Boston, MA 02111
USA

Tel: +1 (0) 617 682-9015
E-mail: cabi-nao@cabi.org

A catalogue record for this book is available from the British Library, London, UK.

Library of Congress Cataloging-in-Publication Data

Names: Shrestha, S. (Shailesh), editor.
Title: Farm-level modelling : techniques, applications and policy / edited by: S. Shrestha, A.P. Barnes, and B. Vosough Ahmadi.
Description: Boston, MA : CABI, [2016]
Identifiers: LCCN 2016002564| ISBN 9781780644288 (hbk : alk. paper) | ISBN 9781786390387 (epub)
Subjects: LCSH: Agriculture--Economic aspects--Mathematical models.
Classification: LCC HD1433 .F37 2016 | DDC 338.101/1--dc23 LC record available at http://lccn.loc.gov/2016002564

ISBN-13: 978 1 78064 428 8

Commissioning editor: Ward Cooper
Editorial assistant: Emma McCann
Production editor: Tim Kapp

Typeset by SPi, Pondicherry, India
Printed and bound in the UK by CPI Group (UK) Ltd, Croydon, CR0 4YY

Contents

Contributors

Patrick Abbot is Managing Director of LTS International Ltd and has a professional and research career in environment and climate change. His more recent research focus has been on applying and measuring open system theories to complex land-use and social change.

Address: LTS International Ltd., Pentlands Science Park, Bush Loan, Penicuik, Edinburgh, EH26 0PL, UK. E-mail: paddy-abbot@ltsi.co.uk

Filippo Arfini is Professor of Agricultural Economics as well as of Rural Policies in the Department of Economics, University of Parma. His fields of research are policy assessment by means of quantitative analysis, agri-food chain management for Geographical Indication (GI) products, quality economics and related policies, and rural development dynamics. His research activity has developed in the framework of national and European research programmes. He is the author of several publications on the issue of GIs and their implication on rural development.

Address: Department of Economics, University of Parma, Via J. F. Kennedy 6 - 43125, Parma, Italy. E-mail: filippo.arfini@unipr.it

Andrew P. Barnes is an agricultural economist and Reader in Innovation and Behavioural Change within the Land Economy, Environment and Society (LEES) Research Group of Scotland's Rural College (SRUC). His main research focus is on capturing the impacts of policy change and farmer behaviour at the farm and catchment level within a modelling framework. He currently leads the Policy, Innovation and Behaviours Team at SRUC.

Address: LEES, Scotland's Rural College, Edinburgh, EH9 3JG, UK, E-mail: andrew.barnes@sruc.ac.uk

Maria Blanco is an Associate Professor of Agricultural and Resource Economics at the Technical University of Madrid. Her research interests are in quantitative analysis of agricultural policies, the development of agro-economic modelling tools for policy impact assessment as well as the integrated assessment of environmental and agricultural policies. She has extensive experience in the linkage of biophysical and economic models, in particular in the fields of water economics and climate change impact assessment.

Address: Department of Agricultural Economics, Universidad Politécnica de Madrid (UPM), Avda. Complutense 3–28040, Madrid, Spain. E-mail: maria.blanco@upm.es

Pavel Ciaian is a senior researcher at the Institute for Prospective Technological Studies (IPTS) of the Joint Research Centre (JRC) of the European Commission. He studied at KU Leuven University in Belgium. Before joining the JRC, he was senior researcher at the LICOS Centre for Institutions and Economic Performance of KU Leuven University and Assistant Professor of Economics at the

Slovak Agricultural University. His research interests focus on agricultural policy, institutional economics, and energy and development economics.
Address: IPTS, JRC – EU Commission, Ed. EXPO-c/Inca Garcilaso 3, 41092 Seville, Spain. E-mail: pavel.ciaian@ec.europa.eu

Jinxiu Ding is an Assistant Professor in the School of Economics, Xiamen University, China. Her recent research focus is on the analysis of the regional impact of increased drought frequency on water management and the regional effects of decadal climate variability on crop yields.
Address: Department of Public Economics, School of Economics, Xiamen University, Xiamen, Fujian, China. E-mail: jinxiuding@gmail.com

Michele Donati is a researcher at the Department of Biosciences, University of Parma, Italy. He has a PhD in Agri-food Economics and teaches Economic Modelling for Environmental Assessments. His main field of research concerns the evaluation of agricultural policies through quantitative models based on mathematical programming. He participated in several European research projects on the evaluation of Common Agricultural Policy (CAP) reforms.
Address: Department of Biosciences, University of Parma, Viale Usberti, 33/A-43124, Parma, Italy. E-mail: michele.donati@unipr.it

Vera Eory is an environmental economist in the Land Economy, Environment and Society (LEES) Research Group of Scotland's Rural College (SRUC). Her research focuses on the economic and environmental implications of greenhouse gas reduction in agriculture, particularly on the cost-effectiveness of mitigation practices, environmental co-effects and uncertainties. She is also interested in farmers' perceptions of greenhouse gas (GHG) mitigation and climate policy design.
Address: LEES, Scotland's Rural College, Edinburgh, EH9 3JG, UK. E-mail: vera.eory@sruc.ac.uk

Maria Espinosa is a senior researcher at the Institute for Prospective Technological Studies (IPTS) of the Joint Research Centre (JRC) of the European Commission. She is an agricultural engineer with a PhD in agricultural economics from the Technical University of Madrid. Before joining the JRC she was a junior researcher at the Andalusian Institute of Agricultural Research and Training (IFAPA). Her main research focus is farm-level modelling, agricultural policy impact analyses and behavioural economics.
Address: IPTS, JRC – EU Commission, Ed. EXPO-c/Inca Garcilaso 3, 41092 Seville, Spain. E-mail: maria.espionsa@ec.europa.eu

Joana Guimarães Ferreira is a researcher within the Policy, Innovation and Behaviours Team of the Land Economy, Environment and Society (LEES) Research Group of Scotland's Rural College (SRUC). Her most recent research has focused on modelling the impacts of farmer decision making on ecosystem services provision and on identifying determinants of conservation behaviour.
Address: LEES, Scotland's Rural College, Edinburgh, EH9 3JG, UK. E-mail: joana.ferreira@sruc.ac.uk

Patrick R. Gillespie is a postdoctoral researcher in the Agricultural Economics and Farm Surveys Department of the Rural Economy and Development Programme (REDP) of Teagasc, the Agriculture and Food Development Authority of Ireland. His main research interests include efficiency and productivity in the Irish dairy sector, economic analysis of greenhouse gas emissions resulting from agricultural activity and analysing the international competitive performance and potential of Irish agriculture.
Address: REDP, Teagasc, Athenry, Co. Galway, Ireland. E-mail: patrick.gillespie@teagasc.ie

Alexander Gocht is a senior researcher at the Thünen Institute of Farm Economics. He studied agricultural economics at Imperial College London and completed his PhD at Bonn University on methods of economic farm modelling. His research interests focus on EU agricultural policy, farm modelling, and environmental protection and sustainability.
Address: Thünen Institute (Institute of Farm Economics), Bundesallee 50, 38116, Braunschweig, Germany. E-mail: alexander.gocht@thuenen.de

Sergio Gomez y Paloma is a senior researcher at the Institute for Prospective Technological Studies (IPTS) of the Joint Research Centre (JRC) of the European Commission. He studied Agricultural Sciences and Agricultural Economics at the Universities of Bologna, Milan and Naples (Italy), and

AgroParisTech (France). Before joining the JRC he was a Lecturer of Development Economics at Roskilde University in Denmark. His main research interests include agricultural policy analysis, agricultural microeconomics and development economics.
Address: IPTS, JRC – EU Commission, Ed. EXPO-c/Inca Garcilaso 3, 41092 Seville, Spain. E-mail: sergio.gomez-y-paloma@ec.europa.eu

Eleonore Guillem is a former postdoctoral research in ecological systems at LPED – UMR (Laboratoire Environnement Populations Développement – Unité Mixte de Recherche), Université Aix-Marseille. She has a PhD in Environmental Sciences from the University of Edinburgh, UK.
Address: LPED – UMR, Université Aix-Marseille, 52 rue du Capitaine Galinat, M2, 13005 Marseille, France. E-mail: eleonore.guillem@gmail.com

Thomas Heckelei is Professor for Economic and Agricultural Policy in the Institute for Food and Resource Economics of the Faculty of Agriculture of the University of Bonn, Germany. His main research work is on quantitative impact analysis of agricultural policies and the development of related methods, specifically the empirical specification of simulation models. More recently, the dynamics of agricultural systems and markets has attracted his attention.
Address: Institute for Food and Resource Economics, University of Bonn, Nußallee 21, 53115 Bonn, Germany. E-mail: thomas.heckelei@ilr.uni-bonn.de

David A. Hennessy is Professor and Elton R. Smith Chair in Food and Agricultural Policy, Department of Agricultural, Food and Resource Economics, Michigan State University. His research interests are in the production economics of animal and crop agriculture, together with their linkages to agricultural, food and environmental policies.
Department of Agricultural, Food and Resource Economics, Michigan State University, 446 W. Circle, Dr., Rm 207, Morrill Hall of Agriculture, East Lansing MI 48824, USA. E-mail: hennes64@anr.msu.edu

Thia Hennessy is an agricultural economist and Head of the Agricultural Economics and Farm Surveys Department of the Rural Economy and Development Programme (REDP) of Teagasc, the Agriculture and Food Development Authority of Ireland. Her main research focus is on modelling the impact of agricultural and environmental policy on the viability and sustainability of farming. She is head of the Teagasc National Farm Survey, the Irish component of the Farm Accountancy Data Network (FADN) of Europe, which produces the official data on output, input and income in farming in Ireland.
Address: REDP, Teagasc, Athenry, Co. Galway, Ireland. E-mail: thia.hennessy@teagasc.ie

Stephen Hynes is a natural resource economist and lecturer at the National University of Ireland, Galway (NUIG). His main research interest is in microeconomic behaviour analysis related to natural resource/environmental policy. He is currently the Scientific Director of the Whitaker Institute of NUIG and is also Head of the Socio-Economic Marine Research Unit of NUIG.
Address: Socio-Economic Marine Research Unit, Whitaker Institute, NUI Galway, Ireland. E-mail: stephen.hynes@nuiglaway.ie

Michael MacLeod is a researcher in climate change mitigation within the Land Economy, Environment and Society (LEES) Research Group of Scotland's Rural College (SRUC). His main research interests are in using life-cycle analysis to quantify greenhouse gas (GHG) emissions arising from the production of food commodities and the identification of cost-effective ways of reducing emissions. He is also engaged in research into the broader regulatory challenges posed by agriculture–environment interactions.
Address: LEES, Scotland's Rural College, Edinburgh, EH9 3JG, UK. E-mail: michael.macleod@sruc.ac.uk

Bruce A. McCarl is University Distinguished Professor and Regents Professor in Agricultural Economics at Texas A&M University. His research focuses on the economic implications of global climate change, greenhouse gas emission reduction and water allocation/policy, as well as on the applications of optimization theory.
Address: Department of Agricultural Economics, Texas A&M University, College Station, TX 77843-2124, USA. E-mail: mccarl@tamu.edu

Dave Murray-Rust is a postdoctoral researcher with the Centre for Intelligent Systems and their Applications within the School of Informatics at the University of Edinburgh. He obtained an MEng in Electrical and Information Systems from Cambridge, an MSc in Informatics from Edinburgh and a PhD in Artificial Intelligence and Music from Edinburgh.

Address: Centre for Intelligent Systems, University of Edinburgh, 10 Crichton Street, Edinburgh EH8 9AB, UK. E-mail: d.murray-rust@ed.ac.uk

Cathal O'Donoghue is Head of the Rural Economy and Development Programme (REDP) of Teagasc (the Agriculture and Food Development Authority of Ireland). He is a graduate of University College Cork (UCC) and a statistician and economist by training, with postgraduate degrees from the University of Oxford, University College Dublin (UCD), the London School of Economics (LSE) and the University of Warwick. His personal research programme involves the development and use of policy simulation models, for which he holds a Chair (extraordinary (adjunct)) at the University of Maastricht, as well as adjunct positions at UCD and National University of Ireland, Galway (NUIG).

Address: REDP, Teagasc, Athenry, Co. Galway, Ireland. E-mail: Cathal.ODonoghue@teagasc.ie

Stephen Ramsden is Associate Professor in Management at the University of Nottingham. He is interested in understanding the complex bioeconomic relationships between: farmer decision makers; farm business resources, inputs and outputs; and the physical and biological environment. Much of his work is necessarily interdisciplinary, spanning economics, the social and natural sciences and bioeconomic modelling. He is also Director of the University of Nottingham Farm.

Address: School of Biosciences, Sutton Bonington Campus, University of Nottingham, Loughborough, Leicestershire, LE12 5RD, UK. E-mail: stephen.ramsden@nottingham.ac.uk

Arnaud Rault is an agricultural economist at the French National Institute of Agricultural Research (INRA). His main research focuses on animal health economics, with emphasis on risk management, collective action and public economics applied to farm animal production.

Address: INRA, Oniris Site del Chantrerie, UMR 1300 BIOEPAR, BP 40706, 44307 Nantes, France. E-mail: arnaud.rault@nantes.inra.fr

Cesar Revoredo-Giha is a Senior Economist and Team Leader of Food Marketing Research within the Land Economy, Environment and Society (LEES) Research Group of Scotland's Rural College (SRUC). His main research focuses on the operation and performance of agri-food supply chains and their implications for the farming sector. Additionally, other research interests are in areas related to development economics, such as food policy, nutrition and poverty.

Address: LEES, Scotland's Rural College, Edinburgh, EH9 3JG, UK. E-mail: cesar.revoredo@sruc.ac.uk

Shailesh Shrestha is an agricultural economist in the Land Economy, Environment and Society (LEES) Research Group of Scotland's Rural College (SRUC). His main research interests are in policy impact assessments, mathematical modelling, farming system analysis, farm-level adaptations and climate change.

Address: LEES, Scotland's Rural College, Edinburgh, EH9 3JG, UK. E-mail: shailesh.shrestha@sruc.ac.uk

Roberto Solazzo is a researcher at the Council for Agricultural Research and Economics (CREA), Italy. His major fields of research are quantitative methodologies for the assessment of agricultural policies, international trade in agri-food products and supply chain analysis. He has participated in several national and European research and evaluation projects on agri-food policies.

Address: CREA – Consiglio per la Ricerca in Agricoltura e l'Analisi dell'Economia Agraria, Via Po, 14, Palazzina B-00198, Roma, Italy. E-mail: roberto.solazzo@crea.gov.it

Fiona Thorne is an agricultural economist in the Agricultural Economics and Farm Surveys Department of the Rural Economy and Development Programme (REDP) of Teagasc (the Agriculture and Food Development Authority of Ireland). Her main research focus is on competitiveness, productivity assessment and microeconomic developments in the crops sector in Ireland. She currently has a number of PhD students and postdocs working in the area of efficiency and competitiveness.

Address: REDP, Teagasc, Ashtown, Dublin 15, Ireland. E-mail: fiona.thorne@teagasc.ie

Mario Veneziani is a research assistant in the Agricultural Economics section of the Department of Economics, University of Parma. His main research focus is on employing quantitative methods to analyse the impact of agricultural policy measures on farmers' decision making as well as investigating topics concerning the food industry and consumer behaviour. He has been involved in two EU FP-7 funded projects centred around analysing the farm sector relying on the individual-level data provided by the Farm Accountancy Data Network (FADN) of Europe.
Address: Department of Economics, University of Parma, Via J.F. Kennedy 6-43125, Parma, Italy. E-mail: mario.veneziani@unipr.it

Bouda Vosough Ahmadi is an agricultural economist with a background in veterinary medicine. Since 2008, he has worked within the Policy Analysis Team of the Land Economy, Environment and Society (LEES) Research Group of Scotland's Rural College (SRUC). His main research focus is on assessing the impacts of policies on farm economics and management, and on livestock health and welfare. Since February 2016, he has been working at the Joint Research Centre (JRC) of the European Commission.
Address: LEES, Scotland's Rural College, Edinburgh, EH9 3JG, UK. E-mail: bouda.v.ahmadi@sruc.ac.uk

Weiwei Wang is a postdoctoral research associate in the Department of Agricultural and Consumer Economics at the University of Illinois Urbana-Champaign. Her research specialties include natural resource and energy economics, climate change economics, food policy and operations research. She currently leads the development of an integrated economic–energy–agriculture model – the Biofuel and Environmental Policy Analysis Model (BEPAM), which also integrates the forestry sector (BEPAM-F).
Address: Department of Agricultural and Consumer Economics, University of Illinois Urbana-Champaign, Mumford Hall, 1301 W Gregory Dr MC-710, Urbana, IL 61801, USA. E-mail: weiwei.wang23@gmail.com

Paul Wilson is Professor of Agricultural Economics at the University of Nottingham and also Chief Executive of Rural Business Research, the consortium of leading agricultural universities and colleges that undertakes the Farm Business Survey research programme in England. His research interests include the analysis of managerial drivers of performance within agriculture, understanding the economics of the food chain and consumer behaviour towards sustainable food consumption, and investigating how managerial objectives influence outcomes for environmental activities and the production of biomass for bioenergy.
Address: School of Biosciences, Sutton Bonington Campus, University of Nottingham, Loughborough, Leicestershire, LE12 5RD, UK. E-mail: paul.wilson@nottingham.ac.uk

Foreword

Computational modelling is playing an ever greater role in policy making. One example is the way that the distribution of greenhouse gas emission reduction targets across EU (European Union) member states in the EU's 2020 climate and energy package was determined largely on the basis of model simulations that suggested where reductions could be achieved most efficiently. The conclusion of international trade agreements is now routinely accompanied by a model analysis showing the expected gains and losses to individual sectors and how overall economic welfare and employment might be affected.

Empirical models are also increasingly used in agricultural policy making. In my own country, Ireland, the decision to go for full decoupling of the Single Farm Payment introduced in the 2003 Common Agricultural Policy (CAP) Mid-Term Review was based on model results showing that farm incomes would be significantly higher under this option rather than retaining the option of coupled payments in the beef and sheep sectors. The annual market outlook projections developed by DG AGRI (the Directorate-General for Agriculture and Rural Development of the EU) make use of sophisticated agricultural sector models, even if expert judgement continues to play an important role.

This trend towards greater use of computational models is driven by a number of factors. The emergence of 'evidence-based policy making', with its emphasis on a more rational, rigorous and systematic approach to the taking of policy decisions, has helped to create a demand for tools and methods that can provide answers to the questions that policy makers pose. What would happen if we chose this policy option rather than that one? Do the observed outcomes justify continuing a policy measure or should the measure be changed? How differently would the world look in the future under one scenario compared with another? The institutionalization of the impact assessment of new policy proposals as well as the requirements of *ex post* 'fitness for purpose' checks add to the demand for evidence on how policies are working or might work in the future. Computational models can help to provide answers to questions of this kind.

There is also a supply side to the growth in interest in models. With advances in computing power, software and data availability, building a useful empirical model is no longer the preserve of large institutes and research teams, even if the appropriate research infrastructure is still an advantage when it comes to updating a model to maintain its usefulness and credibility. Within the area of agricultural policy, many agricultural economics researchers want their work to be relevant for policy advice (Brink, 2013). Applied economic modelling in agriculture thus continues to attract widespread interest and research funding across Europe.

This book of essays eloquently describes how the range of questions being asked of agricultural modellers is rapidly widening. As Maria Blanco argues in her opening chapter, this is partly driven by the evolution of the CAP from a markets policy to greater reliance on non-market measures that directly target farmers, as well as by the broadening in its focus beyond prices and farm incomes to also encompass the impacts of farming on the environment and resource use; then there is also the matter of farming's growing integration with the rest of the food supply chain. This change in the nature of the demand from policy makers has led to a vigorous supply response from the research community, particularly through the development of farm-level modelling to complement the more traditional market models – which continue to play their role. The chapters in this book provide a fine introduction to the techniques used and the issues addressed by farm-level models. They underline the potential that exists to generate new insights and guidance for policy makers as these models come to be more widely used.

The great strengths of farm-level modelling are its capacity to take account of the heterogeneity of individual farms, to more adequately model the interactions between farm production, farm practices, resource use and environmental impacts, and to provide evidence to policy makers not based just on aggregated outcomes, but also on the distributional and spatial impacts of particular policies or policy proposals. These gains come at a cost, of course, and some of the trade-offs are explored in some of the later chapters. Modelling individual farm behaviour requires a better understanding of the constraints faced by and the decision processes used by individual farmers, issues that can be greatly simplified in market models. Scaling up and aggregating the results from farm-level models so that market feedback is taken into account remains a challenge. So does the issue of model validation, as well as the always present issue of data availability and parameterization.

The many potential uses of farm-level models, particularly when integrated with biophysical models to simulate impacts on environmental resources or when used to model the spatial impacts of climate change, can quickly give rise to great complexity. This can be a challenge when communicating model results to policy makers, who may rightly be suspicious of what appears to be a 'black box'. This issue can be addressed in various ways. Modellers must decide on the appropriate degree of realism to be included in a model or suite of models, bearing in mind that going for increased realism with possibly a finer description of the outcomes may add greatly to complexity. More emphasis can be put on open source rather than proprietary models, while extending their use to the wider research community can help to validate relationships and build credibility in interpreting model outcomes. Building user-friendly interfaces to allow policy makers to simulate alternative outcomes to get a feel for the strength of different relationships can also be helpful in building confidence in model results.

This book on farm-level modelling is itself a contribution to the process of dialogue between academic researchers and policy makers. It sets out the potential contributions that farm-level modelling can make. It tries to demystify some of the techniques that are used, both through intuitive introductions and through case study illustrations. It is also upfront on where further progress needs to be made, and from this perspective it provides a challenge to the research community as well. For all of these reasons, it should be warmly welcomed.

Alan Matthews
Professor Emeritus of European Agricultural Policy
Trinity College Dublin, Ireland

Reference

Brink, L. (2013) Making agricultural economics research relevant for policy advice. *Canadian Journal of Agricultural Economics* 61, 15–36.

Preface

Agriculture is the product of a complex mixture of behavioural, biophysical and market drivers. Understanding how these factors interact to produce crops and livestock for food production has been the focus of economic investigation for a number of centuries. However, only in the last 50 years, with the advent of optimization algorithms and the exponential growth in computing technology, have we seen any significant growth in mathematical modelling of the dynamics of agricultural systems. The complexity of approaches has grown in parallel with the availability of data at increasingly finer resolutions and, moreover, in response to the widening demands of agricultural policy. Within Europe and elsewhere, the bulk of agricultural policy has focused on securing and increasing supply. Since the 1990s, these goals have widened to accommodate environmental, social and wider rural issues. Modellers concerned with the agricultural process have responded in kind by developing new approaches to capture these wider complexities and augment classical optimization based approaches to inject new life and accuracy into their estimates.

Ultimately, the purpose of modelling is to understand how farmers, landowners and the industry will respond to market and policy levers. This is why government has and does value the input of, and many man-hours devoted by, agricultural economists to constructing approaches that accurately and adequately reflect response. The purpose of this book is to show the student interested in agriculture the methods that have been employed to inform decision makers how to improve economic, social and environmental outcomes or, at least, to highlight the trade-offs required to achieve the goals of a multidimensional policy.

Agricultural modelling can be as simple or as hard as you wish it to be and this book presents avenues in which the 'farm problem' has been approached by a number of economists working within European, US and developing country contexts. With the advent of Microsoft Excel Solver, mathematical modelling has become a fairly fuss free activity and offers an open invitation to students of agriculture and the environment to explore and develop topics of interest. Some of the models, and indeed the modellers themselves, have, within this book, developed the basics of modelling using Excel Solver before taking the leap to other routes, such as learning GAMS (General Algebraic Modeling System) or Java software approaches. The Internet is now a useful free source of a number of architectures for exploring the models presented within this book, such as R, Java, C++, etc. The purpose of this book is not to give a set of tutorials on how to use these software packages within the context of agricultural modelling, but to provide some 'recipe suggestions' for approaches that students may take in their Masters or doctoral training courses.

We have split this book into two discrete parts based on loosely defined spatial distinctions. Part 1 concerns itself with assessment at the farm level and the various approaches used for this.

Chapter 1 presents a background to farm-level modelling problems on policy impact assessments and Chapter 2 advances the technique of positive mathematical programming as an approach to provide robust calibration of farm-level models. Chapter 3 focuses on a demonstration of the more familiar linear programming-based models in capturing short-term farmer adaptations to meet market, climatic and other demands. Chapters 4 and 5 extend this story further by focusing on the management of risk and how to incorporate this within agricultural models. Chapter 4 examines crop diversification as a risk management strategy, whereas Chapter 5 uses biosecurity and animal health as an illustration of the significant risk that livestock farmers accrue. Chapters 6 and 7 explore various dimensions of farming efficiency and wider climatic measures. Chapter 6 illustrates the process of measurement of resource use efficiency at the farm level, and Chapter 7 underlines efficiency by taking the specific example of management for climate effects and how to account for these within policy and farmer decision-making frameworks.

From the farm-level assessments discussed in the first section of the book we may have indications of how particular farms may respond to changes or challenges, and in some instances these will be either typical or average farms reflective of the study region. Part 2 focuses on how to aggregate these results to reflect a wider set of farming responses at a catchment, national or even supranational level. Chapters 8 and 11 concentrate on the regional and catchment-level approaches needed to accommodate the management of public goods or scarce resources. Chapters 9 and 10 present more novel approaches to accommodate the heterogeneity in farming and also in farmer decision making. Chapter 9 examines spatial disaggregation and utilizes the individual and regional level data sets that are available to modellers using econometric estimation, whereas Chapter 10 uses rule-based approaches as a way to typify and scale complex farmer decision making within a region. Finally, Chapters 12 and 13 give us novel approaches to integrate modelling to higher scales, such as the food supply chain or linking to European general equilibrium models.

While they are diverse in content and approaches, the format of these chapters attempts to be uniform where possible. Thus, each chapter will introduce the modelling or policy problem. The chapter will then review the available methods or past approaches taken to address the recognized problems. The core modelling framework is then presented, and this is followed by presentations of its application to several select case studies. Conclusions are then drawn or a summary of the chapter given.

Shailesh Shrestha
Andrew Barnes
Bouda Vosough Ahmadi

Acknowledgements

We would like to thank the Scottish Government for its support through the 5 year Strategic Research Programme on Food and Economic Adaptation (2011–2016). The funding for this programme allowed us to complete this book and we are grateful for their support. We also thank the contributors for their time, patience and energy in developing their chapters in this book.

1 Policy Impact Assessment

Maria Blanco*
Technical University of Madrid/Universidad Politécnica de Madrid, Madrid

1.1 Introduction

Public intervention in the agricultural sector is common in many developed and developing countries (Gardner, 1992). The rationales for government intervention in the agricultural sector include the need to correct market failures or to provide public goods, redistribution and/or stabilization of income, and concerns about food security and the self-reliance of food supplies. There is a wide cross-country variation in the priority objectives of agricultural policies, which also evolve through time in line with ever-changing societal demands. Whatever the objectives of the agricultural policy, the policy-making process involves the identification and assessment of alternative options to reach those objectives. Any policy change is designed with the expectation of improving the current situation, but the extent to which this can be achieved – even if it can only be known *ex post* – needs to be carefully assessed *ex ante*. The evaluation of public policies prior to their approval is a crucial step in policy design and usually consists of the assessment of the likely impacts of the new policy measures proposed with the final aim of maximizing the benefits to society and avoiding undesirable side effects.

As the assessment of policy impacts has gained focus in policy design, sectoral and/or fragmented studies have been integrated into a systematic process – impact assessment – which addresses impacts across the three dimensions of sustainability in a balanced way and, therefore, contributes to the mainstreaming of sustainability in policy making. According to the Organisation for Economic Co-operation and Development (OECD) (2010), sustainability impact assessment (SIA) has two main functions: (i) it is a tool for developing integrated policies that take full account of the three sustainable development dimensions; and (ii) it is a process for the *ex ante* assessment of the likely economic, social and environmental effects of policies, strategies, plans and programmes. SIA usually combines qualitative and quantitative assessment tools in a systematic framework, and it may involve a wide range of tools, depending on the issue at hand (Ness *et al.*, 2007). This type of assessment is being implemented in many OECD countries and is required for all initiatives of the European Union (EU) that are likely to have significant economic, environmental or social impacts.

In 2002, the EU adopted formal impact assessment procedures to improve the quality

*E-mail: maria.blanco@upm.es

and coherence of the policy-development process (European Commission, 2002). Further, impact assessment aims at contributing to a more coherent implementation of the European Strategy for Sustainable Development (European Commission, 2001). Impact assessment evaluates the likely positive and negative impacts of proposed policy actions, enabling informed political judgements to be made about the proposal and identifying trade-offs in achieving competing objectives. Impact assessment guidelines were first published in 2002 and have been regularly updated since then in response to new developments in both policy and impact assessment areas (European Commission, 2005, 2009, 2015). Impact assessment is now compulsory for major EU policies, including the Common Agricultural Policy (CAP), one of the oldest common policies of the EU.

Since 2003, the EU has conducted *ex ante* impact assessments of the successive reforms of the CAP in order to inform the policy-design process of the economic, social and environmental consequences of changes in policy instruments. For each CAP reform, several impact studies have been conducted using a variety of existing and newly developed tools (European Commission, 2003, 2008, 2011). The EU Impact Assessment Guidelines insist on the need to quantify the relevant impacts of new policies and they promote the use of modelling tools to better inform policy design and improve the performance and coherence of European policies. There is a wide variety of models and tools used for impact studies, involving both qualitative and quantitative tools and both economy-wide and sectoral models. The specificities of the agricultural sector, together with the complexity of the interactions between economy, society and environment, call for specific modelling approaches. In this sense, agro-economic models have always played an important role in the *ex ante* analysis of agricultural policies. Furthermore, as CAP policy has evolved from market intervention towards non-market measures that directly target farmers and sustainability, conventional modelling approaches have been adapted and new models have been developed to better capture the complex economic, social and environmental interactions associated with the changes in CAP policy instruments.

Current agricultural policies put greater emphasis on farm-specific support and environmental performance. As a result, there is an increasing interest in developing farm models capable of properly representing the new CAP instruments and assessing their farm-level effects. Compared with partial and general equilibrium frameworks, farm-level analysis provides greater flexibility for capturing high farm heterogeneity and modelling the multifaceted interactions between farming practices, environmental effects and economic performance. This chapter discusses the increasingly important contribution of farm-level modelling to policy impact assessment. For illustration purposes, it focuses on the EU agricultural sector, where both policies and impact models have experienced a rapid transformation in recent decades.

The chapter is organized as follows. The following section (1.2) highlights how the evolution of agricultural policies over time has shaped the development of agro-economic models, with a focus on European policies. Section 1.3 discusses the role of farm-level approaches in *ex ante* impact assessments, with an emphasis on the diverse modelling approaches used rather than on empirical results obtained. Above all, it focuses on the capabilities of current modelling approaches to properly capture farm heterogeneity, farm-level adaptation strategies, agriculture–environment interactions and market feedback. Finally, the last section (1.4) summarizes the major challenges for farm-level modelling of agricultural policies and presents some concluding remarks.

1.2 Evolution of EU Agricultural Policies and Parallel Development of Impact Models

Early CAP measures were mostly based on price and market support (production subsidies, border protection measures). Accordingly, partial equilibrium models that depict the functioning of agri-food markets were commonly used in policy impact studies. Besides these, general equilibrium models were applied to cope with economy-wide impacts and spillover effects between sectors. As we will see hereafter, the evolving objectives of EU agricultural policy have guided the orientation and design of agro-economic models focused on policy impact assessment.

Since the 1990s, the CAP has been gradually reformed towards stronger market orientation and enhanced agricultural sustainability in response to ever-changing societal demands. Early reforms in the 1990s addressed problems such as overproduction, the high cost of CAP support and international trade tensions. These reforms replaced a large share of the price support in the EU by direct payments coupled to production (defined per hectare of land or per head of livestock), implying a shift from product support to producer support. The use of quantitative tools to analyse policy impacts became widespread in those years (Lansink and Peerlings, 1996; Barkaoui and Butault, 1998). Further, the rapid progress in computer and communication technologies, along with better access to international databases, boosted the development of increasingly sophisticated models. A review of partial and general equilibrium models commonly used for agricultural and trade policy assessment is provided by van Tongeren *et al.* (2001). With subsequent policy initiatives, the weighting of the different objectives of the CAP has changed, as well as the policy instruments used to achieve these objectives. A drastic reform of the CAP was adopted in 2003 to make European agriculture more competitive and market oriented and, at the same time, provide less trade-distorting support to farmers. The main elements of the reform were: (i) the introduction of the Single Payment Scheme that decoupled direct payments from production; and (ii) a greater emphasis on environmental quality, food safety and animal welfare standards.

The 2003 reform represented a major change in the way the EU supports the farm sector. Changes in agricultural policy instruments were accompanied by increased attention to the assessment of policy impacts. Uncertainty of the impacts of the 2003 CAP reform, together with greater sustainability concerns, stressed the need to provide comprehensive impact assessment. Following the recently established Guidelines (European Commission, 2002) – aligned with sustainability goals – a formal impact assessment of the 2003 reform was carried out (European Commission, 2003). This involved a scenario analysis, where a range of policy alternatives were compared against a 'baseline' or reference scenario reflecting the expected developments of EU agriculture in a 'non-policy' situation, while taking into account anticipated technological or societal developments as well as the policies already in place. Parallel to the official requirements from the European Commission on the *ex ante* impact assessment of policy initiatives, considerable progress took place on model development. On the one hand, while existing impact models were suited to analysing production decisions and evaluating the impacts of price support, decoupling was an unprecedented change and represented a great challenge for policy modellers. On the other hand, impact models required in-depth adaptations before being applied to the assessment of environmental effects. In the 2000s, many conventional impact models went through significant improvements that enabled them to better capture the interactions of policy incentives, farmer responses and environmental effects at various spatial and temporal scales. At the same time, new modelling approaches more targeted towards sustainability analysis were developed.

Economic theory suggests that decoupled payments provide a more efficient basis for income transfer and give rise to fewer market distortions than coupled ones. In principle, the decoupling of direct payments from production is expected to make production decisions more market oriented and, therefore, is supposed to lead to the same level of production that would exist without any payments. However, even fully decoupled payments may affect production decisions through: (i) the income effect, which influences farm labour allocation; (ii) the risk-related effect, including a wealth effect and an insurance effect; and (iii) dynamic effects, as payments may influence farmers' investment behaviour (Conforti, 2005). While there is little consensus on the indirect effects of decoupled support, a vast literature exists on the impacts of decoupling on risk behaviour (Sckokai and Moro, 2006; Serra *et al.*, 2006), farm labour allocation (Gohin, 2006), investment decisions (Sckokai and Moro, 2009) and land markets (Guyomard *et al.*, 2004). Bhaskar and Beghin (2009) provide a comprehensive review of the literature on decoupled payments.

Model-based analyses of payment decoupling are numerous. Balkhausen *et al.* (2008) review the results of selected partial and general equilibrium models used for assessing the production and land-use effects of decoupling, finding that, even though results differ across models depending on their specification and assumptions, all models foresee a decline in cereal production and an increase in fodder production as a consequence of decoupling. Also, with decoupled payments, production decisions are more determined by market signals and not by CAP payments. The focus on environmental and sustainability concerns from the 2003 CAP also implied a great challenge for conventional impact models, which were particularly well suited to assessing economic impacts but much less so to accounting for environmental implications. Further, as more disaggregated analysis was needed to quantify the complex interdependencies between agriculture and the environment, models capable of providing farm-level responses became more relevant. Farm-level approaches started to be widely used to assess the environmental implications of agricultural policies and to analyse the impacts of agri-environmental measures (Rohm and Dabbert, 2003; Schmid and Sinabell, 2007).

With the objective of increasing the efficiency and effectiveness of the European agricultural policy, a further step towards decoupled direct payments took place in 2008 with the Health Check package, which led to full decoupling in most countries and sectors (European Commission, 2008). Other measures were the phasing out of milk quotas, the abolition of compulsory set-aside and the increase in modulation; these have been extensively modelled (Britz *et al.*, 2012a). Other environmental concerns have also influenced the development of impact models. Bioenergy is a clear example of how agro-economic models constantly adapt to answer new policy and research questions. The promotion of the use of energy from renewable sources and the rapid development of biofuel markets throughout the 2000s motivated the introduction of new activities in agro-economic models to account for food–energy interactions (Blanco *et al.*, 2010). The 2013 reform further strengthens the environmental objectives of the CAP with the introduction of a Greening Payment, linking the 30% of direct payments to the provision of sustainable farming practices. The reform also aims to move towards a less asymmetric distribution of support, seeking the convergence of payments not only between Member States, but also within Member States. The new policy measures, in particular the greening measures,[1] will have differentiated effects at the regional/farm level. Most of the agro-economic models widely used to conduct impact assessment of previous CAP policies are unable to fully capture the impacts of these new policy measures (Ciaian *et al.*, 2013). The need to develop modelling tools that are able to analyse the socio-economic and environmental impacts of agricultural policies at a much disaggregated level is becoming a crucial issue. Farm-level models play an increasingly prominent role in impact assessment studies. While these models are better suited to assessing the effects of the new farm-specific policy measures, the disaggregated assessment also faces important challenges, namely extensive data requirements and higher complexity in order to extend spatial coverage and to account for market feedback. In brief, agricultural policies are shaped more and more by environmental and sustainability concerns, and policy measures are becoming more territory specific. Climate change and limited resources are identified as future challenges, and will probably be increasingly relevant for policy design. Accordingly, models capable of simulating agricultural policies specified at regional and farm levels become more relevant.

1.3 The Widening Role of Farm-level Modelling in Impact Assessment

As has already been mentioned, recent CAP developments call for more disaggregated assessment of the economic, social and environmental impacts of the increasingly targeted and farm-specific policy measures. As a result, there is a growing interest in developing farm-level models suited to capturing the impacts of agricultural policy instruments at a highly disaggregated level. Better access to databases, developments in computer power and big data tools are also important factors contributing to the advanced development of micro-level modelling tools.

Most farm-level models used for *ex ante* simulation of agricultural policies are built within the framework of mathematical programming (MP). Hazell and Norton (1986) provide a discussion on the standard applications of MP to economic analysis in agriculture. MP is an optimization approach that offers great flexibility: (i) for capturing farm heterogeneity; (ii) for representing the multiple interactions across farming activities and between farming practices and their environmental effects; and (iii) for modelling a wide variety of policy instruments designed at a regional or even a farm level. Broadly speaking, there are several approaches to farm-level modelling, including farm-supply models, bioeconomic models and agricultural household models. While farm-supply models mainly focus on economic objectives, bioeconomic models integrate economic and environmental objectives, and agricultural household models incorporate the social dimension. Furthermore, the models differ in the extent to which they account for spatial and farm heterogeneity, interactions between activities (on the same area or over space), agriculture–environment interactions and intra- and inter-annual variability.

1.3.1 Interactions between activities

Farm models assume that farmers allocate scarce resources among activities so as to obtain the outputs that optimize farm objectives. Agricultural activities are described through input–output coefficients that define the relationship between the amounts of inputs used and the outputs obtained, where outputs involve both agricultural products and externalities. An inherent feature of farm models is that decisions are taken jointly over activities so that interactions between the activities are accounted for. In particular, these models are well suited to account for the interactions between crop and livestock farming through on-farm production and the consumption of animal feed and organic fertilizer.

Farm models are widely used to anticipate the reactions of farmers to alternative policy scenarios. Thanks to their activity-based approach, they are able to capture interactions between activities as well as farm-level adjustments to face new technological, economic or institutional settings. Potential adaptations include changes in cropland allocation, adjustments in production intensity, changes in investment strategies and decisions to expand or exit the farm. While partial equilibrium models are well suited to analysing the effects of market-support measures, farm-level models have some advantages for assessing aid decoupled from production. Thus, since the introduction of decoupled payments in the 2003 CAP, the role of farm-level modelling in policy impact assessment has been steadily increasing. A number of studies have explored the implications of decoupling for the agricultural sector and this is a good example for illustrating the specific contribution of the farm-level approach. Balkhausen *et al.* (2008) discuss the modelling work done with partial and general equilibrium models, drawing attention to the difficulties of aggregated models in accounting for substitution effects between arable land and pasture land, and their implications for livestock production. Farm-level studies are also numerous (Offermann *et al.*, 2005; Buysse *et al.*, 2007; Shrestha *et al.*, 2007), and many of them focus on the substitution effects between crop and fodder areas. For example, Galko and Jayet (2011) used the AROPAj model to assess the effects of several decoupling options, concluding that land used for pasture increases at the expense of land used for cereals and protein crops. Farm-level modelling is a very flexible framework for incorporating a broad range of policy measures, ranging from subsidies to production quotas and conditionality (Mosnier *et al.*, 2009). Compared with more aggregated assessments, farm-level impact analysis is able to capture the interaction between activities in greater detail. Yet, as farm models alone are unable to account for market feedback, these models most likely overestimate the impacts of direct payments on farming practices and agricultural production.

1.3.2 Farm heterogeneity

Growing concern over agricultural sustainability has prompted greater demand for agricultural policy analysis at the local and regional levels. With the introduction of territory-specific measures in recent CAP reforms, the role of impact assessment at the farm level is becoming more relevant. Also, policy makers are increasingly interested in understanding the distributive effects of policies and the differentiated farmer responses when they are faced with various policy and market situations. Overall, farm-supply models applied to the analysis of EU agricultural policies capture farm heterogeneity by modelling a set of farm types, defined through a farm typology. Most of these models have been designed to perform analysis for specific countries or regions, such as FARMIS in Germany (Offermann *et al.*, 2005), AGRISP in Italy (Arfini and Donati, 2011), FAMOS in Austria (Schmid, 2004) or ScotFarm in Scotland (Shrestha *et al.*, 2014). Among the few farm-level models that cover the whole EU, there are CAPRI-FT (Gocht and Britz, 2011) and AROPAj (De Cara and Jayet, 2005).

Farm size and specialization are the main criteria used for defining farm typologies. These typologies greatly depend on data availability and important trade-offs exist between higher detail and the quality of the specification of the corresponding farm model. Within the EU, the main data sources used to build farm typologies are the Farm Accounting Data Network (FADN) and the Farm Structure Survey (FSS). A number of studies have analysed the effects of decoupling on particular countries or regions. Using the FARMIS model, Küpker *et al.* (2006) found that the regional implementation of decoupling in Germany induced a significant redistribution of direct payments and, therefore, led to differences in income effects depending on farm type, location and size. Shrestha *et al.* (2007) used a farm-level approach to estimate the regional effects of decoupling on agricultural production and farm income in Ireland. Their results show that under the historical decoupling scheme, the milk quota would shift from less efficient to larger more efficient farms in all regions. Moreover, switching to a national flat rate form of decoupling would mean that large beef and dairy farmers in the southern regions would lose, while small dairy and sheep farmers in the western and northern regions would be most likely to gain.

Despite evidence of the influence of direct payments on risk aversion, the treatment of risk and uncertainty is not sufficiently covered in farm modelling. Although some farm models incorporate risk, such models have been applied only to a limited number of representative farms (Petsakos *et al.*, 2009; Arata *et al.*, 2014) and their methodologies are not easy to apply when modelling a large set of farms including numerous farm types. Analysing decoupling at the farm level for the whole EU is very challenging, mostly because of data availability and data quality issues. In one of the few attempts that has been made so far, Gocht *et al.* (2013) used the CAPRI-FT model to assess how the harmonization of direct payments in the 2013 CAP reform will affect the distribution of farm income across regions and farm types. With the introduction of greening measures in the 2013 CAP, farm-level modelling becomes crucial. Assessing the effects of capping and greening measures requires the use of models that are able to capture the adaptation strategies followed by farms (Vosough Ahmadi *et al.*, 2015). The diversity of implementation options across Member States, together with the lack of data to adequately define the farm types involved, makes the assessments of greening particularly difficult. The first attempts to model greening measures for particular regions or countries have shown a diversity of effects (Was *et al.*, 2014; Solazzo *et al.*, 2015).

1.3.3 Agriculture–environment interactions

Over the last few decades, there has been increasing concern about the relationship between agriculture and the environment. Since the 1990s, the CAP has introduced agri-environmental measures in order to discourage negative environmental externalities and promote the positive externalities of agricultural activities. Two issues are crucial when addressing environmental issues: (i) the scale of impact assessment should be at the territorial level; and (ii) economic and environmental effects should be jointly assessed. On the one hand, compared with more aggregated approaches, farm-level modelling presents clear advantages for depicting the manifold interactions between agriculture and the environment. On the other hand, the joint *ex ante* assessment of the economic and environmental effects of agricultural policies presents clear advantages because it helps to better target policies towards their intended outcomes. Not only are farmer responses assessed but the environmental consequences of farmer reactions are also accounted for. Compared with more aggregated models, farm-level models are better suited to simulating the interaction of policy incentives, farmer responses and environmental outcomes.

A common approach to assessing the environmental impacts of agricultural policies is to monitor environmental indicators. Several sets of environmental indicators have been developed internationally. The OECD set of agri-environmental indicators (OECD, 2013) has been constantly developed and refined since the 1990s and the modelling of these serves as a basis for informed green growth policies. The European Commission also develops a set of agri-environmental indicators to monitor the integration of environmental concerns into agricultural policy (European Commission, 2006). The integration of environmental aspects into farm-level modelling is straightforward when we can assume a direct link between farm inputs/outputs and some environmental indicator. Based on technical coefficients, the nutrient, energy or carbon balances can be calculated for each activity. Many studies have analysed the impacts of agri-environmental policy measures by translating model outputs into changes in environmental indicators (Buysse *et al.*, 2007; Schmid and Sinabell, 2007). However, due to the complexity of the interactions between farming practices and the environment, even highly disaggregated models may fail to capture some environmental effects. For instance, establishing the relationship between nitrate percolation and groundwater quality, or analysing the impacts of land-use changes on biodiversity and landscape are challenging tasks. Also, data availability and accuracy are often major limitations for depicting the relationship between agricultural activities and the environment.

A more sophisticated way to account for the interactions between agriculture and the environment is the bioeconomic approach, which combines biophysical[2] and economic models (Janssen and van Ittersum, 2007; Flichman, 2011). Belhouchette *et al.* (2011) linked the FSSIM model (Louhichi *et al.*, 2010) to crop growth models to assess environmental externalities. Jayet and Petsakos (2013) coupled the economic model AROPAj with the crop growth model STICS (which enables the derivation of nitrogen-yield response functions) to assess the effects of nitrogen taxes in France under different agricultural policy scenarios. Schönhart *et al.* (2011) coupled the bioeconomic farm optimization model FAMOS, the crop rotation model CropRota and the biophysical process model EPIC to assess the cost-effectiveness of agri-environmental measures in a set of farm types in Austria. As biophysical models operate at a high spatial resolution and are data intensive, increasing the regional coverage of bioeconomic models without loss in accuracy is very challenging. Notwithstanding, the assessment of agri-environmental impacts has experienced major advances in relation to the development of bioeconomic models, the design of interlinked multidisciplinary modelling tools and the variety of environmental impacts considered.

Because decoupled payments encourage more extensive agriculture, they may improve soil and water conditions of the environment (Schmid *et al.*, 2007). None the less, the effects on biodiversity are less clear. To analyse the impacts of decoupling on land use and biodiversity, Brady *et al.* (2009) applied an agent-based approach – which links the dynamics of farm structure to landscape dynamics – to a set of EU regions. They found that decoupled payments may result in further homogenization of land use and loss of biodiversity. The territorial scale of recent policy measures has also motivated the development of impact models that integrate spatial issues. To assess biodiversity, Bamière *et al.* (2011) used a spatially explicit MP farm-based model which accounts for three spatial levels (field, farm and landscape), and showed that valuable insights into agri-environmental programme design are gained through a detailed representation of farming system management.

As already mentioned, models integrating the multiple dimensions of sustainability are crucial tools for assessing current EU policies. Often, the limitation of these approaches is the spatial coverage. As the EU is characterized by high regional variation in agricultural, environmental and socio-economic conditions, quantifying the potential impacts of CAP measures over the whole EU is a challenging task. In recent years, there has been a significant development of bioeconomic approaches, not only to address the multiplicity of objectives in new agricultural policies but also to assess the impacts of climate change in agriculture. Actually, the integration of biophysical and economic models is the most widely used approach to assessing the complex interrelations between climate change, agricultural production and natural resource sustainability (Fernández and Blanco, 2015).

1.3.4 Dynamics and structural change

Sustainable impact assessment involves many dynamic features: investment behaviour, changes in farm structure, the evolution of environmental conditions (soil fertility, water quality, greenhouse gas (GHG) emissions, etc.). Nevertheless, most farm-level models have been developed in a comparatively static framework, mostly because: (i) data on capital assets is not as available as production data; (ii) the dynamic interlinkages between farming practices and environmental quality are highly complex; and (iii) calibrating and solving dynamic models is very challenging. Few attempts to model livestock dynamics are found in the literature and these usually refer to specific regions or countries. For instance, to analyse GHG abatement in German dairy farms, Lengers and Britz (2012) developed a dynamic model that is able to cover a great variety of GHG abatement options and derive farm-specific marginal abatement cost curves.

The potential impacts of decoupled support on investment behaviour and structural change have received great attention in the literature, but current farm modelling approaches are not able to fully cover those impacts (Zimmermann *et al.*, 2009). Accordingly, farm models are often combined with other approaches. Hennessy and Rehman (2006) explored the effect of decoupled payments on production decisions and structural change in Irish farming. To this aim, they combined a multiperiod optimization model with the econometric estimation of farm labour allocation and entry and exit decisions. Similarly, Renwick *et al.* (2013) combined the CAPRI-FT model with a land-use model to analysis the impact of policy changes on land abandonment. The effects of decoupling on structural change vary across regions, farming systems and policy options. While Hennessy and Rehman (2006) found that farm numbers will decline more rapidly under decoupling relative to the baseline situation, Sahrbacher *et al.* (2007) showed that the decoupling scenario slows down structural change, mainly because decoupled payments provide additional income opportunities for farmers with grassland to remain in the sector.

Sahrbacher *et al.* (2007) used an agent-based model, AgriPoliS, in which production and investment decisions are made simultaneously (Balmann, 1997; Happe *et al.*, 2006). This approach is better suited to modelling the dynamics of farm structure. Brady *et al.* (2009) found that decoupled payments will increase land rental prices and that this, in turn, affects future farm income and production decisions. The drawback of this approach is the complexity of extending the analysis to the whole of the EU. Viaggi *et al.* (2010) developed a farm-household dynamic model to evaluate the effects of decoupling on farm investment behaviour and provides an application to northern Italy. These authors argued that, since the introduction of decoupled payments in the 2003 CAP, the use of instruments able to account for multiple objectives, dynamics and investment choices will become even more relevant in the analysis of EU agricultural policy.

1.3.5 Market feedback

While farm-level models are able to capture agriculture–environment interactions at a disaggregated level, these models focus at supply responses and do not take into account market interactions. That is, input and output prices are exogenous in this type of model and, therefore, price effects are not accounted for, at least endogenously. On the contrary, partial equilibrium agro-economic models, which have been standard tools for policy impact assessment to date, are well suited to representing not only the production but also the demand for and trade of agricultural and food products, but they fail to capture the effects of farm-specific policy measures. In the current context of increased globalization, the lack of market feedback is one of the main limitations of farm-level modelling. This limitation can be overcome by linking farm models to partial equilibrium tools. Applications of multi-scale approaches in impact assessment include the AGRISIMU modelling framework (Lehtonen *et al.*, 2010), which integrates a farm-level optimization model, a regional sector model and biophysical models to assess alternative policy options in Finland. The FARMIS model has also been linked to the market model ESIM to measure the impacts of liberalizing European agriculture on farm income distribution in western Germany (Deppermann *et al.*, 2014). In most cases, the link to markets is done through a soft-link approach in which the outcomes from the market model are used as inputs in the farm model. The CAPRI-FT model is a unique case because in this approach, farm-level models and a global market model are fully integrated and solved iteratively (Gocht and Britz, 2011). The distinctive feature of CAPRI-FT is that it enables the assessment of farm-level impacts while covering the whole EU and taking into account market feedback.

1.4 What Models Do We Need to Assess Tomorrow's Agricultural Policies?

The agricultural sector is facing continuous socio-economic and environmental challenges in a rapidly changing economic and institutional environment. In coming years, climate change and environmental concerns will increasingly influence agriculture and shape the design of agricultural policies. Even if current modelling tools designed to assess policy impacts are highly sophisticated, a number of challenges remain in the modelling of agricultural systems. Moreover, models also need to continuously adapt to changing policy and societal concerns. Due to the manifold objectives of agricultural policies and the multitude of environmental impacts associated with agricultural production, sustainability impact assessment requires an integrated approach to account for the interrelated economic, environmental and social impacts at different temporal and spatial scales. Tools for integrated assessment combine models from different disciplines that operate at varying time and spatial scales and provide a multitude of outcomes. Integrated assessments also need methods for scaling up economic, environmental and social variables from the farm level to higher aggregation levels (Britz *et al.*, 2012b). In addition, they require methods for scaling down data and baseline indicators from the administrative level (regional, national) to the farm level.

The development of multi-model multi-scale platforms faces several challenges related to aggregation issues: (i) the diversity of temporal and spatial scales through models and disciplines; (ii) the need to account for cross-effects (i.e. price endogeneity); and (iii) the need to account for the complex and interconnected links between economic and environmental outcomes. Scaling methods become a crucial aspect of sustainability impact assessment tools. Without scaling, the contribution of farm models to the assessment of policy impacts will remain limited. Current agricultural policies require methods that are able to represent farm heterogeneity and to model farm-specific policy measures. Still, these methods should also be able to capture interconnections between farms (i.e. exchange of inputs) as well as market adjustments for inputs and outputs. Relevant trade-offs exist between the benefits of developing more complex models and the costs of providing greater detail (greater data needs, increased complexity and transparency issues).

Environmental factors are another crucial issue. There is a wide diversity of environmental impacts (i.e. impacts on natural resources such as soil and water, biodiversity, landscape) and a wide range of agri-environmental policy measures. Impact models need to identify and measure the causal relationships between policy measures and environmental change. Despite the advances in the integration of environmental variables in impact assessment tools, current farm models are still lacking the capability to properly simulate the environmental impacts of policy measures, in particular those with a strong spatial component (i.e. biodiversity, landscape or hydrology).

Data availability and accuracy are the key limiting factors in model development. Aspects such as the definition of farm types are highly dependent on the available data sources. Even more limitations are found in the case of environmental variables, which also show a high spatial variability. Secondary data sources on the relationship between farming and environmental conditions are lacking, and collecting these data for a large number of regions may become extremely costly. Nevertheless, spatial data are increasingly available and, even if the link to farm practices is still missing, big data tools offer promising opportunities to improve data availability and processing.

The challenges encountered in *ex ante* assessment of recent CAP reforms have led to changes in impact models. Today, the rapid development of sophisticated modelling platforms has been made possible by the collaboration of multidisciplinary teams worldwide. Multi-scale multi-model perspectives require the joint work of research teams from different disciplines. Notable advances have been already achieved in database sharing and in the joint development of model coding. Recent research on climate change and resource scarcity is a clear example and shows how impact assessment will rely more and more on multi-model approaches, which are capable of covering more issues.

Moreover, integrated impact assessment tools also call for joint baselines. Important efforts have been made to homogenize baselines so that model outcomes are comparable when they are used to address the same policy question. Yet, conceptual differences between baselines exist, and it remains difficult to combine baseline indicators across spatial scales. Deficiencies in model validation and uncertainty analysis require special attention. Further efforts are needed in these areas to enhance the contribution of farm models to sustainability impact assessment.

Notes

[1] The three foreseen greening measures are: permanent grassland, crop diversification and ecological focus area.

[2] Biophysical models simulate the interactions between soil, climate, farm management, crop growth and water and nutrient cycles.

References

Arata, L., Donati, M., Sckokai, P. and Arfini, F. (2014) Incorporating risk in a positive mathematical programming framework: a new methodological approach. *Paper prepared for presentation at the EAAE 2014 Congress 'Agri-Food and Rural Innovations for Healthier Societies', August 26–29, 2014, Ljubljana, Slovenia*. Available at: http://ageconsearch.umn.edu/bitstream/182659/2/Arata-Incorporating_risk_in_a_positive_mathematical_programming_framework-387_a.pdf (accessed 17 February 2016).

Arfini, F. and Donati, M. (2011) The impact of the Health Check on structural change and farm efficiency: a comparative assessment of three European agricultural regions. In: OECD (ed.) *Proceedings of an OECD Workshop on Disaggregated Impacts of CAP Reforms*. OECD Publishing, Paris, pp. 75–90.

Balkhausen, O., Banse, M. and Grethe, H. (2008) Modelling CAP decoupling in the EU: a comparison of selected simulation models and results. *Journal of Agricultural Economics* 59, 57–71.

Balmann, A. (1997) Farm-based modelling of regional structural change: a cellular automata approach. *European Review of Agricultural Economics* 24, 85–108.

Bamière, L., Havlík, P., Jacquet, F., Lherm, M., Millet, G. and Bretagnolle, V. (2011) Farming system modelling for agri-environmental policy design: the case of a spatially non-aggregated allocation of conservation measures. *Ecological Economics* 70, 891–899.

Barkaoui, A. and Butault, J.P. (1998) Modélisation de l'agriculture meusienne et «Paquet Santer». *Economie Rurale* 248, 13–20.

Belhouchette, H., Louhichi, K., Therond, O., Mouratiadou, I., Wery, J., van Ittersum, M. and Flichman, G. (2011) Assessing the impact of the Nitrate Directive on farming systems using a bio-economic modelling chain. *Agricultural Systems* 104, 135–145.

Bhaskar, A. and Beghin, J.C. (2009) How coupled are decoupled farm payments? A review of the evidence. *Journal of Agricultural and Resource Economics* 34, 130–153.

Blanco, M., Burrell, A., Gay, S.H., Henseler, M., Kavallari, A., M'Barek, R., Pérez Domínguez, I. and Tonini, A. (2010) *Impacts of the EU Biofuel Target on Agricultural Markets and Land Use: A Comparative Modelling Assessment*. Reference Report EUR 24449, European Commission Joint Research Centre. Publications Office of the European Union, Luxembourg.

Brady, M., Kellermann, K., Sahrbacher, C. and Jelinek, L. (2009) Impacts of decoupled agricultural support on farm structure, biodiversity and landscape mosaic: some EU results. *Journal of Agricultural Economics* 60(3), 563–585.

Britz, W., Gocht, A., Pérez-Domínguez, I., Jansson, T., Grosche, S.C. and Zhao, N. (2012a) EU-wide (regional and farm level) effects of premium decoupling and harmonisation following the Health Check reform. *German Journal of Agricultural Economics* 61, 44–56.

Britz, W., van Ittersum, M., Lansink, A.O. and Heckelei, T. (2012b) Tools for integrated assessment in agriculture. State of the art and challenges. *Bio-based and Applied Economics* 1, 125–150.

Buysse, J., Van Huylenbroeck, G. and Lauwers, L. (2007) Normative, positive and econometric mathematical programming as tools for incorporation of multifunctionality in agricultural policy modelling. *Agriculture, Ecosystems, and Environment* 120, 70–81.

Ciaian, P., Espinosa, M., Gomez y Paloma, S., Heckelei, T., Langrell, S., Louhichi, K., Sckokai, P., Thomas, A. and Vard, T. (2013) *Farm Level Modelling of the CAP: A Methodological Overview*. Publications Office of the European Union, Luxembourg.

Conforti, P. (2005) *The Effect of Direct Payments of the OECD Countries in World of Agricultural Markets: Evidence from Partial and General Equilibrium Frameworks*. Working Paper 12, Commodities and Trade Division, Food and Agriculture Organization of the United Nations, Rome.

De Cara, S., Houzé, M. and Jayet, P.-A. (2005) Methane and nitrous oxide emissions from agriculture in the EU: a spatial assessment of sources and abatement costs. *Environmental and Resource Economics* 32, 551–583.

Deppermann, A., Grethe, H. and Offermann, F. (2014) Distributional effects of CAP liberalisation on western German farm incomes: an ex-ante analysis. *European Review of Agricultural Economics* 41, 605–626.

European Commission (2001) *A Sustainable Europe for a Better World: A European Union Strategy for Sustainable Development (Commission's Proposal to the Gothenburg European Council)*. Communication from the Commission, Document COM(2001) 264 final, Commission of the European Communities, Brussels. Available at: http://ec.europa.eu/regional_policy/archive/innovation/pdf/library/strategy_sustdev_en.pdf (accessed 17 February 2016).

European Commission (2002) *Communication from the Commission on Impact Assessment*. Document COM(2002) 276 final, Commission of the European Communities, Brussels. Available at: http://eur-lex.europa.eu/LexUriServ/LexUriServ.do?uri=COM:2002:0276:FIN:EN:PDF (accessed 17 February 2016).

European Commission (2003) *Mid-term Review of the Common Agricultural Policy – July 2002 Proposals – Impact Analyses*. European Commission, Brussels. Available at: http://ec.europa.eu/agriculture/policy-perspectives/impact-assessment/mid-term-review/rep_en.pdf (accessed 17 February 2016).

European Commission (2005) *Impact Assessment Guidelines*. Document SEC(2005) 791, European Commission, Brussels. Available at: http://ec.europa.eu/agriculture/sfs/documents/documents/sec2005-791_en.pdf (accessed 17 February 2016).

European Commission (2006) *Development of Agri-environmental Indicators for Monitoring the Integration of Environmental Concerns into the Common Agricultural Policy*. Communication from the Commission to the Council and the European Parliament. Document COM(2006) 508 final, Commission of the European Communities, Brussels. Available at: http://eur-lex.europa.eu/legal-content/EN/TXT/PDF/?uri=CELEX:52006DC0508&from=en (accessed 17 February 2016).

European Commission (2008) *Commission Staff Working Document Accompanying the Proposals for [Various Council Regulations and a Council Decision Concerning the 2008 CAP "Health Check"] – Impact Assessment Summary*. Document SEC(2008) 1885. Commission of the European Communities, Brussels. Available at: http://ec.europa.eu/smart-regulation/impact/ia_carried_out/docs/ia_2008/sec_2008_1886_1_en.pdf (accessed 17 February 2016).

European Commission (2009) *Impact Assessment Guidelines*. Document SEC(2009) 92. European Commission, Brussels. Available at: http://ec.europa.eu/smart-regulation/impact/commission_guidelines/docs/iag_2009_en.pdf (accessed 17 February 2016).

European Commission (2011) *Commission Staff Working Paper – Impact Assessment – Common Agricultural Policy Towards 2020*. Document SEC(2011) 1153 final/2, European Commission, Brussels. Available at: http://ec.europa.eu/agriculture/policy-perspectives/impact-assessment/cap-towards-2020/report/full-text_en.pdf (accessed 17 February 2016).

European Commission (2015) *Commission Staff Working Document – Better Regulation Guidelines*. Document SWD(2015) 111 final, European Commission, Strasbourg, France. Available at: http://ec.europa.eu/smart-regulation/guidelines/docs/swd_br_guidelines_en.pdf (accessed 17 February 2016).

Fernández, F.J. and Blanco, M. (2015) Modelling the economic impacts of climate change on global and European agriculture. Review of economic structural approaches. *Economics* 9, 2015–10.

Flichman, G. (ed.) (2011) *Bio-economic Models Applied to Agricultural Systems*. Springer, Dordrecht, The Netherlands.

Galko, E. and Jayet, P.-A. (2011) Economic and environmental effects of decoupled agricultural support in the EU. *Agricultural Economics* 42, 605–618.

Gardner, B.L. (1992) Changing economic perspectives on the farm problem. *Journal of Economic Literature* 30, 62–101.

Gocht, A. and Britz, W. (2011) EU-wide farm type supply models in CAPRI – how to consistently disaggregate sector models into farm type models. *Journal of Policy Modeling* 33, 146–167.

Gocht, A., Britz, W., Ciaian, P. and Gomez y Paloma, S. (2013) Farm type effects of an EU-wide direct payment harmonisation. *Journal of Agricultural Economics* 64, 1–32.

Gohin, A. (2006) Assessing CAP reform: sensitivity of modelling decoupled policies. *Journal of Agricultural Economics* 57, 415–440.

Guyomard, H., Le Mouël, C. and Gohin, A. (2004) Impacts of alternative agricultural income support schemes on multiple policy goals. *European Review of Agricultural Economics* 31, 125–148.

Happe, K., Kellermann, K. and Balmann, A. (2006) Agent-based analysis of agricultural policies: an illustration of the agricultural policy simulator AgriPoliS, its adaptation and behavior. *Ecology and Society* 11(1): 49.

Hazell, P.B. and Norton, R.D. (1986) *Mathematical Programming for Economic Analysis in Agriculture*. Macmillan, New York.

Hennessy, T. and Rehman, T. (2006) Modelling the impact of decoupling on structural change in farming: Integrating econometric estimation and optimisation. *Paper No. 25271, International Association of Agricultural Economists, 2006 Annual Meeting, August 12–18, 2006, Queensland, Australia.* Available at: http://econpapers.repec.org/paper/agsiaae06/25271.htm (accessed 17 February 2016).

Janssen, S. and van Ittersum, M.K. (2007) Assessing farm innovations and responses to policies: a review of bio-economic farm models. *Agricultural Systems* 94, 622–636.

Jayet, P.-A. and Petsakos, A. (2013) Evaluating the efficiency of a uniform N-input tax under different policy scenarios at different scales. *Environmental Modeling and Assessment* 18, 57–72.

Küpker, B., Huttel, S., Kleinhanss, W. and Offermann, F. (2006) Assessing impacts of CAP reform in France and Germany. *German Journal of Agricultural Economics* 55, 227–237.

Lansink, A.O. and Peerlings, J. (1996) Modelling the new EU cereals and oilseeds regime in the Netherlands. *European Review of Agricultural Economics* 23, 161–178.

Lehtonen, H.S., Rötter, R.P., Palosuo, T.I., Salo, T.J., Helin, J.A., Pavlova, Y. and Kahiluoto, H.M. (2010) A modelling framework for assessing adaptive management options of Finnish agrifood systems to climate change. *Journal of Agricultural Science (Toronto)* 2(2), 3–16.

Lengers, B. and Britz, W. (2012) The choice of emission indicators in environmental policy design: an analysis of GHG abatement in different dairy farms based on a bio-economic model approach. *Revue d'Etudes en Agriculture et Environnement* 93, 117–144.

Louhichi, K., Kanellopoulos, A., Janssen, S., Flichman, G., Blanco, M., Hengsdijk, H., Heckelei, T., Berentsen, P., Lansink, A.O. and van Ittersum, M. (2010) FSSIM, a bio-economic farm model for simulating the response of EU farming systems to agricultural and environmental policies. *Agricultural Systems* 103, 585–597.

Mosnier, C., Ridie, A., Képhaliaco, C. and Carpy-Goulard, F. (2009) Economic and environmental impact of the CAP mid-term review on arable crop farming in south-western France. *Ecological Economics* 68, 1408–1416.

Ness, B., Urbel-Piirsalu, E., Anderberg, S. and Olsson, L. (2007) Categorising tools for sustainability assessment. *Ecological Economics* 60, 498–508.

OECD (2010) *Guidance on Sustainability Impact Assessment*. OECD Publishing, Paris.

OECD (2013) *OECD Compendium of Agri-environmental Indicators*. OECD Publishing, Paris.

Offermann, F., Kleinhanss, W., Hüttel, S. and Küpker, B. (2005) Assessing the 2003 CAP reform impacts on German agriculture using the farm group model FARMIS. In: Arfini, F. (ed.) *Modelling Agricultural Policies: State of the Art and New Challenges*. Monte Universita Parma Editore, Parma, Italy, pp. 546–564.

Petsakos, A., Rozakis, S. and Tsiboukas, C. (2009) Risk optimal farm plans in the context of decoupled subsidy payments: the case of cotton production in Thessaly. *Journal of Farm Management* 13(7), 34–50.

Renwick, A., Jansson, T., Verburg, P.H., Revoredo-Giha, C., Britz, W., Goch, A. and McCracken, D. (2013) Policy reform and agricultural land abandonment in the EU. *Land Use Policy* 30, 446–457.

Rohm, O. and Dabbert, S. (2003) Integrating agri-environmental programs into regional production models: an extension of positive mathematical programming. *American Journal of Agricultural Economics* 85, 254–265.

Sahrbacher, C., Schnicke, H., Kellermann, K., Happe, K. and Brady, M. (2007) *Impacts of Decoupling Policies in Selected Regions of Europe*. Deliverable 23 of the IDEMA [The Impact of Decoupling and Modulation in the European Union: A Sectoral and Farm Level Assessment] Project. IAMO (Leibniz Institute of Agricultural Development in Transition Economies), Halle, Germany.

Schmid, E. (2004) *Das Betriebsoptimierungssystem FAMOS – FArM Optimization System*. Diskussionpapier DP-09, Institut für nachhaltige Wirtschaftsentwicklung, Universität für Bodenkultur, Vienna.

Schmid, E. and Sinabell, F. (2007) On the choice of farm management practices after the reform of the Common Agricultural Policy in 2003. *Journal of Environmental Management* 82, 332–340.

Schmid, E., Sinabell, F. and Hofreither, M.F. (2007) Phasing out of environmentally harmful subsidies: consequences of the 2003 CAP reform. *Ecological Economics* 60, 596–604.
Schönhart, M., Schauppenlehner, T., Schmid, E. and Muhar, A. (2011) Integration of bio-physical and economic models to analyze management intensity and landscape structure effects at farm and landscape level. *Agricultural Systems* 104, 122–134.
Sckokai, P. and Moro, D. (2006) Modeling the reforms of the common agricultural policy for arable crops under uncertainty. *American Journal of Agricultural Economics* 88, 43–56.
Sckokai, P. and Moro, D. (2009) Modelling the impact of the CAP Single Farm Payment on farm investment and output. *European Review of Agricultural Economics* 36, 395–423.
Serra, T., Zilberman, D., Goodwin, B.K. and Featherstone, A. (2006) Effects of decoupling on the mean and variability of output. *European Review of Agricultural Economics* 33, 269–288.
Shrestha, S., Hennessy, T. and Hynes, S. (2007) The effect of decoupling on farming in Ireland: a regional analysis. *Irish Journal of Agricultural and Food Research* 46, 1–13.
Shrestha, S., Ahmadi, B.V., Thomson, S.G. and Barnes, A.P. (2014) An assessment of the post 2015 CAP reforms: winners and losers in Scottish farming. *Contributed Paper prepared for presentation at the 88th Annual Conference of the Agricultural Economics Society, April 9–11, 2014, AgroParisTech, Paris.* Available at: http://ageconsearch.umn.edu/bitstream/170536/2/Shailesh_Shrestha_CAP%20post%20 2015_AES%20full%20paper_Shailesh.pdf (accessed 17 February 2016).
Solazzo, R., Donati, M. and Arfini, F. (2015) CAP towards 2020 and the cost of political choices: the case of Emilia-Romagna region. *Land Use Policy* 48, 575–587.
van Tongeren, F., van Meijl, H. and Surry, Y. (2001) Global models applied to agricultural and trade policies: a review and assessment. *Agricultural Economics* 26, 149–172.
Viaggi, D., Raggi, M. and Gómez y Paloma, S. (2010) An integer programming dynamic farm-household model to evaluate the impact of agricultural policy reforms on farm investment behaviour. *European Journal of Operational Research* 207, 1130–1139.
Vosough Ahmadi, B., Shrestha, S., Thomson, S.G., Barnes, A.P. and Stott, A.W. (2015) Impacts of greening measures and flat rate regional payments of the Common Agricultural Policy on Scottish beef and sheep farms. *The Journal of Agricultural Science* 153, 676–688.
Was, A., Zawalinska, K. and Britz, W. (2014) Impact of “greening” the Common Agricultural Policy on sustainability of European agriculture: from perspective of the Baltic Sea countries. *Journal of Agribusiness and Rural Development* 34, 191–212.
Zimmermann, A., Heckelei, T. and Pèrez-Dominguez, I. (2009) Modelling farm structural change for integrated ex-ante assessment: review of methods and determinants. *Environmental Science and Policy* 12, 601–618.

2 Positive Mathematical Programming

Filippo Arfini,[1]* Michele Donati,[1] Roberto Solazzo[2] and Mario Veneziani[1]
[1]University of Parma, Parma, Italy; [2]Council for Agricultural Research and Economics, Roma, Italy

2.1 Introduction

Agricultural economists have always considered farm models a useful tool for analysing and simulating farmer decisions about technological change, market dynamics and agricultural policy scenarios. The most frequently used quantitative tools are based on mathematical programming and econometrics. Both of these methodologies benefit from developments in information technology which simplify the computing process, thus freeing researchers to develop models that are able to respond to policy makers' needs and use large amounts of data. Particularly useful nowadays is the capacity of models to use micro-information in order to produce macro-level solutions.

This chapter describes the use of mathematical programming for developing farm-level models that are useful for agricultural policy assessment. It focuses on the technique most frequently used in agricultural policy evaluation in the European Union (EU), positive mathematical programming (PMP). Unlike other mathematical programming (MP) tools, PMP can be applied at different levels of aggregation, such as individual farms, farm types and regions.

MP for agriculture was traditionally developed through linear programming (LP) models using information obtained directly from farms (Hazell and Norton, 1986). The main characteristic of LP models is that they describe primal and dual information in detail for each farm. They can take into account revenues, costs and farming practices, as well as relationships between farm activities, and relationships with product markets and input factors. They can also take into account limiting factors and their dual prices. LP aims to provide support for the production decisions of individual farmers and is mainly intended for application to the area of technical assistance. It is, however, less useful for policy makers. One limitation is the difficulty and cost of obtaining statistically representative economic and technical information about farmers' production choices directly. A second limitation is that databases such as the EU Farm Accountancy Data Network (FADN) do not supply information about the technology in use – specifically, technical coefficients – or about production process costs. Another problem is that where farms have specialized to a low number of production activities, or operate under many constraints, the output of LP can be made unacceptable by misspecified production problems (Paris and Howitt, 1998). Further typical disadvantages of LP are corner solutions and the difficulty of calibrating the

*E-mail: filippo.arfini@unipr.it

model to the observed production situation. Also, LP often has a normative aim and the distance between observed reality and the model is frequently explained by the well-known inefficiency of farmers. For all these reasons, LP is unsatisfactory for the analysis of farm policy.

The shift from LP to PMP has taken place for two main reasons: the availability of regional farm data databases; and the need to provide sound information, mainly to EU policy makers, on impact assessment of agricultural policies. Agricultural policy introduces constraints and restrictions (e.g. milk production quotas and the Nitrates Directive), provides coupled and decoupled payments, affects technological decisions and pushes farmers towards environmentally friendly practices and multifunctional farming. Evaluating the effect of agricultural policies allows for the identifying of weaknesses, corrective actions and instruments to achieve sector, regional and local objectives.

A methodology featuring the advantages of LP but without its disadvantages was therefore required. The new methodology needed to have the same capacity as LP to model the future behaviour of farmers from economic and technological viewpoints. A key assumption is that the configuration of economic variables can be explained as a result of the actions of rational actors (agricultural producers) making decisions that maximize their utility, taking into account market prices and subsidies – coupled to specific crops, decoupled from production or linked to the adoption of specific farming methods – as well as farm and family constraints, and relative costs not visible to external experts but well known (or perceived) by farmers.

In order to consider these aspects, some researchers (Howitt, 1995; Paris and Howitt, 1998) have moved from a normative to a positive approach and from LP to non-linear programming models, which are able to represent observed entrepreneurial behaviour. The new methodology is known as PMP, and is one of the most important innovations in the field of mathematical programming applied to the agricultural sector. However, the PMP models described in the literature tend to vary, and in fact tend to evolve towards econometrics (Buysse *et al.*, 2004).

This chapter thus outlines PMP farm models and how they can be used for research involving databases, cost function estimation and key issues in agricultural policy.

2.2 Review of Existing Methods

Work on PMP was started in the early 1990s by two researchers from the University of California, Davis, initially on the basis of early research by Heady and Egbert (1964) and Meister *et al.* (1978). There were two different approaches to implementation. The first (Howitt, 1995) focused on the capacity to calibrate farm decisions, the second on the capacity to estimate variable costs for each farm process. Very soon though, these approaches were combined into a much more detailed and powerful model (Paris and Howitt, 1998) with calibration and variable cost estimation at its centre. The idea underpinning PMP models is that it is a more transparent and manageable practice to collect data about output production level per activity, rather than data on production costs. Such data are fairly similar to those in existing farm databases, which show the structural characteristics of individual farms: land area, production plan and, especially, levels of production for each agricultural process. PMP models take as their starting point the production levels for each process.

The development of PMP also made it necessary to overcome the limitations of databases, problems in the representation of the cost function in models and problems posed by new farm policy scenarios.

The limitations of databases were as follows:

1. Shortage of information on variable costs by process: these can be considered as explicit accounting costs, and are important in assessing total variable costs and farm profit function. Researchers are often obliged to develop methodologies to integrate PMP models. For the cost function, researchers were obliged to choose the functional form of the variable cost function from, for example, quadratic, logarithmic or semi-logarithmic forms, as well as choosing between the diagonal elements or full form of the cost matrix and choosing the method of estimating the function. As described below, the most frequently used methodology is maximum entropy (ME), which is particularly useful when little information is available. The disadvantage is that results vary according to the decisions made by researchers.

2. Lack of balanced sample of farms: farms are often treated as 'single observations', and this has repercussions on the time periods used for certain parameters, particularly yields. Farm data may be not truly representative of an area. Researchers are, therefore, faced with important decisions when trying to ensure the representativeness of models of a production sector or a region – for example, NUTS2 (Level 2 Nomenclature of Units for Territorial Statistics) regions.

3. Analysing new policy scenarios: the first aspect of this is the problem of self-selection by farmers in choosing new production processes; the second aspect is the forecasting of scenarios involving the possibility of new activities or technologies.

2.2.1 PMP, calibration and cost estimates

The PMP models described in the literature show how researchers have interpreted and developed the seminal work of Howitt (1995), Paris (1993) and Paris and Howitt (1998), and attempted to overcome the problem of lack of information about variable costs by process in farm databases, particularly FADN. There were two clear strands in addressing this challenge. One strand was that the key aspect of PMP was taken to be calibration, and the other was the estimation of costs, with calibration being used for verification.

In the original formulation put forward by Howitt (1995), Paris and Arfini (1995) and Paris and Howitt (1998), the word 'positive' in PMP implies increasing the reliability of a constrained optimization model by using observed behaviour in the specification phase. So, the general idea of PMP is to add to a very simple LP model one additional constraint which binds the LP problem to observed activity levels (calibration constraints). This constraint allows the use of the associated dual information (Phase 1) to specify a non-linear objective function (Phase 2), such that the observed activity levels are reproduced by the optimal solution of a new programming problem without calibrating bounds (Phase 3).

Phase 1 of this procedure is formally described by extending a classical LP model, in which a farm gross margin function, Z, is maximized by also taking into account one additional constraint on the observed production level:

$$\max Z = \mathbf{p}'\mathbf{x} - \mathbf{c}'\mathbf{x} \tag{1}$$

subject to

$$\mathbf{Ax} \leq \mathbf{b} \quad [\mathbf{y}] \tag{2}$$

$$\mathbf{x} \leq (\bar{\mathbf{x}} + \boldsymbol{\varepsilon}) \quad [\boldsymbol{\lambda}] \tag{3}$$

$$\mathbf{x} \geq 0 \tag{4}$$

where: $\mathbf{p}$ is the vector of output prices; $\mathbf{x}$ is the vector of activity levels; $\mathbf{c}$ is the vector of accounting costs associated with inputs; $\bar{\mathbf{x}}$ is the vector of observed activity levels; ε is the vector of small positive numbers and serves to prevent linear dependency between the structural constraint (2) and the calibrating constraint (3) which is binding to the observed production level; $\mathbf{y}$ is the vector of the dual variables associated with the structural (or resource) constraints; $\boldsymbol{\lambda}$ is the vector of the dual variables associated with the calibration constraints. The addition of the calibration constraints forces the optimal solution of the LP model represented by equations (1)–(4) to reproduce the observed base year activity levels, $\bar{\mathbf{x}}$, exactly. 'Exactly' is understood to mean within the range of the positive perturbations of the calibration constraints, $\boldsymbol{\varepsilon}$, which are included to guarantee that all binding resource constraints of model (1)–(4) remain binding here and thus avoid a degenerate dual solution (Heckelei and Britz, 2005). The known non-negativity constraint placed on the primary variables for the problem is defined in relation (4). The dual values occur in two vectors ($\mathbf{y}$ and $\boldsymbol{\lambda}$): the first is the vector of resources dual values, which are associated with the resource constraints, while the second vector includes the dual costs of the production activities, which are associated with the calibration constraints. PMP, unlike the LP model, takes into account only one resource, usually land. The $\mathbf{A}$ matrix of technical coefficients is defined as the ratio between the limiting input ($\mathbf{h}$) and the observed production level ($\bar{\mathbf{x}}$).

The papers by Howitt (1995), Paris and Howitt (1998) and Heckelei and Wolff (2003) provide a complete description of the structure of both the primal and dual models. The economic interpretation of model (1)–(4) is important. The dual of the model (1)–(4) provides information about the cost associated with the resource (land) and the production. Howitt (1995) and Paris and Howitt (1998) interpret the dual variable vector $\boldsymbol{\lambda}$ associated with the calibration constraints as capturing any type of information that is insufficiently known to the researcher, such as risk perception, technological knowledge and personal preferences.

In Phase 2 of the procedure, the dual values of the calibration constraints related to realized production levels are used to specify a non-linear cost function. The variable cost function must be convex in output levels.

PMP does not require researchers to use any specific functional form. The next section (2.2.2) discusses this, together with the method of estimation, but at this point of the description, note that the most frequent form is quadratic, which is shown as follows:

$$\text{TVC} = \mathbf{d}'\overline{\mathbf{x}} + \frac{1}{2}\overline{\mathbf{x}}'\mathbf{Q}\overline{\mathbf{x}} = (\mathbf{c}+\boldsymbol{\lambda})'\overline{\mathbf{x}} \quad (5)$$

where TVC represents total variable costs; $\mathbf{d}$ is the vector linking parameters associated with linear terms (and is not binding); $\overline{\mathbf{x}}$ represents the $\mathbf{x}$ calibrated from the LP model; $\mathbf{Q}$ is the symmetric, positive semi-definite matrix which is the function of marginal costs (**mc**) for the observed farm on the basis of the following equation:

$$\mathbf{mc} = \mathbf{c} + \boldsymbol{\lambda} = \mathbf{d} + \mathbf{Q}\overline{\mathbf{x}} \quad (6)$$

Note that the derivatives (6) of this variable cost function do not incorporate the opportunity cost of fixed resources ($\mathbf{A'y}$) which are still captured in the model (1)–(4) by the dual values of the resource constraints.

Phase 3 of the procedure combines information from Phase 1 (price, technological matrix, resources vector) and information from Phase 2 (cost function), and gives rise to a new non-linear problem that reproduces observed activity levels as:

$$\max Z = \mathbf{p}'\mathbf{x} - \mathbf{d}'\mathbf{x} - \frac{1}{2}\mathbf{x}'\mathbf{Q}\mathbf{x} \quad (7)$$

subject to

$$\mathbf{Ax} \leq \mathbf{b} \quad [\mathbf{y}] \quad (8)$$

$$\mathbf{x} \geq 0 \quad (9)$$

This non-linear model gives a new calibration for the observed production level ($\overline{\mathbf{x}}$) and farmland shadow price without the calibration constraints of Phase 1. Once the new calibration is complete, the model is fit to implement different policy scenarios taking account of price interventions, public payments (coupled, partially coupled or fully decoupled), quota systems and environmental interventions.

The model represented by equations (7)–(9) reveals certain inconsistencies in PMP (Heckelei, 2002).[1] The first is that for a farm under the constraint of calibration in Phase 1, which happens for farm activities characterized by the lowest marginal profit, the value of $\boldsymbol{\lambda}$ is zero. The second limitation is that the FADN does not supply information on variable costs for every agricultural activity (**c**). These two problems significantly limit the possibility of including all processes and all costs in a non-linear model.

In other words, the absence of specific costs at farm level means that it is not possible to derive cost function parameters for the least profitable product, because its marginal cost value is null. The literature offers a number of solutions that modify the traditional formulation of PMP. In particular, Heckelei (2002) identifies a large set of instruments permitting estimation of the cost function starting from the observed production level.

One solution consists of implementing an alternative Phase 1 in which shadow prices associated with the resource constraints and with the calibrating constraint are derived using the equilibrium conditions of the problem (1)–(4). This implies that Phase 1 of the PMP methodology changes in such a way that all the marginal costs can be recovered from the observed production information values and so directly by Phase 2 of PMP, which becomes the first PMP step. This procedure directly defines the first-order conditions of the problem (1)–(4), and optimizes the problem by deriving the shadow price values and the production levels, thus minimizing the slackness variables associated with dual endogenous variables.

On the basis of this consideration, Heckelei (2002) suggests avoiding the first phase of PMP and replacing it by two steps in which the first-order optimality conditions are imposed in the first phase of PMP.[2] The procedure consists of a maximization model that overcomes the problem encountered when the data source transmits poor-quality information about farm behaviour. This is possible by estimating the shadow prices of the resource constraints simultaneously with the other parameters of the model; the methodology can provide a general and flexible tool for estimating parameters of duality-based behavioural functions with the explicit allocation of fixed factors. In this context, the only difference between programming and econometric models is the form of the model used for simulation purposes. Hence, the approach proposed by Heckelei (2002) and Heckelei and Wolff (2003) does not require the first phase of PMP to calibrate, even for more complex constraint structures. At this level, two types of solution can be adopted:

1. Derive the shadow prices emerging from the first phase of the traditional PMP model using specific econometric tools, such as generalized least-squares or Bayesian methods, directly applied to the Karush–Kuhn–Tucker (KKT) conditions.
2. Derive the shadow prices of the outputs from exogenous information related to the land market (this reflects the marginal value of the land).

Júdez *et al.* (2001) and Henry de Frahan *et al.* (2007) also suggest directly using first-order conditions without the first phase of PMP, using the observed rental prices of resources and solving for the parameters under selected additional restrictions on the **Q** matrix parameters. Heckelei and Wolff (2003) opened up a new research area: they introduced error terms allowing for a deviation of estimated from observed values and, using the ME estimator, investigated the impact of additional prior information on the estimation results. In the first phase, their approach is limited to three production activities and one constraint, but further developments by Gocht (2005) and Polomé *et al.* (2005) resulted in PMP being used to estimate farm-level cost functions based on FADN data. Moreover, many papers (Gocht, 2005; Kanellopoulos *et al.*, 2010; Gocht and Britz, 2011) and, in particular, many European researchers (Kanellopoulos *et al.*, 2010; Gocht and Britz, 2011), have adopted Heckelei's suggestions for developing PMP models capable of assessing the impact of policies of the Common Agricultural Policy (CAP). These studies use aggregated FADN information for different years. Only aggregate farm information allows the collection of consistent input market prices (e.g. land price) suitable for regional and sectorial models. At the same time, farms are considered in panel data containing several years of observations. This strategy makes it possible to model variation in yield, and thus characteristics of the agricultural technologies, more closely (Heckelei and Britz, 2005).

The method suggested by Heckelei and Wolff (2003) has many advantages, but it is necessary to take account of its capacity to represent observed reality. This is especially important if policy is to be assessed at local level and if each region is characterized by a stratification of different farm types. In PMP based on the work of Heckelei, each model considers as representative one farm identified by production specialization, or farm type (FT) at regional level. In other words, FADN observations on similar farms are used to represent the economic and production characteristics of n regional FTs. This limits the capacity to observe variation between farms in the same region belonging to the same FT.

However, the literature contains other and different methodologies in which farm cost function is estimated using the opportunity given by FADN data to construct as many models as there are farms observed. It is particularly useful to construct farm models to show the behaviour and strategy of individual farms in a region. This enables data to be aggregated at NUTS2 or FT level in policy simulations, which provides a basis for territorial and sector analysis (Arfini and Donati, 2011).

The development of models for all farms in a region entails adding a 'benchmark farm' to the sample of similar farms. This benchmark farm carries out all the activities actually observed in the sample and furthermore:

- The surface area and total output of each agricultural process observed is obtained, as is the total number of hectares and output from all farms carrying out the process in their production plan.
- Prices and yields are obtained as average price and yield for similar farms in the sample.

- Costs are represented by a benchmark cost function and each farm in the sample is characterized by the same cost function, which is indexed by a vector of deviations to account for that farm's environment and for farmer preferences.

Although the sample farms are considered homogeneous from a technical, income and environmental viewpoint, not every farm cultivates every crop occurring in that FT. This aspect is defined as the self-selection problem (Paris and Arfini, 2000), and it characterizes the behaviour of each farmer. The assumption is that each farmer whose farm is classified within a given FT could have produced every crop in the given FT, but did not do so for reasons of preference. In other words, the cost function of every individual farm in the given sample needs to measure a cost level for any crop that is greater than or equal to the analogous cost level for the homogenous farm (i.e. FT). In order to account for this self-selection process, the marginal cost relations of each farm must be further characterized to distinguish between the activities that are actually carried out and the potential activities. This is done by postulating two sets of constraints for the *n*th farm.

The first constraint is connected with the existing crops, which will have a marginal cost given by the **Q** matrix specified for the frontier cost multiplied by the observed productions, and by a (positive or negative) deviation component compared with the frontier itself (Paris and Arfini, 2000). The second set of constraints concerns the activities not implemented by the *n*th farm. In this case, the marginal cost of the non-implemented process(es) could be less than or equal to that specified for the frontier. In turn, the relationship between the marginal costs can be represented as a weak inequality compared with the level of average marginal cost calculated for each activity at the sample level. Hence, a proper methodology for estimating a valid cost function for the given FT must incorporate a self-selection rule (Paris and Arfini, 2000; Arfini and Donati, 2013).

In this methodological setting, an alternative PMP approach was developed for estimating the production cost for each activity. This approach was developed in the framework of the EU-FP7 FACEPA (the Farm Accountancy Cost Estimation and Policy Analysis of European Agriculture project of the European Community's Seventh Framework Programme; Deliverable 6.2, Donati *et al.*, 2013) and is characterized by using only endogenous information available from the FADN.

The model merges the first phase with the second phase through the dual properties of the PMP approach. It considers a sample composed of *n* farms where information about the production plan, prices and technical coefficients is known at the farm level and where only one limiting factor, land available at the farm level, b_n, is considered. The matrix $\mathbf{A}_n$ is composed of the coefficients $\mathbf{a}_n$, which link the production factors to yields. The observed production levels for each farm are indicated by vector $\overline{\mathbf{x}}_n$, and output market prices are represented by vector $\mathbf{p}_n$. The model assumes that for each farm activity, marginal costs consist of two components: vector $\mathbf{c}_n$, which represents the explicit marginal costs related to each activity (explicit marginal costs can be seen as the exogenous cost in farm accountancy and can be observed),[3] and $\boldsymbol{\lambda}_n$, which represents the implicit marginal cost. $\boldsymbol{\lambda}_n$ provides the measure of additional costs, which can consist of price expectations, specific production preferences and the technological skills of the individual farmers. The latter costs do not appear in the FADN, but can be derived through PMP as follows:

$$\min_{\mathbf{u}_n, y_n, \boldsymbol{\lambda}_n, \mathbf{Q}} \left\{ \sum_{n=1}^{N} \frac{1}{2} \mathbf{u}_n' \mathbf{u}_n + \sum_{n=1}^{N} \begin{pmatrix} b_n y_n + \boldsymbol{\lambda}_n' \overline{\mathbf{x}}_n \\ + \mathbf{c}_n' \overline{\mathbf{x}}_n - \mathbf{p}_n' \overline{\mathbf{x}}_n \end{pmatrix} \right\} \quad (10)$$

subject to

$$\mathbf{A}_n' y_n + \boldsymbol{\lambda}_n + \mathbf{c}_n \geq \mathbf{p}_n \quad (\mathbf{w}_n) \quad (11)$$

$$\mathbf{c}_n + \boldsymbol{\lambda}_n = \mathbf{Q}\overline{\mathbf{x}}_n + \mathbf{u}_n \quad (\mathbf{z}_n) \quad (12)$$

where: $y_n \geq 0$, $\boldsymbol{\lambda}_n \geq 0$ and **Q** is a symmetric positive semi-definite matrix, as stated by Paris and Howitt (1998) and Paris (2011); $\mathbf{w}_n$ and $\mathbf{z}_n$ are the shadow prices associated with equations (11) and (12) respectively; $\mathbf{u}_n$ is the vector of marginal cost deviations per farm, that is, the distance between the marginal cost $\mathbf{c}_n + \boldsymbol{\lambda}_n$ and the marginal cost $\mathbf{Q}\overline{\mathbf{x}}_n$ of a non-linear cost function, such that $\mathbf{c}_n + \boldsymbol{\lambda}_n - \mathbf{Q}\overline{\mathbf{x}}_n = \mathbf{u}_n$. The estimated parameters of **Q** are part of a quadratic cost function aiming to provide flexibility to model responses towards farm simulations. The model is optimized by a combined objective function, equation (10), which adopts a least-squares (LS) technique and minimizes the difference between the total revenue, $\mathbf{p}'_n\bar{\mathbf{x}}_n$, and the total cost, $b_n y_n + \boldsymbol{\lambda}'_n\bar{\mathbf{x}}_n + \mathbf{c}'_n\bar{\mathbf{x}}_n$. This latter expression identifies the optimal conditions for the standard PMP approach or, in general terms, states that under optimal conditions, the primal objective function should be equal to the dual function.

The above model as represented by equations (10)–(12) integrates the first and second phases of the standard PMP approach using PMP dual properties (Paris, 2011). In this model, there is no explicit trace of either the calibrating constraints or the epsilon ($\boldsymbol{\varepsilon}$) terms that help to break the linear dependence between structural and calibration constraints. The constraints of the model shown in equations (11)–(12) concern the equilibrium conditions with marginal costs greater than or equal to marginal revenue, and the relationship by which a linear cost function is transformed into a quadratic cost function. The model does not repeat the tautological procedure of the standard approach, but rather reveals hidden information about the differential marginal costs embedded in the production levels and makes this information available for the simulation phase.

The economic meaning and the properties of the solution of the problem represented by equations (10)–(12) is clarified by the Lagrangian (L) representation, as follows:

$$L = \sum_{n=1}^{N}\frac{1}{2}\mathbf{u}'_n\mathbf{u}_n + \sum_{n=1}^{N}\left(b_n y_n + \boldsymbol{\lambda}'_n\bar{\mathbf{x}}_n + \mathbf{c}'_n\bar{\mathbf{x}}_n - \mathbf{p}'_n\bar{\mathbf{x}}_n\right)$$
$$+\sum_{n=1}^{N}\mathbf{w}'_n\left(\mathbf{p}_n - \mathbf{A}'_n y_n - \boldsymbol{\lambda}_n - \mathbf{c}_n\right)$$
$$+\sum_{n=1}^{N}\mathbf{z}'_n\left(\boldsymbol{\lambda}_n + \mathbf{c}_n - \mathbf{Q}\bar{\mathbf{x}}_n - \mathbf{u}_n\right) \quad (13)$$

From the Lagrangian function, we can obtain the following relevant KKT conditions:

$$\frac{\partial L}{\partial \mathbf{u}_n} = \mathbf{u}_n - \mathbf{z}_n = \mathbf{0} \quad (14)$$

$$\frac{\partial L}{\partial \boldsymbol{\lambda}_n} = \bar{\mathbf{x}}_n - \mathbf{w}_n + \mathbf{z}_n \geq \mathbf{0} \quad (15)$$

$$\frac{\partial L}{\partial y_n} = b_n - \mathbf{A}_n\mathbf{w}_n \geq \mathbf{0} \quad (16)$$

The partial derivatives (14) indicate that the deviation terms, $\mathbf{u}_n$, are equal to the dual values, $\mathbf{z}_n$, linked to equation (12). Because the problem attempts to minimize the squares of the farm cost, the deviations $\mathbf{u}_n$ and $\mathbf{z}_n$ should take very low values close to zero. The KKT condition (15) can be rewritten as $\mathbf{w}_n - \mathbf{z}_n \leq \bar{\mathbf{x}}_n$, showing that the difference between the two shadow prices associated with equations (11) and (12) should be less than or equal to the outputs realized. In this respect, if we consider that the shadow price of the equation representing the equilibrium condition can be interpreted as the shadow output quantity, we can state that $\mathbf{w}_n \approx \bar{\mathbf{x}}_n$. Furthermore, as for the KKT condition (11), $\mathbf{z}_n$ can be viewed as a small term close to zero, and thus, we can state that $\mathbf{z}_n \approx \boldsymbol{\varepsilon}$. Rearranging this information, the KKT condition (11) becomes $\mathbf{w}_n \leq \bar{\mathbf{x}}_n + \mathbf{z}_n$, corresponding to the calibration constraint of the standard approach, which implies that the model (10)–(12) correctly replicates the standard PMP specification without the explicit calibration constraints. Taking this into account, the KKT condition (16) can be interpreted as the structural constraint related to land use, and by moving b_n to the right-hand side of equation (16) and changing the sign, we obtain $\mathbf{A}_n\mathbf{w}_n \leq b_n$, corresponding to equation (8).

This PMP approach overcomes the tautological procedure of the standard PMP approach and provides all the necessary information on the total marginal cost needed to recover a cost function for all of the farms belonging to the FADN sample and to implement the simulation phase. According to these objectives, model (10)–(12) takes the information related to the total variable costs available at the farm level in the European FADN as a guide for the total accounting cost estimation in which the squares of marginal cost deviations ($\mathbf{u}_n$) are minimized (min LS) in the following manner:

$$\min_{\mathbf{u}_n} \text{LS} = \sum_{n=1}^{N}\frac{1}{2}\mathbf{u}'_n\mathbf{u}_n \quad (17)$$

subject to

$$\boldsymbol{\alpha}_n + \boldsymbol{\lambda}_n = \mathbf{R}'\mathbf{R}\bar{\mathbf{x}}_n + \mathbf{u}_n \qquad \text{if } \bar{\mathbf{x}}_n > 0 \quad (18)$$

$$\boldsymbol{\alpha}_n + \boldsymbol{\lambda}_n \leq \mathbf{R}'\mathbf{R}\bar{\mathbf{x}}_n + \mathbf{u}_n \qquad \text{if } \bar{\mathbf{x}}_n = 0 \quad (19)$$

$$\boldsymbol{\alpha}'_n\bar{\mathbf{x}}_{nn} \leq \text{TVC}_n \quad (20)$$

$$\mathbf{u}'_n\bar{\mathbf{x}}_n + \frac{1}{2}\bar{\mathbf{x}}'_n(\mathbf{R}'\mathbf{R})\bar{\mathbf{x}}_n \geq \text{TVC}_n \quad (21)$$

$$\boldsymbol{\alpha}_n + \boldsymbol{\lambda}_n + \mathbf{A}'_n \mathbf{y}_n \geq \mathbf{p}_n \quad (22)$$

$$\mathbf{b}'_n \mathbf{y}_n + \boldsymbol{\lambda}'_n \bar{\mathbf{x}}_n = (\mathbf{p}_n - \boldsymbol{\alpha}_n)' \bar{\mathbf{x}}_n \quad (23)$$

$$\mathbf{R} = \mathbf{L}\mathbf{D}^{1/2} \quad (24)$$

$$\sum_{n=1}^{N} \mathbf{u}_n = 0 \quad (25)$$

The objective of the model represented by equations (17)–(25) is to estimate a non-linear cost function, including the unknown accounting variable cost $\boldsymbol{\alpha}_n$. Restrictions (18) and (19) define the relationship between marginal costs derived from a linear function and marginal costs derived from a quadratic cost function; $\boldsymbol{\alpha}_n + \boldsymbol{\lambda}_n$ defines the sum of the unknown (or estimated) accounting variable costs and the differential variable marginal costs.

Differential variable marginal costs are implicit in the decision-making process of the farmer, but they do not appear in farm accounts. Both components are endogenous variables in the minimization problem. Restrictions (18) and (19) also guarantee that the self-selection rule is followed, enabling farmers to select possible production activities from all activities present in the region (represented by the sample dimension), but restricting activities to those observed in the first phase of the PMP methodology (Paris and Arfini, 2000).

Moreover, to ensure consistency between the estimated accounting variable costs and those effectively recorded by the farm accounting system, constraint (20) requires that the total estimated variable cost be not greater than the total variable cost observed in the FADN data at farm level. Equation (21) states that the costs estimated by the model by means of a non-linear cost function must be at least equal to the value of the observed TVC.

To ensure consistency between the estimation process and the optimal conditions, constraint (22) introduces the traditional condition of economic equilibrium, in which total marginal costs must be greater than or equal to marginal revenues. Total marginal costs also include the cost of binding resources as defined by the product of the technical coefficients matrix, $\mathbf{A}_n'$, and the shadow price of the restricting factors, $\mathbf{y}_n$. Marginal revenues are defined by the sum of the selling prices of the products, $\mathbf{p}_n$, and any associated public coupled subsidy. The additional constraint (23) defines the optimal condition in which the value of the primal function corresponds exactly to the value of the objective function of the dual problem. To ensure that the matrix of the quadratic cost function is symmetric positive semi-definite, the model uses Cholesky factorization (24). Finally, constraint (25) establishes that the sum of the errors, $\mathbf{u}_n$, must be equal to zero.

The cost function estimated with model (17)–(25) can be used in a model of maximization of the farm gross margin, ignoring the calibration restrictions imposed during the first phase of the standard PMP approach. In this case, the dual relations entered into the cost estimation model above ensure that the observed situation is reproduced. The model for the gross margin (GM) of the n-farm, therefore, appears as follows:

$$\max_{\mathbf{x}_n \geq 0} \mathrm{GM} = \mathbf{p}'_n \mathbf{x}_n - \left\{ \frac{1}{2} \mathbf{x}'_n \hat{\mathbf{Q}} \mathbf{x}_n + \hat{\mathbf{u}}'_n \mathbf{x}_n \right\} \quad (26)$$

subject to

$$\mathbf{A}_n \mathbf{x}_n \leq \mathbf{b}_n \quad (27)$$

The model represented by equations (26)–(27) calibrates the observed farming system, thanks to the non-linear objective function which preserves the (economic) information on the actual levels of production. The estimated matrix $\mathbf{Q}$ is reconstructed using Cholesky's decomposition: $\hat{\mathbf{Q}} = \hat{\mathbf{R}}'\hat{\mathbf{R}} = \hat{\mathbf{L}}\hat{\mathbf{D}}\hat{\mathbf{L}}'$. Constraint (27) represents the restriction on the structural capacity of the farm. Once the initial data have been calibrated through the maximization of the farm gross margin, it is possible to introduce variations into the public aid mechanisms and/or in the market price levels to evaluate the farm reactions to various policy conditions. The reaction of the farm production plan will take into account the information used during the cost function estimation phase.

The use of the LS method to estimate the cost function as an alternative to the ME method has several advantages, such as avoiding the unsolved problems of the arbitrary use of support values (Golan *et al.*, 1996; Lence and Miller, 1998; Henry de Frahan *et al.*, 2007).

The PMP model that has been described can be used in three different ways: (i) to estimate accounting variable costs (α_n) related to each activity with data from the FADN (output price, yields, level of farm production, land use and total variable cost at farm level) (Donati *et al.*, 2013); (ii) to estimate the total variable cost of each crop ($\alpha_n + \lambda_n$), which makes it possible to evaluate farm behaviour by defining a new profit function for a given set of information (Donati *et al.*, 2013); and (iii) to model farm behaviour under different market and policy pressures as a simulation of the impact of CAP reforms (Arfini and Donati, 2008, 2011).

An alternative approach developed by Heckelei (2002) estimates input costs assuming Leontief technology for intermediate inputs. The use of Leontief technology has been widely assumed in the literature (Léon *et al.*, 1999) because it provides an explicit link between production activities and total physical input use. However, Heckelei and Wolff (2003) note an absence of rationalization as the intermediate input uses are assumed to be independent of unknown marginal costs captured by the quadratic behavioural function. In fact, Heckelei *et al.* (2008) have proposed the use of a Bayesian model to recover the true disaggregated crop-specific input costs as an alternative to entropy methods for deriving solutions to underdetermined systems of equations. The main advantage is a more direct and straightforwardly interpretable formulation of information available *a priori* and a clearly defined estimation objective.

2.2.2 PMP and cost function characteristics

One of the key characteristics of PMP is the construction of the marginal cost function in its functional form and estimation method. Theoretically, any functional form could be used – quadratic, logarithmic or semi-logarithmic but, in practice, the quadratic form is most widely used. For its estimation, various approaches have been proposed on the basis of the objectives and the restraints in research.

Heckelei and Britz (2005) summarizes the various approaches to PMP starting from the very first models (Howitt and Mean, 1983; Bauer and Kasnakoglu, 1990; Paris and Arfini, 1995), and emphasizes the limitations of realistically reproducing farmer behaviour in the presence of one constraint. Röhm and Dabbert (2003) propose a dual calibration that takes account of endogenous and exogenous information on input prices.

Heckelei and colleagues make further important contributions to PMP in two papers. The first, Heckelei and Britz (2000), obtained an estimate of diagonal and off-diagonal elements in matrix $\mathbf{Q}$ using the average cost approach. This was also suggested in an earlier paper by Gohin and Chantreuil (1999). Heckelei and Britz (2000) assume that the observed vector of accounting cost per activity unit, $\boldsymbol{c}$, is equal to the average cost of the crop-specific variable cost function. The second contribution (Heckelei, 2002) proposed incorporating exogenous own-price elasticities into the model. This is based on the fact that PMP uses calibration, especially when dealing with single observations, but it also makes it possible to insert earlier information. According to this approach, the off-diagonal elements of matrix $\mathbf{Q}$ are set at zero. This method has proved particularly efficacious in the calibration phase in the presence of external elasticities.

Paris and Arfini (2000) put forward an alternative to Heckelei and colleagues' suggestions and estimated the marginal cost function for all of the *n* farms belonging to a sample of homogenous farms when the accounting costs, $\boldsymbol{c}$, are known. Because the linearity aspect of the model is important when the number of farms is large, they proposed a specification that is linear in its parameters. The desired total variable cost function is obtained by integrating the marginal cost function with respect to the output variables. Input prices are not available in the FADN database and are assumed to be fixed. The cost function is usually hypothesized to be a quadratic functional form in output quantities, $C(\mathbf{x}) = \mathbf{x}'\mathbf{Q}\mathbf{x}/2$, where the $\mathbf{Q}$ matrix is symmetric and positive semi-definite. The requirement that the $\mathbf{Q}$ matrix of the quadratic cost function be symmetric positive semi-definite is ensured by Cholesky factorization:

$$\mathbf{Q} = \mathbf{LDL}' \tag{28}$$

where $\mathbf{L}$ is a unit lower triangular matrix, $\mathbf{L}'$ is its transpose and $\mathbf{D}$ is a diagonal matrix whose elements are non-negative. It can be shown that $\mathbf{LDL}'$ is a positive semi-definite matrix if and only if all the diagonal elements of $\mathbf{D}$ are non-negative (Lau, 1978).

For the recovery of the marginal cost function, the literature (Paris and Howitt, 1998; Paris and Arfini, 2000) describes two approaches. The first employs the generalized maximum entropy (GME) estimator, and the second the least-squares estimator (LSE).

According to the ME principle, the Cholesky parameters of the $\mathbf{L}$ and $\mathbf{D}$ matrices are regarded as expected values of associated probability distributions defined over a set of known discrete support values, $\mathbf{S}$ (Golan *et al.*, 1996). In matrix notation, equations linking the diagonal matrix $\mathbf{L}$ and $\mathbf{D}$ are represented as $\mathbf{L}=\mathbf{Z}_L\mathbf{P}_L$ and $\mathbf{D}=\mathbf{Z}_D\mathbf{P}_D$, respectively, where the multiplication is performed only for index $s=1,\ldots,S$. $\mathbf{Z}_L$ and $\mathbf{Z}_D$ are the matrices of the known support values for the $\mathbf{L}$ and $\mathbf{D}$ matrices, respectively, while $\mathbf{P}_L$ and $\mathbf{P}_D$ are the corresponding probability matrices.

The ME problem consists of recovering the probability matrices $\mathbf{P}_L$, $\mathbf{P}_D$ and $\mathbf{P}_{un}$, which maximize the entropy function subject to the appropriate constraints. Formally, the attempt is to find matrices $\mathbf{P}_L$, $\mathbf{P}_D$ and $\mathbf{P}_{un}$ with elements $P_L(j,j',s)\gg 0$, $P_D(j,j',s)\gg 0$ and $P_{un}(j,s)\gg 0$, such that:

$$\max_{\mathbf{P}_L,\mathbf{P}_D,\mathbf{P}_{un}} H(\mathbf{P}_L,\mathbf{P}_D,\mathbf{P}_{un}) = -\sum_{j,j',s} P_L(j,j',s)\log\left[P_L(j,j',s)\right] - \sum_{j,j,s} P_D(j,j,s)\log\left[P_D(j,j,s)\right] \sum_{n,j,s} P_{un}(j,s)\log\left[P_{un}(j,s)\right] \quad (29)$$

subject to

$$\bar{\lambda}_{LP}+\bar{\mathbf{c}}=\mathbf{LDL}'\bar{\mathbf{x}}_R=(\mathbf{Z}_L\mathbf{P}_L)(\mathbf{Z}_D\mathbf{P}_D)(\mathbf{Z}_L\mathbf{P}_L)'\bar{\mathbf{x}}_R \quad (30)$$

$$\lambda_{nk}+c_{nk}=L_k\mathbf{DL}'\mathbf{x}_{Rn}+u_{nk}=\mathbf{Z}_L(k)\mathbf{P}_L(k)(\mathbf{Z}_D\mathbf{P}_D)(\mathbf{Z}_L\mathbf{P}_L)'\mathbf{x}_{Rn}+\mathbf{Z}_{un}(k)\mathbf{P}_{un}(k), \text{if } x_{Rnk}>0 \quad (31)$$

$$\bar{\lambda}_{LPk}+\bar{\mathbf{c}}_k\le L_k\mathbf{DL}'\mathbf{x}_{Rn}+u_{nk}\le\mathbf{Z}_L(k)\mathbf{P}_L(k)(\mathbf{Z}_D\mathbf{P}_D)(\mathbf{Z}_L\mathbf{P}_L)'\mathbf{x}_{Rn}+\mathbf{Z}_{un}(k)\mathbf{P}_{un}(k), \text{if } x_{Rnk}=0 \quad (32)$$

$$1=\sum_s P_L(j,j',s) \quad (33)$$

$$1=\sum_s P_D(j,j',s) \quad (34)$$

$$1=\sum_s P_{un}(j,s),\ j,j'=1,\ldots,J \quad (35)$$

where, constraints (33), (34) and (35) express the adding-up conditions of probabilities. The expression $\log\left[P_L(j,j',s)\right]$ indicates the logarithm of the individual probabilities. The $H(\mathbf{P}_L,\mathbf{P}_D,\mathbf{P}_{un})$ function is strictly concave in the probabilities. After the solution of the ME problem, with $\hat{\mathbf{P}}_L$, $\hat{\mathbf{P}}_D$, and $\hat{\mathbf{P}}_{un}$ indicating the solution probabilities, the matrix $\mathbf{Q}$ and the vectors $\mathbf{u}_n$ can be calculated as $\hat{\mathbf{Q}}=\hat{\mathbf{L}}\hat{\mathbf{D}}\hat{\mathbf{L}}'=(\mathbf{Z}_L\hat{\mathbf{P}}_L)(\mathbf{Z}_D\hat{\mathbf{P}}_D)(\mathbf{Z}_L\hat{\mathbf{P}}_L)'$ and $\hat{\mathbf{u}}_n=\mathbf{Z}_{un}\hat{\mathbf{P}}_{un}$. Hence, the total variable cost function for the nth farm is stated as $C(\mathbf{x}_n)=\hat{\mathbf{u}}_n'\mathbf{x}_n+\mathbf{x}_n'\hat{\mathbf{Q}}\mathbf{x}_n/2$.

ME has proved to be very efficient in recovering the probability matrices when the number of observations is low. Applying it to a sample of n farms allows us to recover an $\mathbf{LDL}'$ for all the observed activities, which solves the problem of inconsistency noted by Heckelei and Britz (2005).

However, the use of ME is criticized because of the relationships between support values and output of the model. Support values are chosen arbitrarily and variation in magnitude modifies the results of the model estimates. Data management is the key to solving this issue: a scaling procedure needs to be implemented in order to minimize the size of the figures related to the model data (price, hectares and production level).

Considering the LSE for a population of n farms implies using the vector $\mathbf{u}_n$, n = 1, ..., N, as a vector of deviations from the frontier cost function (Paris and Arfini, 2000). In this case, the explicit model can be stated as:

$$\min_{\mathbf{u}_n}\sum_{n=1}^{N}\mathbf{u}'\mathbf{u}_n/2 \quad (36)$$

subject to

$$\bar{\lambda}_{LP}+\bar{\mathbf{c}}=\mathbf{LDL}'\bar{\mathbf{x}}_R \quad (37)$$

$$\lambda_{nk}+c_{nk}=L_k\mathbf{DL}'\mathbf{x}_{Rn}+u_{nk}, \text{if } x_{Rnk}>0, k=1,\ldots,J_n \quad (38)$$

$$\bar{\lambda}_{LPk} + \bar{\mathbf{c}}_k \leq L_k \mathbf{DL}'\mathbf{x}_{Rn} + u_{nk} \text{ if } x_{Rnk} = 0,\ k = 1, \ldots, J - J_n \tag{39}$$

The elements of the diagonal matrix **D** are restricted to being non-negative.

Both approaches outlined above have the great advantage that they generate an **LDL′** matrix that has positive and negative elements off the diagonal, thus generating a process of substitution between activities.

Moreover, when the model is calibrated, the model is suitable for policy analysis and ensures that the marginal value of land y is sensitive to variation in output prices. The matrices given in Tables 2.1 and 2.2 show the **LDL′** formulation obtained by GME and LS for n farms drawn from a sample of 549 farms in the Emilia Romagna FADN 2011 database.

2.2.3 PMP models and latent activities

A big limitation on the use of PMP in farm policy analysis to date has been that it is not possible to add new processes or technologies to those actually observed for the sample. In some contexts, however, it can be useful to know how farmers will adopt new technologies when there are changes in public policy and market scenarios. There have been numerous attempts to fill this gap. The first was made by Röhm and Dabbert (2003), who included 'variant activities' in the model, so that a given crop can be divided into several activities. Each activity is subject to two types of calibrating constraints: one for each crop and one for the different variants of the same crop. The Röhm–Dabbert calibrating model produces shadow prices associated with calibrating constraints greater than those obtained by adopting the standard PMP approach in which the calibrating constraint is the same for all farm activities. The consequence is that the dual values associated with structural constraints might be underestimated.

An alternative approach uses all the information from the FADN, and takes advantage of the possibility of recovering a benchmark cost function for homogenous farms and using the 'self-selection' rule. The idea is to use 'latent information' as a key element in the farmer's decision process. Latent information is all the information relating to production activities (including production costs) that enables farmers to change production. It is possible to differentiate between: (i) 'latent technology', when the information is related to a production technology option available within the investigated sample of farms; and (ii) a 'latent crop', when the information is related to a crop that does not exist in the farm production plan for any farms in the investigated sample.

The use of latent information enables all the activities taking place in the territory to be considered in the farmer's production plan even if they are not in fact practised by the farm in question at the time of reporting. The marginal costs of each farm are thus specified as distinguishing between activities actually practised and those that are not.

Consider the case, for the sake of simplicity, of a sample of four farms, of which two practise conventional farming and two practise organic farming, with two conventional crops and two organic crops, where the subscripts r and s represent the conventional product and the organic product, respectively. The **Q** matrix estimated for the benchmark farm is as follows:

$$\hat{\mathbf{Q}} = \begin{bmatrix} q_{r1,r1} & q_{r1,r2} & q_{r1,s1} & q_{r1,s2} \\ q_{r2,r1} & q_{r2,r2} & q_{r2,s1} & q_{r2,s2} \\ q_{s1,r1} & q_{s1,r2} & q_{s1,s1} & q_{s1,s2} \\ q_{s2,r1} & q_{s2,r2} & q_{s2,s1} & q_{s2,s2} \end{bmatrix} \tag{40}$$

The **Q** matrix retains all the information on substitution and complementarity relationships between conventional and organic production processes, even though organic production is not yet present in the farm plan. Therefore, during the simulation phase, modifications to the initial production can be shown, and the new processes, which exist in latent form, can be added if they give higher profits than existing processes.

The new model with the information related to cost of production estimated appears as follows:

$$\begin{aligned} \max_{\mathbf{x}_r, \mathbf{x}_s} \quad & \mathbf{p}_r'\mathbf{x}_r + \mathbf{p}_s'\mathbf{x}_s - \frac{1}{2}[\mathbf{x}_r \quad \mathbf{x}_s]\mathbf{Q}\begin{bmatrix} \mathbf{x}_r \\ \mathbf{x}_s \end{bmatrix} \\ & \mathbf{Bx}_r + \mathbf{Mx}_s \leq \mathbf{b} \\ & \mathbf{x}_r \geq 0 \\ & \mathbf{x}_s \geq 0 \end{aligned} \tag{41}$$

Table 2.1. **Q** matrix estimated through generalized maximum entropy (GME) for a sample of 549 farms in the Emilia Romagna region, Italy.

Product	Durum wheat	Common wheat	Maize	Alfalfa	Barley	Tomato	Rice	Soya	Milk
Durum wheat	0.120185	0.001754	−0.00583	−0.00489	0.002548	0.012773	−0.0023	−0.0017	−0.01173
Common wheat	0.001754	0.021045	−0.00069	−0.00052	0.000483	0.002713	−0.0004	−0.00028	−0.00198
Maize	−0.00583	−0.00069	0.01516	0.000804	−0.00061	−0.00314	0.00054	0.000365	0.002514
Alfalfa	−0.00489	−0.00052	0.000804	0.018923	−0.00047	−0.00242	0.00042	0.000283	0.00195
Barley	0.002548	0.000483	−0.00061	−0.00047	1.208391	−0.04204	0.00262	0.000969	0.004338
Tomato	0.012773	0.002713	−0.00314	−0.00242	−0.04204	0.07242	−0.0028	−0.00135	−0.00945
Rice	−0.00239	−0.00043	0.000549	0.000423	0.002626	−0.00284	0.02125	−0.07208	−0.01583
Soya	−0.00173	−0.00028	0.000365	0.000283	0.000969	−0.00135	−0.07208	0.627339	−0.06007
Milk	−0.01173	−0.00198	0.002514	0.00195	0.004338	−0.00945	−0.01583	−0.06007	0.056291

Table 2.2. **Q** matrix estimated through least squares (LS) for a sample of 549 farms in the Emilia Romagna region, Italy.

Product	Durum wheat	Common wheat	Maize	Alfalfa	Barley	Tomato	Rice	Soya	Milk
Durum wheat	0.013701	–0.00446	–0.00141	–0.00319	0.004477	0.000906	0.004309	0.001103	0.025989
Common wheat	–0.00446	0.015126	0.013709	0.027606	0.004336	0.012389	0.008474	0.005522	–0.08496
Maize	–0.00141	0.013709	0.017532	0.031253	–0.00225	0.013487	0.007589	0.003528	–0.09554
Alfalfa	–0.00319	0.027606	0.031253	0.061376	0.002728	0.028469	0.012312	0.005429	–0.19964
Barley	0.004477	0.004336	–0.00225	0.002728	0.018408	0.003785	0.006232	0.00465	–0.0033
Tomato	0.000906	0.012389	0.013487	0.028469	0.003785	0.016054	0.005622	0.003042	–0.09268
Rice	0.004309	0.008474	0.007589	0.012312	0.006232	0.005622	0.021232	0.01152	0.000213
Soya	0.001103	0.005522	0.003528	0.005429	0.00465	0.003042	0.01152	0.00957	0.008966
Milk	0.025989	–0.08496	–0.09554	–0.19964	–0.0033	–0.09268	0.000213	0.008966	0.755972

2.3 PMP Application for Policy Assessments

Over time, PMP models have been developed in response to the requirements for policy assessment and, particularly, policy making, rather than for the purposes of farm planning. Policy makers (of the European Commission and European Parliament) have required the EU Directorate-General (DG) for Research to provide modelling of the evolving CAP.[4] In recent years, the CAP has been reformed in various aspects: tariff barriers, subsidies and payments (from coupled to decoupled aid), quotas and environmental schemes. These have all been successfully modelled by PMP.

Faced with these scenarios, policy makers have tried to assess, *ex ante*, the potential impact of various reforms on agricultural production (cereals, milk, meat, etc.), agricultural inputs (particularly land), the income and efficiency variation of agricultural farmers, environmental implications and, more generally, the impacts on rural development. Impact assessment models have gradually moved from macroeconomic sector models to microeconomic models evaluating production and economic effects at the local level, focusing on farmer strategies (Langrell *et al.*, 2012; Solazzo *et al.*, 2014).

PMP models have, consequently, been widely used for impact assessment in most areas of European agricultural policy, as shown by the large amount of research that has resulted and has been reported through the proceedings of the European Association of Agricultural Economists (EAAE) Congresses held in Ghent, Zurich and Ljubljana, and through EAAE Seminars held in Bonn (Heckelei *et al.*, 2001), Parma (Arfini, 2005), Seville (Bartova *et al.*, 2008) and Viterbo (Sorrentino *et al.*, 2011).[5] PMP research has also been widely published in leading agricultural economics journals. See Heckelei and Britz (2005) for a summary of PMP models developed in Europe.

Thanks to its flexibility, and the fact that it has similar advantages to MP and particularly LP, PMP can be applied to a wide variety of topics and policy areas. PMP-based models can accommodate different types of linear constraints (e.g. agronomic and environmental constraints) using techniques described in standard agricultural economics textbooks (Hazell and Norton, 1986; McCarl and Spreen, 2011; Paris, 2013). However, once the model is calibrated and contains all the economic and technical information available (output prices, input costs, payments, tax, technology levels and relationships between products and factors), it is possible, by means of linear constraints, to generate complex models specifically designed to meet research objectives.

2.4 Conclusions

The literature on PMP shows that there is keen interest on the part of agricultural economists in a methodology that overcomes many of the limitations of LP and provides useful input for policy makers in this extremely complex field. Compared with the very early work by Howitt and Paris (Howitt, 1995; Paris, 1993; Paris and Arfini, 1995; Paris and Howitt, 1998), PMP has now become a powerful instrument to model production choices and structural constraints and hence to assess *ex ante* the impact of future market scenarios and public policy on farms, sectors and regions of Europe. Ongoing developments, principally the higher risks that farmers will face in future, will make further modifications to the model necessary, including the insertion of risk as a variable. However, PMP will remain an essential instrument as new scenarios appear on the horizon of agricultural policy.

Notes

1 '… the determination of shadow prices in Phase 1 of PMP creates a fundamental inconsistency between parameter specification and the resulting quadratic optimisation model, which renders any estimation of parameters using multiple observations inconsistent …' (Heckelei and Britz, 2005, p. 62).

2 'The general alternative to the PMP with respect to calibrating or estimating a programming model is a simple methodological principle: always to directly use the first order condition of the optimisation model

that is assumed to represent or approximate producer behaviour and is suitable to the simulation needs of the analysts.' (Heckelei and Britz, 2005, p. 62).

[3] Unlike the European FADN, data-gathering protocols in some individual countries (e.g. Italy and Belgium) allow for variable costs per activity.

[4] Projects on the impact assessment of CAP reforms financed by the EU Directorate-General for Research to date are: CAPRI, EUROTOOLS, GENEDEC, IDEMA, CARERA, SEAMLESS, FACEPA and FADNTOOL. See the CORDIS website, www.cordis.europa.eu, or individual project websites.

[5] Congress and Seminar papers as grey literature are available at http://ageconsearch.umn.edu/.

References

Arfini, F. (ed.) (2005) *Proceedings of the 89th European Seminar of the European Association of Agricultural Economics. Modelling Agricultural Policies: State of the Art and New Challenges, 3–5 February, 2005, MUP, Parma*. Monte Università Parma, Parma, Italy.

Arfini, F. and Donati, M. (2008) Health Check ed efficienza delle aziende agricole: una valutazione comparativa su quattro regioni agricole europee. *Rivista di Economia Agraria* 1, 65–92.

Arfini, F. and Donati, M. (2011) Impact of the Health Check on structural change and farm efficiency: a comparative assessment of three European agricultural regions. In: Moreddu, C. (ed.) Disaggregated Impacts of CAP Reforms: Proceedings of an OECD Workshop, 10–11 March 2010, Paris OECD Conference Centre. OECD Publishing, Paris, pp. 75–90.

Arfini, F. and Donati, M. (2013) Organic production and the capacity to respond to market signals and policies: an empirical analysis of a sample of FADN farms. *Agroecology and Sustainable Food Systems* 37, 149–171.

Bartova, L., Gil, J.M., M'Barerek, R. and Ratinger, T. (2008) *Modelling Agricultural and Rural Development Policies. Proceedings of 107th EAAE Seminar. Selected Papers, 29th January–1st February, 2008, Sevilla, Spain*. Office for Official Publications of the European Communities, Luxembourg.

Bauer, S. and Kasnakoglu, H. (1990) Nonlinear programming models for sector policy analysis. *Economic Modelling* 7, 275–290.

Buysse, J., Fernagut, B., Harmignie, O., Henry de Frahan, B., Lauwers, L., Polomé, P., Van Huylenbroeck, G. and Van Meensel, J. (2004) Modelling the impact of sugar reform on Belgian agriculture. *Selected paper prepared for presentation at the International Conference on Policy Modelling EcoMod2004, Paris, 30 June–2 July, 2004*. Available at: http://ecomod.net/sites/default/files/document-conference/ecomod2004/808.pdf (accessed 19 February 2016).

Donati, M., Arfini, F. and Paris, Q. (2013) Positive mathematical programming to estimate specific costs of production. In: Cesaro, L. and Marongiu, S. (eds) *The Use of Rica to Estimate the Cost of Production in Agriculture: Application of Econometric and Mathematical Programming Methodologies*. Collana Studi e Ricerche, Istituto Nazionale di Economia Agraria (INEA), Rome, pp. 89–109.

Gocht, A. (2005) Assessment of simulation behaviour of different mathematical programming approaches. In: Arfini, F. (ed.) *Proceedings of the 89th European Seminar of the European Association of Agricultural Economics. Modelling Agricultural Policies: State of the Art and New Challenges, 3–5 February, 2005, MUP, Parma*. Monte Università Parma, Parma, Italy, pp. 166–187.

Gocht, A. and Britz, W. (2011) EU-wide farm type supply models in CAPRI – how to consistently disaggregate sector models into farm type models. *Journal of Policy Modeling* 33, 146–167.

Gohin, A. and Chantreuil, F. (1999) La programmation mathématique positive dans les modèles d'exploitation agricole. Principes et importance du calibrage. *Cahiers d'Economie et de Sociologie Rurales* 52, 59–79.

Golan, A., Judge, G.G. and Miller, D. (1996) *Maximum Entropy Econometrics: Robust Estimation with Limited Data*. Wiley, New York.

Hazell, P.B.R. and Norton, R.D. (1986) *Mathematical Programing for Economic Analysis in Agriculture*. Macmillan, New York.

Heady, E.O. and Egbert, A.C. (1964) Regional programming of efficient agricultural production patterns. *Econometrica* 32, 374–386.

Heckelei, T. (2002) Calibration and estimation of programming models for agricultural supply analysis. Habilitation thesis, University of Bonn, Bonn, Germany.

Heckelei, T. and Britz, W. (2000) Positive mathematical programming with multiple data points: a cross-sectional estimation procedure. *Cahiers d'Economie et Sociologie Rurales* 57, 27–50.

Heckelei, T. and Britz, W. (2005) Models based on positive mathematical programming: state of the art and further extensions. In: Arfini, F. (ed.) *Proceedings of the 89th European Seminar of the European Association of Agricultural Economics. Modelling Agricultural Policies: State of the Art and New Challenges, 3–5 February, 2005, MUP, Parma*. Monte Università Parma, Parma, Italy, pp. 48–73.

Heckelei, T. and Wolff, H. (2003) Estimation of constrained optimisation models for agricultural supply analysis based on generalised maximum entropy. *European Review of Agricultural Economics* 30, 27–50.

Heckelei, T., Witzke, H.P. and Henrichsmeyer, W. (eds) (2001) *Proceedings of the 65th EAAE Seminar on Agricultural Sector Modelling and Policy Information Systems, Bonn (Germany), March 29–31, 2000.* Vauk Verlag, Kiel, Germany.

Heckelei, T., Mittelhammer, R. and Jansson, T. (2008) *A Bayesian Alternative to Generalized Cross Entropy Solutions for Underdetermined Econometric Models*. Agricultural and Resource Economics, Discussion Paper 2008:2, Institute for Food and Resource Economics, Bonn, Germany. Available at: http://www.ilr.uni-bonn.de/agpo/publ/dispap/download/dispap08_02.pdf (accessed 19 February 2016).

Henry de Frahan, B., Buysse, J., Polomé, P., Fernagut, B., Harmignie, O., Lauwers, L., Van Huylenbroeck, G. and Van Mensel, J. (2007) Positive mathematical programming for agricultural and environmental policy analysis: review and practice. In: Weintraub, A., Romero, C., Bjørndal, T. and Epstein, R. (eds) *Handbook of Operations Research in Natural Resources.* International Series in Operations Research and Management Science, Volume 99, Springer, New York, pp. 129–154.

Howitt, R.E. (1995) Positive mathematical programming. *American Journal of Agricultural Economics* 77, 329–342.

Howitt, R.E. and Mean, P. (1983) *A Positive Approach to Microeconomic Programming Models*. Working Paper 6, Department of Agricultural Economics, University of California, Davis, California.

Júdez, L., Chaya, C., Martínez, S. and González, A.A. (2001) Effects of the measures envisaged in Agenda 2000 on arable crop producers and beef and veal producers: an application of positive mathematical programming to representative farms of a Spanish region. *Agricultural Systems* 67, 121–138.

Kanellopoulos, A., Berentsen, P., Heckelei, T., van Ittersum, M. and Oude Lansink, A. (2010) Assessing the forecasting performance of a generic bio-economic farm model calibrated with two different PMP variants. *Journal of Agricultural Economics* 61, 274–294.

Langrell, S., Ciaian, P., Gomezy Paloma, S., Cunningham, D.L., Garnier, J.-F., Isermeyer, F. and Mishra, A.K. (2012) *Sustainability and Production Costs in the Global Farming Sector: Comparative Analysis and Methodologies*. JRS Scientific and Technical Reports, Report EUR 25436 EN, European Commission Joint Research Centre. Publications Office of the European Union, Luxembourg and Food and Agriculture Organization of the United Nations, Rome. Available at: http://ftp.jrc.es/EURdoc/JRC70116.pdf (accessed 19 February 2016).

Lau, L.J. (1978) Testing and imposing monotonicity, convexity and quasi-convexity constraints. In: Fuss, M. and McFadden, D. (eds) *Production Economics: A Dual Approach to Theory and Applications. Volume I: The Theory of Production*. North Holland, Amsterdam, pp. 409–453.

Lence, S.H. and Miller, D.J. (1998) Estimation of multi-output production functions with incomplete data: a generalised maximum entropy approach. *European Review of Agricultural Economics* 25, 188–209.

Léon, Y., Peeters, L., Quinqu, M. and Surry, Y. (1999) The use of maximum entropy to estimate input–output coefficients from regional farm accounting data. *Journal of Agricultural Economics* 50, 425–439.

McCarl, B.A. and Spreen, T.H. (2011) *Applied Mathematical Programming Using Algebraic Systems*. Texas A&M University, College Station, Texas.

Meister, A.D., Chen, C.C. and Heady, E. (1978) *Quadratic Programming Models Applied to Agricultural Policies*. Iowa State University Press, Ames, Iowa.

Paris, Q. (1993) *PQP, PMP, parametric programming, and comparative statics*. Chapter 11 in: *Notes for AE253, November 1993*. Department of Agricultural Economics, University of California, Davis, California.

Paris, Q. (2011) *Economic Foundations of Symmetric Programming*. Cambridge University Press, New York.

Paris, Q. (2013) *An Economic Interpretation of Linear Programming*. Iowa State University Press, Ames, Iowa.

Paris, Q. and Arfini, F. (1995) A positive mathematical programming model for regional analysis of agricultural policies. In Sotte, F. (ed.) *Proceedings of the 40th EAAE Seminar on the Regional Dimension in Agricultural Economics and Policies, 26–28th June 1995, Ancona*. Università di Ancona, Ancona, Italy.

Paris, Q. and Arfini, F. (2000) Funzioni di costo di frontiera, auto-selezione, rischio di prezzo, PMP e Agenda 2000. *Rivista di Economia Agraria* 55, 211–242.

Paris, Q. and Howitt, R. (1998) An analysis of ill-posed production problems using maximum entropy. *American Journal of Agricultural Economics* 80, 124–138.

Polomé, P., Fernagut, B., Harmignie, O. and Henry de Frahan, B. (2005) Multi-input multi-output farm-level cost function: a comparison of least squares and entropy estimators. *Contributed paper, 2005 EAAE [European Association of Agricultural Economists] International Congress, August 23–27, 2005, Copenhagen, Denmark*. Available at: http://ageconsearch.umn.edu/handle/24727 (accessed 19 February 2016)

Röhm, O. and Dabbert, S. (2003) Integrating agri-environmental programs into regional production models: an extension of positive mathematical programming. *American Journal of Agricultural Economics* 85), 254–265.

Solazzo, R., Donati, M., Arfini, F. and Petriccione, G. (2014) A PMP model for the impact assessment of the Common Agricultural Policy reform 2014–2020 on the Italian tomato sector. *New Medit* 2, 9–19.

Sorrentino, A., Henke, R. and Severini, S. (eds) (2011) *The Common Agricultural Policy After the Fisher Reform*. Ashgate, Burlington, Vermont.

3 Modelling Farm-level Adaptations Under External Shocks

Shailesh Shrestha* and Bouda Vosough Ahmadi
Land Economy, Environment and Society (LEES) Research Group, Scotland's Rural College (SRUC), Edinburgh, UK

3.1 Introduction

Farmers make decisions regularly on managing their resources to achieve the best possible outcome from their activities, either to improve profits or to fulfil other objectives such as improving environmental benefits. They consider a number of 'farm adjustment strategies' and select the best strategies to achieve their goals. These can be represented by a number of pathways taken by a farm over a period of time, as suggested by Bowler (1992). This decision making is more frequent if there is an expectation of certain external shocks in farming conditions such as changes in agricultural policies, market prices, new technologies, livestock/crop diseases or climate. Farmers often make rapid adjustments to their management practices in response to these changes. For example, dairy farmers may increase herd size and production if changes in future trade policies lead them to expect a higher milk price in the market; or they may try to the reduce cost of production to improve farm margins if they believe they will be receiving lesser payment support under a change in payment schemes. Such adjustments made by farmers as a result of certain changes in the farming environment can be termed short-term farm-level adaptation and are the subject of this chapter.

Examining farm-level adaptation is an important part of farm-level modelling, especially if a study is undertaken to analyse the impacts of external shocks on farms. Without considering farmers' responses in farm-level modelling studies, the impact of a shock on a farming system would be widely exaggerated. This is illustrated in Fig. 3.1, which shows that the impact on crop productivity under a climate change scenario is very different with and without the inclusion of adaptations made by farmers on the farms in the study. Similar kinds of work have been presented by many researchers to show the importance of farm adaptations in a farm-level economic analysis (Mendelsohn *et al.*, 1996; Schimmelpfennig *et al.*, 1996; Rosenzweig and Tubiello, 1997; Reilly and Schimmelpfennig, 1999; Tan and Shibasaki, 2003).

Adaptations can be categorized into long term adaptations and short-term adaptations. Long-term adaptation generally includes the changes that require higher investments and institutional involvement. These adaptations also take a longer time to implement and are expected to have an impact for a longer duration of time (>5 years). Short-term adaptations, in contrast, are relatively small adjustments to farm management that do not incur a great deal of additional costs. These are the changes that farmers generally

*E-mail: shailesh.shrestha@sruc.ac.uk

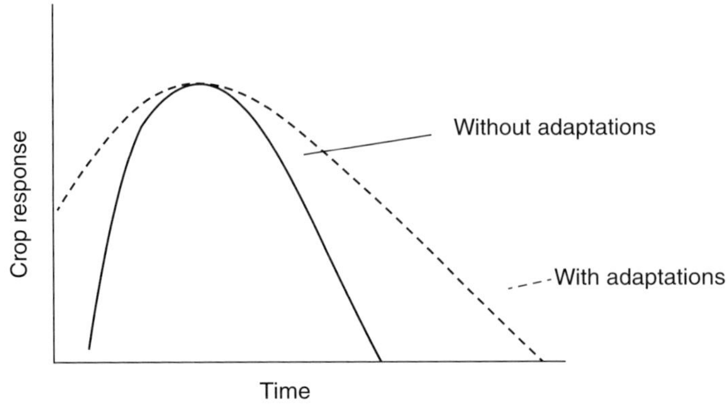

Fig. 3.1. Impact of climate change on crop productivity under a climate change scenario with and without adaptation by farmers. Adapted from Mendelsohn *et al*., 1996.

take in their day-to-day business and can be considered to be farm-level adaptations. In the past, both of these types of adaptations have been modelled in impact assessment studies. The main difference between them is that long-term adaptation would need more information, such as the cost of implementation, the lag period after implementation and information on the costs and benefits of the measures implemented in addition to detailed farm-level data (Bosello *et al.*, 2011; Mandryk *et al.*, 2012). Modelling short-term adaptations, however, does not need this additional information on the implementation of the adaptive measures. Instead, the model can use available farm-level data and examine adjustments within existing farm-management practices. As mentioned earlier, this chapter focuses only on such short-term farm-level adaptations.

3.2 Review of Existing Methodologies

There are a number of ways in which farm adaptations can be modelled. The most commonly used methods are statistical and mathematical models.

A range of studies used various statistical models to make decisions on different adaptations. These include the Ricardian method (Mendelsohn *et al.*, 1996; Van Passel *et al.*, 2012), which is a simple empirical cross-sectional analysis technique that has been used to estimate the relationships between the adaptation variables and a fixed variable such as land use; and multinomial logit models (Deressa *et al.*, 2009; Gbetibouo, 2009) and multivariate probit models (Nhemachena and Hassan, 2007), which have been used to estimate the probability of certain adaptation measures taking place under a studied condition. These models require preset information on adaptation measures, which means that the adaptation options are first identified and their impacts determined before their inclusion in the model. In many such studies, the adaptation measures are identified beforehand, based either on a survey (Chiotti *et al.*, 1997; Bryant *et al.*, 2000), or on past experience and experiments (Mendelsohn, 2000; Mortimore and Adams, 2001), or on ‘expert opinions’ (Easterling *et al.*, 1992, 1993). This is a useful approach for generating likely outcomes based on the relationships between a few variables and in comparing these outcomes to decide the best option for a farmer. Statistical models are built to use empirical data and aim to discover relationships between an adaptation variable and a dependent variable. However, they ignore the fact that a farm is a combination of multiple farming activities in which relationships between different farming activities are interlinked and have compounding effects upon one another. This means that farmers generally make decisions based on the compounding effects of adjusting different activities on the whole farming system, and not on the effect of the one-to-one basis of adjusting a single variable on a dependent variable.

The second method, mathematical modelling, is widely used to study farm-level adaptations. There are numerous examples of different types of mathematical models used to identify various aspects of farm adaptations. Examples of these modelling techniques are: decision models such as the agent-based model (ABM) and the decision support model (DSM); and optimization models such as linear programming (LP), non-linear programming (NLP) and mixed-integer programming models (MIP). The ABM and DSM are useful tools for examining adaptation measures on farms (Matthews *et al.*, 2007; Gallerani *et al.*, 2008; Guillem *et al.*, 2011, 2015; Vosough Ahmadi *et al.*, 2015a,b). There are a number of ways in which adaptation measures are explored in these models. The most frequently used method is to combine the models with a statistical approach to select the adaptation measures that are most appropriate under specific changes. This is useful for examining the impacts of adaptation that are already identified, but it can also be considered to be a limitation if the adaptation measures are not known or if additional measures need to be explored. In a manner similar to that employed in statistical approaches, the modeller needs to fix a set of adaptation measures before running ABM and DSM, and the models are restricted to use within those prefixed measures.

Optimization models (LP, NLP, MIP), though, are useful tools in identifying adaptation measures and hence very popular among agricultural economists (Berntsen *et al.*, 2003; Veysset *et al.*, 2005; Kerselaers *et al.*, 2007; Shrestha *et al.*, 2007, 2014, 2015; Gibbons and Ramsden, 2008; Louhichi *et al.*, 2010; Farquharson *et al.*, 2013). These models mimic farmers' decision making by adjusting farm management activities to optimize their objectives, for example, maximizing profits or minimizing production costs. Optimizing farm models have been used by various authors to analyse farm adaptations under different farming conditions, e.g. MIP (Dogliotti *et al.*, 2005; Gibbons and Ramsden, 2008), NLP (Louhichi *et al.*, 2010), static linear programming (Ramsden *et al.*, 1999; Berntsen *et al.*, 2003; Calker *et al.*, 2004) and dynamic linear programming (Beukes *et al.*, 2002; Acs *et al.*, 2007; Shrestha *et al.*, 2007). Most of the analyses of adaptation measures in agricultural production are focused on the impact assessment of policy changes, technological changes and climate changes, but optimization models have also been used to provide decision support for farmers and farm managers in tackling management problems, for example with respect to animal health and welfare. Some examples of livestock health management issues that have been explored using optimization models are: optimal replacement policies for dairy and beef cattle (Galligan *et al.*, 1993; Frasier and Pfeiffer, 1994; Heikkilä *et al.*, 2008), the optimization of profit and animal welfare (Stott *et al.*, 2012; Vosough Ahmadi *et al.*, 2015b), dairy cattle replacement in the case of failures in fertility and reproduction (Dijkhuizen *et al.*, 1985) and animal disease management (Stott *et al.*, 2003, 2005).

In this chapter, we describe the LP modelling technique, which is one of the most widely used optimization modelling techniques used to determine farm-level adaptations. The two examples provided in Section 3.5 near the end of the chapter illustrate the applicability of the technique to two different kinds of farm-level studies, namely, adaptation to climate change and adaptation to regulatory and consumer demand for higher animal welfare.

3.3 LP Modelling

A linear programming (LP) model optimizes farm activities to maximize or minimize an objective function (z), such as farm net margin or costs, within a set of limiting farm resources. A general form of the LP model is shown below:

$$\max\ z = \sum \left(p_{f,t} \times x_{f,t}\right) - \left(c_{f,t} \times x_{f,t}\right)$$

subject to

$$A_{f,t} \times x_{f,t} \leq R_{f,t}$$

$$x_{f,t} \geq 0$$

where: $x_{f,t}$ represents the farm activities for each farm type (f) and year (t); $p_{f,t}$ is a measure of returns and $c_{f,t}$ denotes the costs procured for $x_{f,t}$ activity; $A_{f,t}$ denotes an input–output coefficient for activity $x_{f,t}$; and $R_{f,t}$ is a limiting resource such as land or labour.

The LP model presented here is based on optimizing farming system analysis. A farming system analysis includes all existing farm activities on a study farm (Fig. 3.2). The optimizing nature of the technique provides flexibility in the model

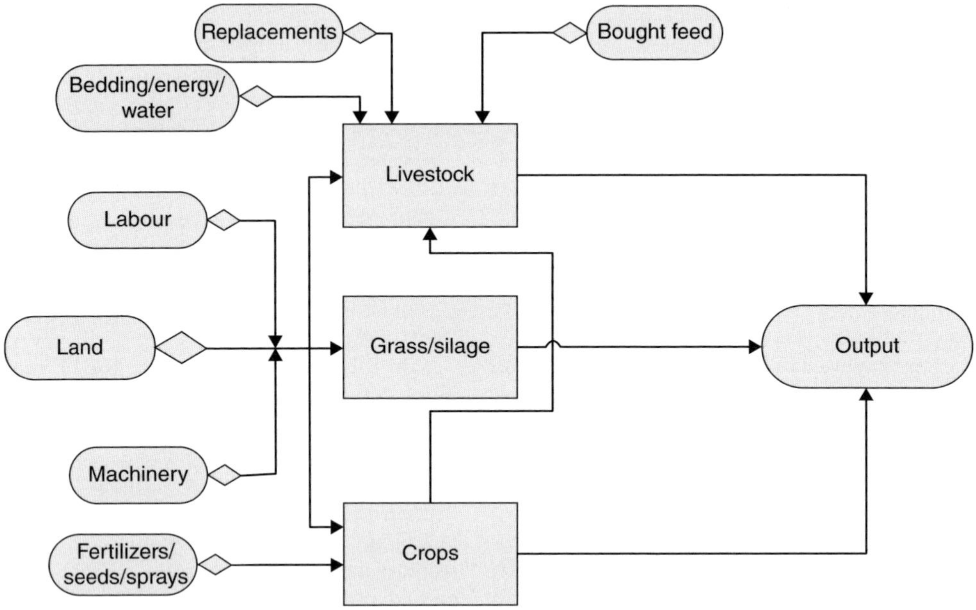

Fig. 3.2. A graphical representation of a farming system model.

to adjust management activities on a farm. All farm activities are replicated in the model by linking them together mathematically. Thus, the model selects the optimal pathway within the system by adjusting the number of activities as required to achieve farm objectives.

Based on the farming system (Fig. 3.2), mathematical relationships are formulated for different components to make decisions in the model. All of the components are constrained within a set of limiting farm resources. The model finds adjustment within these constraints based on the prices and revenues they generate, which ultimately enter into the objective function. The optimization process of the objective function therefore dictates the adjustments of activities within each of the components. A detailed description of some of these components is provided below.

3.3.1 Land

Land is one of a farm's limiting resources and is used by all farming activities. Land in our described example of an LP model is designated as arable land (ALand) and grassland (GLand). The model allows land to be transferred between grassland and arable land on farm (f) in year (t). There is also flexibility for farms to rent in (RentLand) or to let out (LetLand) farmland if required.

$$\text{Land}_{f,t} = \text{ALand}_{f,t} + \text{GLand}_{f,t} + \text{RentLand}_{f,t} - \text{LetLand}_{f,t}$$

The number of animals (Livestock) in each category/type of animal (a) is restricted to the grassland available through the stocking rate (STR), which is given as livestock units (LU) per hectare of land for each type of animal (a) on farm (f) in a year (t). The grassland is further divided into grazing land (GRAZLand) and silage land (SILLand).

$$\text{GLand}_{f,t} \times STR_f \geq \text{Livestock}_{f,a,t} \times \text{LU}_a$$

$$\text{GLand}_{f,t} = \text{GRAZLand}_{f,t} + \text{SILLand}_{f,t}$$

Arable land is the total area covered by all crops (c) grown on farm (f) in year (t). The farm is allowed to reallocate land under different crops on a yearly basis.

$$\text{ALand}_{f,t} = \sum crop_{f,c,t}$$

These equations provide the flexibility in land use by reallocating land under different farm components and adjusting corresponding farm activities accordingly.

3.3.2 Feed

A number of possible feed regimes can be included in the LP model based on availability and the requirement for feed on a farm. The feed included in the model can be farm-grown feed, such as grass, grass silage, feed crops (e.g. maize silage and cereal grains) and bought-in concentrate or other forms of feed. The feed constraint in the model is achieved by balancing the feed availability and feed requirement by each individual animal type (*a*) in each month (*m*) in each year (*t*). The feed requirement can either be determined by calculating the energy and protein requirements for each type of animal based on species, age and production level of that animal, or it can be taken from the literature.

$$\text{Livestock}_{f,a,t} \times \text{FEEDREQ}_{a,i,m} \le \sum \text{FeedAvail}_{f,t,m,b} \times \text{FEEDCONT}_{i,b}$$

where: Livestock is the number of animals of type (*a*) on farm (*f*) in year (*t*); FEEDREQ is the requirement (*i* = energy, protein) for each type of animal (*a*) in a month (*m*); FEEDAvail is the availability of feed types (*b*) for each farm (*f*) in a month (*m*) and in a year (*t*); and FEEDCONT is the availability of *i* (energy, protein) in each feed type (*b*).

The model provides flexibility in adjusting the rationing regime by changing to different feeds under the energy and protein requirement constraints in operation. This constraint is useful in identifying farm responses under certain changes as it can directly affect the production, conservation and the availability of grass and other feeds on the farm. The model can also use supplement or additive feeds as adaptive measures for the instant reduction of greenhouse gas (GHG) emissions on farms.

3.3.3 Labour

The total labour required (TOTLab) on a farm is determined by the labour required (LABREQ) for all the farm activities, such as the labour required for livestock (Livestock) or crop (crop) production. The labour requirement for each of the activities is based on either farm-level data or data from the literature:

$$\text{TOTLab}_{f,t} = \text{Livestock}_{f,a,t} \times \text{LABREQ}_{a} + crop_{f,c,t} \times \text{LABREQ}_{c}$$

where *f* is farm, *t* is year, *a* is type of animal and *c* is type of crop.

The total labour (TOTLab) required on a farm is balanced over the types of labour available to a farm, such as family labour (FAMLab) and paid or hired labour (PAIDLab). There is no price associated with family labour so the model will first select family labour and then allocate paid labour only if the family labour is not sufficient:

$$\text{TOTLab}_{f,t} = \text{FAMLab}_{f} + \text{PAIDLab}_{f,t}$$

$$\text{LABCost}_{f,t} = \text{PAIDLab}_{f,t} \times \text{LABPRICE}$$

The cost of paid labour (LABCost) is determined by the labour cost (rate per hour) and deducted from the objective function of the model. The model is thus structured to adopt a suitable type of labour based on activities as well as on the costs of labour.

3.3.4 Herd size

The size of the herd is also one of the main adjustments that a farmer can make under changes in management and in response to policies and market structure or to environmental protection concerns. The herd size can be increased if the market and relevant policies are suitable for doing so; if they are not, the herd size can be decreased by selling and/or culling the herd. In addition, farmers can choose replacement animals from their own replacement stocks or by buying from the market. Replacements for dairy cattle, breeding beef cattle, sows and sheep on the farm can be based on preset replacement rates. For example, lactating dairy animals have a minimum 25% replacement rate to represent a 4 year dairy cycle, which is normal practice in dairy production. If herd size needs increasing, animals are first replaced by on-farm replacement stocks and then by off-farm replacement stocks.

$$\text{Livestock}_{f,a,t} = \text{Livestock}_{f,a,t-1} + \text{REPStock}_{f,t} - \text{CULL}_{f,t}$$

$$\text{CULL}_{f,t} = \text{Livestock}_{f,a,t} \times 0.25$$

$$\text{REPStock}_{f,t} \geq \text{CULL}_{f,t}$$

where: Livestock is the number of animals of type (a) on farm (f) in year (t) or year $t - 1$; REPStock is the off-farm replacement stock for each farm (f) and year (t); and CULL is the number of animals replaced for each farm (f) and year (t).

3.3.5 Exogenous adaptation measures

Besides adjustments within the existing management practices, additional farm-level adaptive measures that are new to a farm can also be included in the model. These exogenous measures can be added under different constraints or included within the existing constraints. An important point to remember is to structure the exogenous measure similarly to the other activities in the model, such as the costs, labour requirement and revenues associated with the measure. These need to be incorporated into the model in the same format as the activities that are already included. In the equation provided below, for example, pig numbers on the farm can be increased beyond the existing farm capabilities only if it is profitable to incur a starting investment cost of additional building and variable costs for each of the additional pigs. The investment cost of the building and additional variable costs are added to the input costs of pig production in the model:

$$\text{Livestock}_{f,pig,t} = \text{Livestock}_{f,pig,t-1} + \text{REPStock}_{f,t} + \text{Livestock\_}add_{f,pig,t-1} - \text{CULL}_{f,t}$$

where: $\text{Livestock}_{f,pig,t}$ = total pig number on farm (f) in year (t); $\text{Livestock}_{f,pig,t-1}$ = total pig number on farm (f) in year $t - 1$; $\text{REPStock}_{f,t}$ = replacement of pig stock on farm (f) in year (t); $\text{Livestock\_add}_{f,pig,t-1}$ = additional pig numbers beyond farm's existing capacity on farm (f) at in year ($t-1$) and $\text{CULL}_{f,t}$ = culled pig on farm (f) in year (t).

The total pig number on the farm, $\text{Livestock}_{f,pig,t}$, is restricted by the farm capacity to accommodate a fixed number of pigs. This also depends on the criteria of animal welfare that the farm wishes to include in its farming objectives. Any additional animal beyond that fixed number is categorized under $\text{Livestock\_add}_{f,pig,t-1}$. The model can add on $\text{Livestock\_add}_{f,pig,t-1}$ only if pig revenues are greater than production costs plus the investment cost of extra building, and additional pig numbers still fulfil the welfare criteria set by the farmer. For greater revenue, the following equation needs to be positive:

$$\text{PigGm}_f = \text{sellani}_{f,pig} \times \text{anirev}_{f,pig} - \text{totani}_{f,pig} \times \text{LIV\_VAR}_{f,pig} - \text{feed}_{f,pig,feed} \times \text{FEED\_PRICE}_f - \text{BUILDCOST}_f$$

where: PigGM_f = pig gross margin for farm (f); $\text{sellani}_{f,pig}$ = number of pigs sold on farm (f); $\text{anirev}_{f,pig}$ = revenues per pig on farm (f); $\text{LIV\_VAR}_{f,pig}$ = variable costs per pig; $\text{feed}_{f,pig,feed}$ = total feed used; FEED_PRICE_f = price of feed; and BUILDCOST_f = investment cost for additional building on farm (f).

If the above equation is negative then the model will not select $\text{Livestock\_add}_{f,pig,t-1}$ and will remove BUILDCOST from the equation.

Later on in this chapter, an example is provided in Section 3.5 (Application) which explores the use of exogenous adaptation measures to optimize the profitability of breeding sow pig farms in the UK.

3.3.6 Modelling runs

In the modelling process, as shown in Fig. 3.3, the LP model is first run under a baseline scenario where all farm activities are maintained as on the study farm. The model is then run under a shock scenario in which changes in farm components and parameters due to external shocks (such as changes in payments, market prices

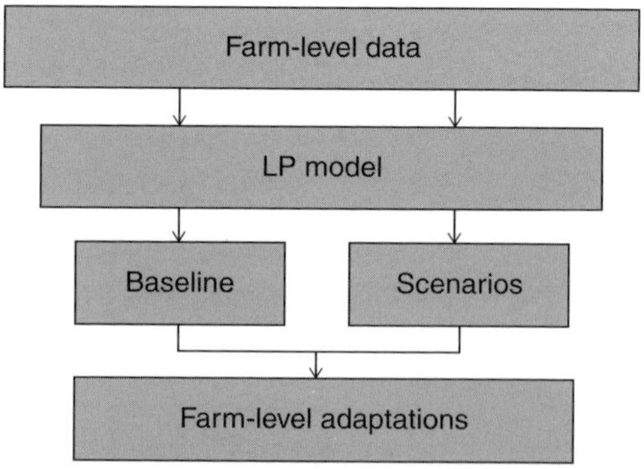

Fig. 3.3. Schematic diagram of the linear programming (LP) modelling process to identify farm adaptations.

and crop yields) are incorporated into the model. A comparison between the farm resources used by the model under the shock scenario and the baseline scenario will identify changes that have occurred in the management, activities or resources of the farm under the shock conditions. These changes are considered to be an adaptation for that particular farm to optimize its resources.

3.3.7 Input data

The model requires farm-level data that include physical as well as financial information. It can be run on individual farms or on a representative farm type, the choice being based upon the nature of the study and the research objectives. Farm survey data acquired from services such as the Farm Accountancy Data Network (FADN) for European farms, the National Agricultural Statistics Service (NASS) for US farms and the Australian Bureau of Agricultural and Resources Economics and Sciences (ABARES) surveys for Australian farms fit this type of modelling work very well. At the national level, these data vary in terms of the depth and details of the information collected. The quality of analysis depends on the availability of accurate data on indicators and parameters such as physical quantities of feed requirements or market prices of individual commodities. Information that is not available from the survey data set can be based on the literature or on expert opinion.

3.4 Limitations

The LP modelling technique has some limitations. The first is the data requirement for this kind of study. The model requires very detailed farm-level data for a study to be as close to reality as possible. The more detailed the data that are available, the more robust the model results will be. In cases where data from FADN-type surveys are not available, a farm survey needs to be carried out to provide enough information for the model, which could prove both time-consuming and costly.

Another limitation is the fact that the model is based on the optimizing technique and does not consider any other factors that do not contribute to optimizing farm activities. In other words, traditional LP optimization does not adequately capture the full behavioural drivers and motivations behind decision making that diverges from full optimization of the system. The consequences could be failure to select the adaptive measures that are not linked to the optimizing process of the model. For instance, in the case of identifying farm adaptations under the decoupling of Common Agricultural Policy (CAP) payments, an LP model (Teagasc_Mod), showed that the adaptive measure that would maximize farm profits was to destock beef animals from the farm (Shrestha *et al.*, 2007). However, in reality, these farms kept beef animals on the farm – the farmers decided to remain within beef farming even if they were making a loss compared with other activities that they could adopt, simply because that was what they had been doing in the past.

In many cases, LP models also assume that farms are operating at the maximum efficiency level, which is not always possible for a real farm. This is a difficult assumption to apply to medium and small-scale farms that have neither the capital nor the resources to achieve optimal efficiency levels. The model for these farms may select the optimal adaptive measures, but these might not be practical on the farm.

3.5 Application

Two examples are provided in this section to show the application of optimizing farm-level models to identifying adaptation measures on the farm. These examples focus on adaptation under: (i) climate change; and (ii) regulatory and consumer demand to improve the welfare of breeding sows. The first example demonstrates the use of endogenous variables within the model that had been adjusted to farm conditions under future climate change; the second example describes the effects of exogenous factors and possible trade-offs between profit and animal welfare.

3.5.1 Irish farms under climate change

This work examined the effect of a regional variation of the impact of climate change on Irish agricultural farms and determined the optimal farm-level adaptation in response to this climate change (Shrestha *et al.*, 2014, 2015). The study used an optimizing farm-level LP programming dynamic model that included all of the existing farm activities and allowed all possible interactions between those activities on the farm to maximize farm profits. Any adjustments to these activities and relationships under the climate change scenario are considered to be farmers' adaptive measures in response to climate change.

The methodology conceived for this study is represented in Fig. 3.4. The study was divided into two phases. The first phase determined the effects of future climate on yields of crops and grass, and on soil trafficability (not shown in Fig. 3.4) in Irish regions using a number of biophysical and simulation models: the Crop Environment REsource Synthesis model, CERES (Ritchie *et al.*, 1998); the Johnstown Grass Model, JGM (Brereton *et al.*, 1996); and the Johnstown Soil Moisture Deficit simulator, JSMD (Schulte *et al.*, 2005). The second phase of the study then used these model outputs in a farm-level LP model (FLLP) to identify possible farm-level adaptations made to maximize farm profits. Family farm income provided a measure of profitability for each farm group and hence was used as an indicator to determine the effects of climate change in this study.

A list of possible adaptation measures used in the model is presented in Table 3.1. This provides an idea of how the adaptation measures are identified. As shown in the table, a possible adaptive measure can be anything that is adjusted in the model in response to climate change.

The study showed that there was a regional variation in the impact of climate change on Irish farms and that they adapted differently under climate change based on the type and location of the farm. Farms in the northern regions showed some negative impact on their farm profits under the climate change scenario. Farms in other regions, however, benefited financially under climate change. The model suggested that farms in all regions tend to adopt a more grass-based feeding system to exploit the increase in grass yield under the climate change scenario and minimize the cost of buying concentrate feed. These farms were projected to decrease the use of concentrate feed on farms to a very varying degree – from less than 2% on dairy farms in the border region to 100% replacement of concentrate feed by grass feed in beef farms in the mid-eastern region. There was some adjustment to land use observed in the mid-eastern region,

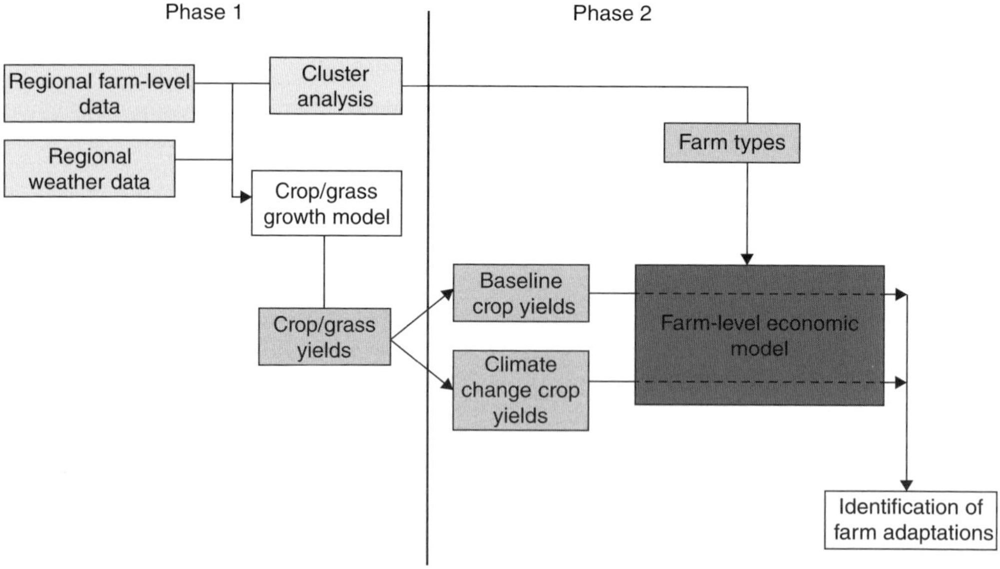

Fig. 3.4. Schematic diagram of the methodology used to model regional variation of the impact of climate change on Irish agricultural farms and farm-level adaptations to this. Adapted from Shrestha *et al.*, 2015.

Table 3.1. A selection of possible adaptation measures used in a farm-level linear programming model (FLLP) model.

Adaptation measures	Variable in the model	Description
Land use	Endogenous	Land use is changed by reallocating land between activities
Production system	Endogenous	The model can select between crop, dairy, beef, sheep or no production system
Number of animals	Endogenous	The model optimizes the animal numbers on the farm
Feeding system	Endogenous	The model chooses the most cost-effective feeding system
Labour	Endogenous	The model optimizes family and paid labour

where tillage farms replaced the arable land to produce more grass and grass silage, and moved from an arable to a livestock production system. For these farms, arable farming was not profitable any more because of a decrease in crop yields under climate change.

The model also selected herd size on farms as one of the major adaptive measures under climate change. Dairy farms in the south-east region reduced herd size by 2%. A higher cost of production and a restriction on the use of concentrate to maintain milk production limited the benefit of increase in grass production under climate change for these farms. In contrast, beef farms in other regions increased their herd sizes by up to 20%. The costs of production on these farms are comparatively lower than on other farms and their farm margins could be further improved by moving from concentrate feed to grass-based feed and increasing herd size.

Farms in all regions also opted for a one-cut grass silage production system to minimize production costs. As the quality of grass silage was assumed to be same in all types of grass conservation methods, the results suggested that the higher production costs for two-cut silage systems outweigh the benefits of the increase in grass silage compared with a one-cut silage system. It has been suggested that cost-saving strategies such as lowering production costs would be preferred by farmers in the future (Ramsden *et al.*, 1999; Rötter and Van de Geijn, 1999).

3.5.2 The case of alternative farrowing housing for sows in the UK

The development of an alternative and economical farrowing system that provides high sow and piglet welfare is regarded as beneficial for the pig industry in response to consumer demand for higher animal welfare (exogenous to farming). Although many alternative housing systems have been developed in different countries, no commercially viable indoor option has emerged for large-scale units. In this work, a farm-level LP model was developed and used to examine possible trade-offs between profits and animal welfare by comparing alternative farrowing housing systems under a scenario of higher consumer demand for improved animal welfare (Vosough Ahmadi *et al.*, 2011).

Three systems were studied. These included: conventional crates that restrict the sow within a metal frame; simple pens that provide more space for the sow and a separate creep area for piglets; and 'designed' pens, i.e. pens that had been modified and enhanced with pen 'furniture', such as sloped walls to protect the piglets from crushing by the sow (Baxter *et al.*, 2011; Vosough Ahmadi *et al.*, 2011). Switching to alternative systems may impose costs on farmers and alter the production functions. The objective of the LP was therefore to optimize the economic performance of the housing systems subject to both managerial and animal welfare constraints. Quantitative values for the constraints were derived from the literature. The potential effects of each welfare component on productivity were assessed by a group of animal welfare scientists and used in the model.

The welfare components (WC) modelled were: space, substrate (litter/bedding) and temperature. Baseline WC constraints and coefficients characteristic of each system were included in the LP model. Extra labour, extra space, increased electricity usage and substrate were introduced at some additional expense to allow the WC constraints to be relaxed (within the limits of a given system), thus improving welfare, and possibly productivity, by reducing piglet mortality. The LP model would then enhance welfare within any system provided that the extra WC raised net profit. Three production functions determining the relationships between inputs (WC) and outputs were provided by a group of five animal welfare scientists, and polynomial curves were fitted to the assessed production functions. In parallel to the optimization model, welfare scores for each system were assessed by a group of welfare scientists. Figure 3.5 illustrates the flow of information and the relationships used in the model.

The results showed that when the piglet survival rates were used in the LP model, the crates gave the highest annual net profit and the designed pens and simple pens were the less favoured options, in that order (Fig. 3.6). However, the annual net profit from the newly designed pens that provided higher welfare, and also from the simple pens, were improved once alternative reference points that followed expert-derived production functions were used to adjust piglet survival rates in response to the extra space, extra substrate and modified pen heating that were provided. The non-crate systems then provided higher welfare and higher net profit for both sows and piglets than the crates, implying the possibility of a win–win situation. The LP model in this example was used to assess the expected profitability of non-crate housing systems as a new technology used as an adaptation measure by farmers in response to exogenous regulatory changes and consumer demand.

3.6 Summary

The chapter has described the use of the LP technique to model farm-level adaptations under external shocks. The technique replicates farmers' responses under a changed farming environment to optimize farm objectives. The chapter focuses on short-term farmers' adaptations only, as these are the adaptations that farmers implement promptly without the need for capital investment under an expectation of change(s). The technique is widely used in economic impact assessment studies. Limitations of the technique are the general assumption that all farmers are profit oriented and also the requirement for large amounts of farm-level data. Two diverse research examples presented in the chapter show the strength and flexibility of the technique for modelling farm-level adaptations.

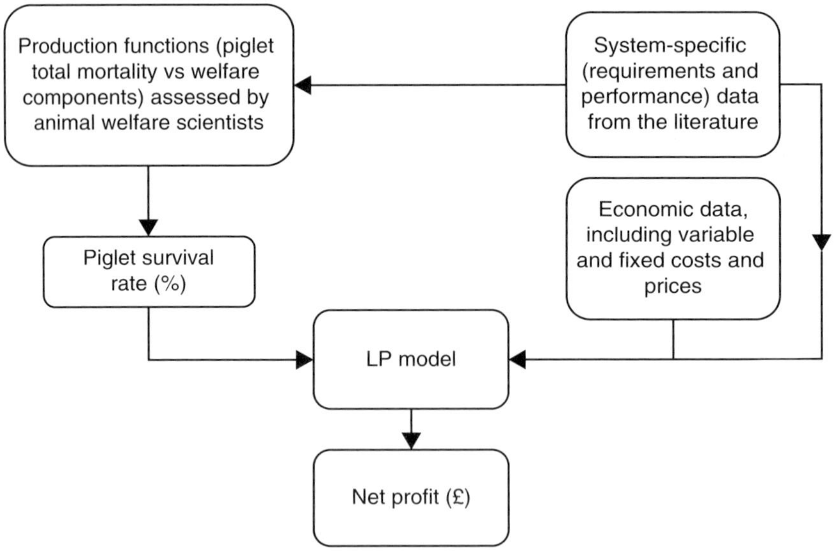

Fig. 3.5. A schematic diagram of the input/output relationships in a linear programming (LP) model used to optimize alternative farrowing housing for sow/piglet welfare in relation to production. Adapted from Vosough Ahmadi *et al.*, 2011.

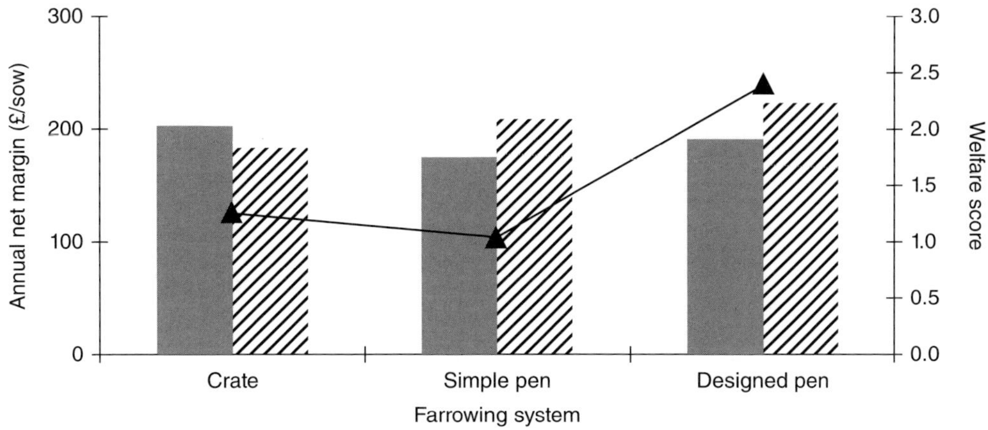

Fig. 3.6. Annual net margins as £/sow (bars) and welfare scores (the line) for three sow farrowing systems (a crate and two pens) estimated for two runs of the linear programming (LP) model developed to optimize them: RUN1 (grey bars) and RUN2 (dashed bars). The costs of the extra labour, substrate, space and power inputs for improved welfare were used in the LP model for both runs. Adapted from Vosough Ahmadi *et al.*, 2011.

References

Acs, S., Berentsen, P.B.M. and Huirne, R.B.M. (2007) Conversion to organic arable farming in the Netherlands: a dynamic linear programming analysis. *Agricultural Systems* 94, 405–415.

Baxter, E.M., Lawrence, A.B. and Edwards, S.A. (2011) Alternative farrowing systems: design criteria for farrowing systems based on the biological needs of sows and piglets. *Animal* 5, 580–600.

Berntsen, J., Petersen, B.M., Jacobsen, B.H., Olesen, J.E. and Hutchings, N.J. (2003) Evaluating nitrogen taxation scenarios using the dynamic whole farm simulation model FASSET. *Agricultural Systems* 76, 817–839.

Beukes, P., Cowling, R.M. and Higgins, S.I. (2002) An ecological economic simulation model of a non-selective grazing system in the Nama Karoo. *Ecological Economics* 42, 221–242.

Bosello, F., Nicholls, R.J., Richards, J., Roson, R. and Tol, R.S.J. (2011) Economic impacts of climate change in Europe: sea-level rise. *Climatic Change* 112, 63–81.

Bowler, I. (1992) 'Sustainable agriculture' as an alternative path of farm business development. In: Bowler, I.R., Bryant, C.R. and Nellis, M.D. (eds) *Contemporary Rural Systems in Transition – Volume* 1 *Agriculture and Environment*. CAB International, Wallingford, UK, pp. 237–253.

Brereton, A.J., Danielov, S.A. and Scott, D. (1996) *Agrometeorology of Grass and Grasslands for Middle Latitudes*. Technical Note No. 107, World Meteorological Organization, Geneva, Switzerland.

Bryant, C.R., Smit, B., Brklacich, M., Johnston, T.R., Smithers, J., Chiotti, Q. and Singh, B. (2000) Adaptation in Canadian agriculture to climatic variability and change. *Climatic Change* 45, 181–201.

Calker, K.J.V., Berentsen, P.B.M., de Boer, I.M.J., Giesen, G.W.J. and Huirne, R.B.M. (2004) An LP model to analyse the economic and ecological sustainability on Dutch dairy farms: model presentation and application for experimental farm. *Agricultural Systems* 82, 139–160.

Chiotti, Q., Johnston, T.R.R., Smit, B. and Ebel, B. (1997) Agricultural response to climate change: a preliminary investigation of farm-level adaptation in southern Alberta. In: Ilbery, B., Chiotti, Q. and Rickard, T. (eds) *Agricultural Restructuring and Sustainability: A Geographical Perspective*. CAB International, Wallingford, UK, pp. 167–183.

Deressa, T.T., Hassan, R.M., Ringler, C., Alemu, T. and Yesuf, M. (2009) Determinants of farmers' choice of adaptation methods to climate change in the Nile Basin of Ethiopia. *Global Environmental Change* 19, 248–255.

Dijkhuizen, A.A., Renkema, J.A. and Stelwagen, J. (1985) Economic aspects of reproductive failure in dairy cattle. II. The decision to replace animals. *Preventive Veterinary Medicine* 3, 265–276.

Dogliotti, S., van Ittersum, M.K. and Rossing, W.A.H. (2005) A method for exploring sustainable development options at farm scale: a case for vegetable farms in south Uruguay. *Agricultural Systems* 86, 29–51.

Easterling, W.E., Rosenberg, N.J., Lemon, K.M. and McKenny, M.S. (1992) Simulations of crop responses to climate change: effects with present technology and currently available adjustments (the 'smart farmer' scenario). *Agricultural and Forestry Meteorology* 59, 75–102.

Easterling, W.E., Crosson, P.R., Rosenberg, N.J., McKenney, M.S., Katz, L.A. and Lemon, K.M. (1993) Agricultural impacts of and responses to climate change in the Missouri–Iowa–Nebraska–Kansas (MINK) region. *Climatic Change* 24, 23–61.

Farquharson, R., Abadi, A., Finlayson, J., Ramilan, T., Liu, D.L., Anwar, M. and Clark, S. (2013) Modelling to evaluate agricultural adaptation to climate change in southern Australia. In: Piantadosi, J., Anderssen, R.S. and Boland J. (eds) MODSIM2013, *20th International Congress on Modelling and Simulation, Adelaide, Australia, 1–6 December 2013.* Modelling and Simulation Society of Australia and New Zealand Inc. (MSSANZ). Available at: http://www.mssanz.org.au/modsim2013/L2/farquharson.pdf (accessed 22 February 2016).

Frasier, W.M. and George H. Pfeiffer, G.H. (1994) Optimal replacement and management policies for beef cows. *American Journal of Agricultural Economics* 76, 847–858.

Gallerani, V., Gomez y Paloma, S., Raggi, M. and Viaggi, D. (2008) Modelling the effect of EU policy reforms on farm investment behaviour. *Paper prepared for poster presentation at the 107th EAAE Seminar "Modelling of Agricultural and Rural Development Policies", Sevilla, Spain, January 29th–February 1st, 2008*. Available at: http://ageconsearch.umn.edu/bitstream/6444/2/pp08ga19.pdf (accessed 23 February 2016).

Galligan, D.T., Ramberg, C., Curtis, C. and Johnstone, C. (1993) Adjustment for the competitive risk of culling in financial evaluation of health programs. *Preventive Veterinary Medicine* 16, 15–20.

Gbetibouo, G.A. (2009) *Understanding Farmers' Perceptions and Adaptations to Climate Change and Variability: The Case of the Limpopo Basin, South Africa*. IFPRI Discussion Paper 00849, International Food Policy Research Institute, Washington, DC.

Gibbons, J.M. and Ramsden, S.J. (2008) Integrated modeling of farm adaptation to climate change in East Anglia, UK: scaling and farm decision making. *Agriculture, Ecosystems and Environment* 127, 126–134.

Guillem, E.E., Murray-Rust, D., Robinson, D.T., Barnes, A.P. and Rounsevell, M.D.A. (2011) The effects of farmers perceptions and objectives on land use change and ecosystem services: the case of a Scottish arable catchment. In: Young, C. (ed.) *Landscape Ecology and Ecosystem Services. Proceedings of the eighteenth annual ialeUK conference, held at The University of Wolverhampton, Telford Campus, 6th–8th September 2011*. ialeUK (International Association for Landscape Ecology) Book Sales, Department of Geography, The University of Reading, Whiteknights, Reading, UK. Available at: https://www.researchgate.net/publication/260420783_The_effects_of_farmers_perceptions_and_objectives_on_land_use_change_and_ecosystem_services_the_case_of_a_Scottish_arable_catchment (accessed 22 February 2016).

Guillem, E.E., Murray-Rust, D., Robinson, D.T., Barnes, A. and Rounsevell, M.D.A. (2015) Modelling farmer decision-making to anticipate tradeoffs between provisioning ecosystem services and biodiversity. *Agricultural Systems* 137, 12–23.

Heikkilä, A.M., Nousiainen, J.I. and Jauhiainen, L. (2008) Optimal replacement policy and economic value of dairy cows with diverse health status and production capacity. *Journal of Dairy Science* 91, 2342–2352.

Kerselaers, E., De Cock, L., Lauwers, L. and Van Huylenbroeck, G. (2007) Modelling farm-level economic potential for conversion to organic farming. *Agricultural Systems* 94, 671–682.

Louhichi, K., Kanellopoulos, A., Janssen, S., Flichman, G., Blanco, M., Hengsdijk, H., Heckelei, T., Berentsen, P., Oude Lansink, A. and van Ittersum, M. (2010) FSSIM, a bio-economic farm model for simulating the response of EU farming systems to agricultural and environmental policies. *Agricultural Systems* 103, 585–597.

Mandryk, M., Reidsma, P. and van Ittersum, M.K. (2012) Scenarios of long-term farm structural change for application in climate change impact assessment. *Landscape Ecology* 27, 509–527.

Matthews, R.B., Gilbert, N.G., Roach, A., Polhill, J.G. and Gotts, N.M. (2007) Agent-based land-use models: a review of applications. *Landscape Ecology* 22, 1447–1459.

Mendelsohn, R. (2000) Efficient adaptations to climate change. *Climatic Change* 45, 583–600.

Mendelsohn, R., Nordhaus, W. and Shaw, P. (1996) Climate impacts on aggregate farm value: accounting for adaptation. *Agricultural and Forest Meteorology* 80, 55–66.

Mortimore, M.J. and Adams, W.M. (2001) Farmer adaptation, change and 'crisis' in the Sahel. *Global Environmental Change* 11, 49–57.

Nhemachena, C. and Hassan, R. (2007) *Micro-level Analysis of Farmers' Adaptation to Climate Change in Southern Africa*. Centre for Environmental Economics and Policy in Africa (CEEPA), IFPRI Discussion Paper 00714, International Food Policy Research Institute, Washington, DC.

Ramsden, S., Gibbons, J. and Wilson, P. (1999) Impacts of changing relative prices on farms level dairy production in the UK. *Agricultural Systems* 62, 201–215.

Reilly, J.M. and Schimmelpfennig, D. (1999) Agricultural impact assessment, vulnerability, and the scope for adaptation. *Climatic Change* 43, 745–788.

Ritchie, J.T., Singh, U., Godwin, D. and Bowen, W.T. (1998) Cereal growth, development and yield. In: Tsuji, G.Y., Hoogenboom, G. and Thornton, P.K. (eds) *Understanding Options for Agricultural Production*. Kluwer, Dordrecht, The Netherlands.

Rosenzweig, C. and Tubiello, F.N. (1997) Impacts of global climate change on Mediterranean agriculture: current methodologies and future directions. An introductory essay. *Mitigation and Adaptation Strategies for Global Change* 1, 219–232.

Rötter, R. and van de Geijn, S.C. (1999) Climate change effects on plant growth, crop yield and livestock. *Climatic Change* 43, 651–681.

Schimmelpfennig, D., Lewandrowski, J., Reilly, J., Tsigas, M. and Parry, I. (1996) Agricultural adaptation to climate change. Agricultural Economic Report 740, Economic Research Service, US Department of Agriculture, Washington, DC. Available at: http://www.ers.usda.gov/media/490977/aer740a_1_.pdf (accessed 22 February 2016).

Schulte, R.P.O., Diamond, J., Finkele, K., Holden, N.M. and Brereton, A.J. (2005) Predicting the soil moisture conditions on Irish grasslands. *Irish Journal of Agricultural and Food Research* 44, 95–110.

Shrestha, S., Hennessy, T. and Hynes, S. (2007) The effect of decoupling on farming in Ireland: a regional analysis. *Irish Journal of Agricultural and Food Research* 46, 1–13.

Shrestha, S., Hennessy, T., Abdalla, M., Forristal, D. and Jones, M. (2014) Determining responses of Irish dairy farms under climate change. *German Journal of Agricultural Economics* 63, 143–155.

Shrestha, S., Abdalla, M., Hennessy, T., Forristal, D. and Jones, M. (2015) Irish farms under climate change – is there a regional variation on farm responses? *Journal of Agricultural Science* 153, 385–398.

Stott, A.W., Lloyd, J., Humphry, R.W. and Gunn, G.J. (2003) A linear programming approach to estimate the economic impact of bovine viral diarrhoea (BVD) at the whole-farm level in Scotland. *Preventive Veterinary Medicine* 59, 51–66.

Stott, A.W., Jones, G.M., Humphry, R.W. and Gunn, G.J. (2005) The financial incentive to control paratuberculosis (Johne's disease) on UK dairy farms. *Veterinary Record* 156, 825–831.

Stott, A.W., Vosough Ahmadi, B., Dwyer, C.M., Kupiec, B., Morgan-Davies, C., Milne, C.E., Ringrose, S., Goddard, P., Phillips, K. and Waterhouse, A. (2012) Interactions between profit and welfare on extensive sheep farms. *Animal Welfare* 21(Supplement 1), 57–64.

Tan, G. and Shibasaki, R. (2003) Global estimation of crop productivity and the impacts of global warming by GIS and EPIC integration. *Ecological Modelling* 168, 357–370.

Van Passel, S., Massetti, E. and Mendelsohn, R. (2012) *A Ricardian Analysis of the Impact of Climate Change on European Agriculture*. Nota di Lavoro [Working Paper] 83.2012, Fondazione Eni Enrico Mattei (FEEM), Milan, Italy. Available at: http://www.feem.it/userfiles/attach/201464173554NDL2012-083.pdf (accessed 22 February 2016).

Veysset, P., Bebin, D. and Lherm, M. (2005) Adaptation to Agenda 2000 (CAP reform) and optimisation of the farming systems of French suckler cattle farms in the Charolais area: a model-based study. *Agricultural Systems* 83, 179–202.

Vosough Ahmadi, B., Stott, A.W., Baxter, E.M., Lawrence, A.B. and Edwards, S.A. (2011) Animal welfare and economic optimisation of farrowing systems. *Animal Welfare* 20, 57–67.

Vosough Ahmadi, B., Moran, D., Barnes, A.P. and Baret, P.V. (2015a) Comparing decision-support systems in adopting sustainable intensification criteria. *Frontiers in Genetics* 6 (23). Available at: http://journal.frontiersin.org/article/10.3389/fgene.2015.00023/full (accessed 22 February 2016).

Vosough Ahmadi, B., Shrestha, S., Thomson, S.G., Barnes, A.P. and Stott, A.W. (2015b) Impacts of greening measures and flat rate regional payments of the Common Agricultural Policy on Scottish beef and sheep farms. *The Journal of Agricultural Science* 153, 676–688.

4 Farm-level Modelling, Risk and Uncertainty

Stephen Ramsden* **and Paul Wilson**
School of Biosciences, University of Nottingham, Loughborough, UK

4.1 Introduction

Farmers and others who manage our land resources face a degree of uncertainty as to what the outcomes of their decisions will be. We should not overstate this relative to other areas of human activity – farming outputs usually have some form of market in developed economies – but, as is well documented, variable supply and price inelastic demand combine to generate high levels of variability in product sales, particularly for agricultural field crops. Even in the livestock sector, where, depending on location and type of system, supply is less prone to weather-induced variability, prices vary considerably over time as a result of production cycles, demand shocks (for example, 'food scares' such as the bovine spongiform encephalopathy, BSE), currency movements, policy changes and indirect weather effects on the supply of animal feedstuffs.

Agricultural decision makers therefore face risk and, if, as is often assumed, farmers are risk averse, they will want to take action. This may be action to reduce the variability associated with a given risk (for example, using insurance); or to reduce the consequences of a given level of variability by building up resilience (for example, a cash reserve), or some combination of these two approaches. A large amount of research has focused on the first option, particularly on the potential for using market instruments such as futures trading to reduce exposure to price variability. This work would typically consider the 'optimal hedge' – the proportion of the output of a single farm-produced commodity that should be protected, through the use of futures trading, from falling prices (Tomek and Peterson, 2001). In contrast, farm-level models, by design, are concerned with the whole farm system. If we think of a farm as having a mix of enterprises (different crops and animals and other productive outputs), it is a short step to thinking of these enterprises as being the farmer's 'portfolio', which can be manipulated to achieve both profitability and risk management objectives. Starting with portfolio analyses that draw on Markowitz's 1952 paper in *The Journal of Finance*, a wide range of farm models have been used to capture elements of both the farm system and the variability of farm outputs. In principle, variability can be extended to other parts of the farm: resource use–output relationships (for example, uncertainty in response to fertilizer use), or constraint variability (for example, weather-induced time availability for field operations), though in practice, models have tended to focus on output variability.

In many developed economies, risk management, either explicitly or implicitly, has also

*E-mail: stephen.ramsden@nottingham.ac.uk

involved state intervention. One of the five original objectives of the Common Agricultural Policy (CAP) of the European Economic Community (EEC), outlined in the 1957 Treaty of Rome, was to stabilize, as well as support, agricultural prices. However, from the 1980s, successive reforms of the CAP have reduced end-price support for many agricultural commodities. Throughout most of the 1970s and 1980s, European cereal prices, protected by CAP intervention buying, were above world price levels, often substantially so. A feature of the pre-1980s-reformed CAP was that it also had a price-reducing mechanism – export levies rather than the more usual export refunds – that prevented prices from rising above a certain level if world prices were higher than the European threshold price. Although this mechanism was only used once, during the early 1970s, the pre-reform CAP, in line with the original Treaty of Rome objectives, did attempt to stabilize prices, both 'high' and 'low'. In the UK, it was quite typical in the 1980s for farmers producing wheat to think in terms of £100/t, with movement around this level being due to normal seasonal supply effects. This level of predictability has now largely been removed; tighter quality standards and restrictions on the type of cereals eligible for intervention have also reduced the amount of grain eligible for support. In addition, import liberalization measures, introduced at the conclusion of the Uruguay Round of the General Agreement on Tariffs and Trade in 1994, and continued in subsequent reforms of the CAP, have exposed European Union (EU) farmers to world price variability. Within some EU countries (e.g. the UK), further instability occurs as a result of fluctuating exchange rates.

European farmers are compensated for reduced historical end-price support through the Single Payment Scheme (SPS), which was replaced in 2015 by the Basic Payment Scheme (BPS). These payments are designed to fit the 'green box' requirements of international trade agreements and are relatively decoupled from production and increasingly tied to on-farm environmental obligations. Consequently, they offer a distinct income stream for a farm, one that is relatively risk free (with the caveat that the payment is affected by exchange rate movements for countries outside the 'Eurozone'). One effect is that there is a reduced need for other approaches to managing risk, through either the management of price variability or increased financial resilience. It is, therefore, not clear whether risk management has become more important under a CAP with both decoupled – but relatively low-risk – support payments and increased price instability compared with higher price support and lower price variability.

Changes to European agricultural policies, largely driven by objectives unrelated to concerns about uncertainty and risk management, have thus led to changes in the farm decision-making environment. The extent to which these changes influence production choices, the way that farmers manage land and wider concerns about, for example, the sustainability of farming, is an important area for research. In this chapter, we review whole farm-modelling approaches that attempt to capture: (i) the stochastic nature of the outcomes of farm decision making; (ii) farmers' attitudes to risk; and (iii) ways that risk-averse farmers can manage variability. As an example, we employ a utility efficient programming (UEP), whole-farm modelling approach to address some of the issues that arise from policy reform, in this case, from the fundamental reforms of the CAP in the mid-2000s. In particular, we explore the expectation that decoupled support payments create an incentive to grow a riskier and less diverse mix of crops and other farm enterprises, and the extent to which policy reform has led to changes in the incentive for farmers to use private methods for managing risk – the diversification of enterprises and use of futures markets.

4.2 Review of Existing Models

Good summaries of the different farm-level approaches to modelling risk can be found in Hardaker *et al.* (2004) and Hoag (2010). A fascinating and more general overview of production economics, including risk analysis, can be found in Chavas *et al.* (2010). As noted by Dorward (1999), we can think of risk as being either embedded, when there is some scope for the decision maker to make tactical adjustments that will affect outcomes as new information arises over time; or non-embedded, when the decision maker has no scope, once an initial decision has been made, of influencing the final outcome. Studies of non-embedded risk have a long history. Markowitz (1952) introduced the idea of the expected returns (E)–variance of returns (V) rule, to establish an efficient set of E–V combinations associated with different share portfolios; from this an investor could select a preferred portfolio from the efficient set. Starting with Freund in 1956, authors have integrated E–V analysis into a farm modelling framework, using quadratic programming (QP). In a recent example, Chavez *et al.* (2014) employed the technique normatively to assess the diversification possibilities on tobacco farms in Argentina. In a final run of the model, the removal of government subsidy improved environmental performance (a positive carbon balance is possible) but substantially reduced overall farm profitability. As is generally the case with farm-level models, a standard approach, using a well-known technique, can produce results that are of profound importance for policy makers.

To ease computation at a time when computer memory was the binding constraint on research efforts, Hazell, in 1971, introduced the idea of MOTAD (minimization of total absolute deviation) as a way of approximating the QP approach. MOTAD and its extension, Target MOTAD (minimization of total absolute deviation below a target income level; Tauer, 1983), has been employed extensively in the agricultural economics literature. Hope and Lingard (1992) used the technique to explore whether early land-idling policy (set-aside), part of the CAP in various guises until 2008, would be attractive to specialized and mixed farms of differing risk aversion. Stott *et al.* (2003) used MOTAD in a livestock disease context, with results that suggested a positive relationship between maintaining a disease-free herd and both farm income improvement and risk reduction. MOTAD was integrated with an ecological modelling system, VEMM (the Vegetation Environment Management Model) by Oglethorpe and Sanderson (1999) to assess the effect of changes in farm management on the diversity of plant species on pasture land in the north-east of England. Trade-offs between farmer utility and environmental benefits are captured. A calibration phase is employed to simulate initial levels of land use and intensity of production; in principle, this could involve adjusting the level of risk aversion assumed to 'fit' model results to observed data, though this was not used by the authors.

Calibration is quite unusual in farm-level modelling outside the field of positive mathematical programming (PMP). From a normative perspective, as noted by Janssen and van Ittersum in their 2007 overview of what they term 'bioeconomic' farm models, modellers are trying to determine what farmers should do: in an optimization framework, by definition, there is no better solution than the optimum. Calibration, from this perspective, only takes us back to what is suboptimal. Of course, this assumes that models capture the richness and complexity that is inherent in farm systems. If variability and attitude to risk are included in the model, for a known level of risk aversion, and model results still do not calibrate, is this because the decision maker is in some sense deficient in management ability – or could it be that the model is deficient in some way?

Pannell *et al.* (2000), in a much cited paper, argue that researchers have paid insufficient attention to: (i) building better, more realistic models that capture the complexity of the real world situation; (ii) the effects on risk of large investment decisions, such as investment in land and capital; and (iii) capturing the benefits of tactical decision making over the course of the farming year. With respect to the latter, discrete stochastic programming (DSP) models have been used to capture the benefit of tactical adjustments over the course of the production cycle. Dorward (1999) illustrated the use of a DSP model in Malawi, in which tactical responses were found to be important for labour- and capital-scarce households.

Kingwell and other authors, starting with a 1993 paper in *Agricultural Economics*, have built up a body of work on the benefits of tactical decision making, in an Australian context, using the MUDAS (Model of an Uncertain Dryland Agricultural System) model. For example, the benefits of flexibly adjusting crop or pasture areas during the production cycle in response to weather variability were estimated to be up to 25% greater than a less flexible strategy (Kingwell *et al.*, 1993). The greatest benefits arise during the 'best and worst' seasons and thus, a priori, we would expect tactical adjustment to become more important under climate change, for which predictions suggest that extremes will become more common. Notwithstanding, in an application of MUDAS, John *et al.* (2005)

provided compelling evidence that climate-induced shifts in variability may reduce the benefit of tactical adjustments if the frequency of favourable weather years is reduced.

More recently, authors using DSP have included Mosnier *et al.* (2009), who considered tactical adjustments to production and price shocks on French suckler cow (for beef production) farms, and Dono and Mazzapicchio (2010), who aimed (again) to improve the ability of the farming sector to adapt to climate change. Despite the attractiveness of DSP for modelling real farm problems and potential decision maker responses, publications using the technique are relatively scarce, particularly in Europe. The technique itself is data intensive and, as argued by Janssen and van Ittersum (2007) and others, this can act as a deterrent to its use. However, this is true of mechanistic crop models as well, and the use of these has been reported extensively in the literature; the availability of improved computing power has also reduced model runtimes. The problem is probably more related to availability of suitable data, particularly the production responses of what are often subtly different management interventions.

Studies that consider the risk of larger scale investment over time are also, perhaps surprisingly, thin on the ground, despite calls for greater attention in this area (e.g. Pannell *et al.*, 2000; Janssen and van Ittersum, 2007). In their 2007 review of bioeconomic models, Janssen and van Ittersum found only two papers (out of 48 studies considered in the period 1982–2006) that considered longer term strategy. One of these, by Wallace and Moss (2002), used a dynamic farm-level optimization framework with multiple objectives, including farm investment objectives, over a 7 year time horizon. More recently, the model has been used to address the structural adjustment objectives of the CAP, with a resulting policy recommendation to encourage younger people into farming rather than to use early retirement schemes – largely because younger people have a more strategic (longer term) approach to investment and business growth (Davis *et al.*, 2013).

Gibbons and Ramsden (2008), in a dynamic mixed integer programming model of a catchment in the East Anglia region of England, considered the effect of long-term investment decisions on the variability of the net margin of farms across the catchment in: (i) reservoir capacity (for irrigation); (ii) animal housing; and (iii) crop storage. Variability is a function of climate-model generated yields under different climate change scenarios and farmer decision making, which, in this 2008 paper was assumed to follow naïve, 'adaptive expectations', based on averages in the preceding 4 years. The results indicated little evidence either for increased investment or of a 'tipping point' shift to 'exotic' crops (sunflowers). However, neither Wallace and Moss (2002) nor Gibbons and Ramsden (2008) considered strategic decision making within a risk modelling framework; for example, although covariances between yields are conveniently retained in the Gibbons and Ramsden framework, the benefits that this confers, for instance through enterprise diversification, are not captured by the model. The arguments put forward by Pannell *et al.* (2000) and Pannell (2006) are relevant here: if net benefits to a farmer are relatively insensitive to changes in risk preference over a wide range of a particular decision variable, such as crop area, why complicate the model by including attitude to risk? It should be noted though that this margin for error is informative, as while the benefits of a more diversified portfolio of enterprises may be small for the individual farmer, as measured by utility or certainty equivalent, they may be important for other reasons, such as improved farmland habitats (Benton *et al.*, 2003) and landscape aesthetics (Frank *et al.*, 2013).

Returning to embedded risk, UEP, which is described in more detail in Section 4.3 below, is the technique that most closely satisfies the assumptions of subjective expected utility theory (Hardaker *et al.*, 2004), without the limitations of the QP or MOTAD approaches. Developed by Hardaker *et al.* (1988) from earlier work by Lambert and McCarl (1985), UEP can be used to generate an efficient set of farm plans (efficient in that there is no better set) for decision makers with different degrees of risk aversion, for a given utility function. Flaten and Lien (2007) combined the approach with a stochastic programming model and found that, relative to production and institutional constraints, the degree of risk aversion is relatively unimportant as a determinant of resource use. Neither the form of the utility function (in this case utility of income) nor the level of risk aversion were as important as a change in policy regime in Lien and Hardaker's 2001 study of Norwegian agriculture, in which a relatively risk-free decoupled subsidy payment led to improved utility and reduced livestock production. More recent applications of UEP include those of Ogurtsov *et al.* (2015) who considered 'catastrophic risk', and Lien *et al.* (2009), who found that both UEP and QP approaches were wanting in situations in which data on enterprise net incomes were sparse.

4.3 The Utility Efficient Programming Model

The UEP model combines the standard constraints of a farm business (land, labour, capital) with an objective function that maximizes expected utility for a given utility function and degree of risk aversion. UEP maintains the advantages of the farm-system approach and avoids the restrictive assumptions of other techniques for including risk, such as QP (Hardaker *et al.*, 2004). The model incorporates both yield and price instability, for representative 'mainly cereal' farms in the East Midlands region of England. Here we present results for a hypothetical case study farm as an example. Following Hardaker *et al.* (2004), a UEP takes the form:

$$\max E[U] = PU(z, r), r$$

subject to

$$\mathbf{A}x \le b$$

$$\mathbf{C}x - \mathbf{I}z = f$$

$$x \ge 0$$

where the expected utility, $E[U]$, is equal to U, the utility of net incomes, z, for a given level of risk aversion, r, weighted by the probabilities, P, of these net incomes occurring under different states of nature. Net income is often the gross margin of a farm enterprise, and that is the convention used here. The matrix **A** consists of the standard programming technical coefficients; x is a vector of activity levels and b is a vector of resource availabilities. Activity gross margins form the matrix **C**; a gross margin is assigned for each probability. Thus, if there are m activities and n probabilities, **C** will be an $m \times n$ matrix. The identity matrix, **I**, transfers the activity gross margins to the objective function; setting this constraint equal to f (the fixed costs) ensures that the optimal solution is net of fixed costs.

For the maximand of a UEP model, we need a suitable (concave) utility function, information on activity gross margins and the probability that different gross margins will occur. To find the efficient set of plans, the model can be run for different levels of absolute risk aversion, r, if we adjust r to reflect the farmer's level of wealth. For computational ease, we use the negative exponential function:

$$U = 1 - \exp[-\{(1 - a)\, r_{\min} + a r_{\max}\} z]$$

The parameter a is varied over a range of zero to 1, with r_{min} when a is zero and r_{max} when a is 1. Following Hardaker *et al.* (2004), we take account of the effect of wealth (i.e. Net Worth) on risk aversion by calculating absolute risk aversion values from a reasonable range of relative values, r_r, with 0.5 representing 'hardly risk averse' and 4 representing 'very risk averse'. The minimum and maximum values for the above function are then calculated as: r_r/Net Worth.

For the hypothetical case study farm, resource availability and technical coefficients were based on a simplified version of the Farm-adapt model (see, for example, Gibbons and Ramsden, 2008). It was assumed that labour and machinery jointly supply time available for field operations and that this time was determined for the farm by the number of labourers, including the farmer's own labour, reported in the Farm Business Survey (a survey conducted by Rural Business Research, RBR, on the performance of farm businesses in England on behalf of the UK Department for Environment, Food and Rural Affairs, Defra). Constraints were divided into periods when specific operations (harvesting, cultivations, etc.) normally take place in the East Midlands; the time available was adjusted by standard figures (Nix, 2013) to account for holidays, illness and weather; field work rates were also taken from Nix. Casual labour was made available for all operations. Rotational constraints limited the amount of break crops (oilseed rape, pulses and set-aside[1]) to be less than 50% of the cropping area. Total cereal area was limited by total cropping area, though the area of first-crop winter wheat was also restricted to be less than or equal to the area of break crops. Following Nix (2013), the first wheat crop was assumed to yield 0.85 t/ha more than wheat following other cereal crops in the rotation. Winter wheat was also separated into three sowing date activities, September, October and November, with yield penalties (Spink *et al.*, 2000) imposed for sowing in October and November. Two activities were specified for winter barley: an early crop, sown in August, constrained to follow early-sown (and therefore harvested) winter wheat, and a later crop, sown in the September sowing period. Winter oilseed rape, field peas and field beans were each represented by one activity. No sowing date yield penalties were imposed for barley or break crops. Renting out land was included as a risk-free alternative activity. Fixed costs included the opportunity cost of farmer and spouse labour and were assumed to remain unchanged across all model runs.

Model results were expressed as certainty equivalents. A certainty equivalent is the value of a guaranteed sum of money that would make the decision maker indifferent between this guaranteed or 'certain' return and the risky alternative, for example, growing crops on the farm (Hardaker *et al.*, 2004). The certainty equivalent captures the combined weighted benefits and costs (in a monetary sense) of the expected financial returns and the amount of risk that corresponds to these expected returns. For farmers who are risk averse, the utility (certainty equivalent) from a given cropping plan with uncertain returns is lower than a farmer who puts less weight on the negative impact that risk has on utility. Thus, the certainty equivalent is an objective measure of the welfare that a cropping plan generates as a function of the amount of risk a farmer is willing to accept. An improvement in the certainty equivalent represents an improvement in the value of a risky farm plan and, as a result, an improvement in farmer welfare.

4.3.1 Gross margin distributions

The net incomes, and the risk associated with these net incomes, were captured through gross margins for different crop and marketing decisions. Gross margins were available at an aggregate level from Farm Business Survey publications such as the *Report on Farming in the Eastern Counties of England* (e.g. Lang, 2005). However, as reported figures are averages, the variability of gross margin over time will be less than that experienced by individual farmers (Tomek and Peterson, 2001, made a similar point in relation to US data). Accordingly, the risk associated with growing crops will be understated if average figures are used. It is possible to construct gross margins for individual farms given access to relevant data. For the East Midlands region, yield data for winter wheat, winter barley and oilseed rape were available over the 8 year period (1997–2004) preceding the Mid Term Review of reforms of the CAP on Mainly Cereal farms that have been in the Farm Business Survey for these 8 years. As farm-level yield data for field pea and bean crops were not available from the Survey, field pea and bean gross margins were calculated using average yield figures reported by Lang (2005). As the information was limited to 8 years, it was assumed that each yield had an equal chance of occurring.

A farmer using the spot market (the market for goods for immediate delivery) faces daily price uncertainty. Daily spot prices (i.e. the prices for immediate delivery) were available for feed wheat, feed barley and rapeseed from the Home Grown Cereals Authority (HGCA) for the period 8 July 1999–1 July 2004. Field bean and pea prices were available on a weekly basis, for the same period, from the UK trade journal *Farmers Weekly*. Data were only included where there were observations on the same date, for each of the five crops. The data were adjusted for inflation using the retail price index, and summary statistics for each crop are shown in Table 4.1. The distribution for all crop prices, apart from those for oilseed rape, is positively skewed, indicating that although high prices do occur, these prices are only achieved relatively infrequently over the domain of the data used. Feed wheat has the greatest mean-weighted variability as measured by the coefficient of variation.

4.3.2 Calculation of spot price gross margins

The full range of feed wheat prices (£54/t) was divided to give five classes, each with a range of £10.80/t. Adding this value to the lowest recorded feed wheat price gave the first class of prices; this includes all prices less than £64.6/t.

The next class contains all prices greater than £64.60/t and less than £75.40/t, and so forth. The associated probability of prices falling within each class was then calculated (see Fig. 4.1); the price associated with each probability was taken as the average price for each class, giving five possible outcomes and five associated probabilities. It was assumed that the prices for other crops were jointly determined with the wheat prices. For example, the average price for barley in the less than £64.60/t class was calculated by averaging all barley prices that occurred on the same days as the wheat prices that were less than £64.60/t. The same procedure was used for other crops.

For each crop, there are therefore eight yields (derived from yearly yield data for each farm) and five price (as outlined above) outcomes, giving 40 possible output value combinations in total. Variable costs were taken from

Table 4.1. Summary statistics for crop prices (in £) from 8 July 1999 to 1 July 2004 in the East Midlands of England.

Statistic/Crop	Feed wheat[a]	Feed barley[a]	Oilseed rape[a]	Field peas[b]	Field beans[b]
No. observations	444	444	444	444	444
Mean	76	72	152	93	92
Median	75	73	155	89	88
Mode	57	58	116	82	69
Sample variance	165	89	494	111	190
Coefficient of variation	0.17	0.13	0.15	0.11	0.15
Kurtosis	−0.11	−0.22	−1.09	−0.77	−0.65
Skewness	0.53	0.31	−0.10	0.71	0.35
Range	54	43	91	43	57
Minimum	54	52	111	74	68
Maximum	108	96	202	118	125

[a]Data from the Home-Grown Cereals Authority (HGCA), UK.
[b]Data from the UK trade journal *Farmers Weekly*.

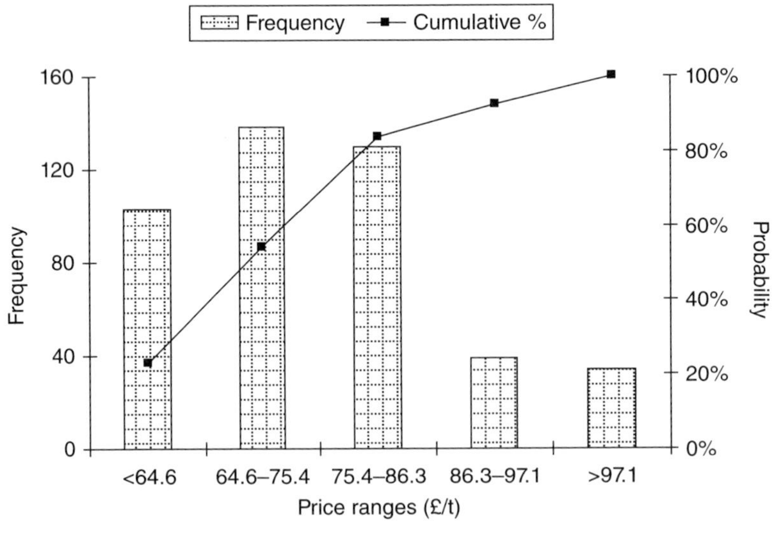

Fig. 4.1. Histogram of feed wheat prices (by range) in the UK from 8 July 1999 to 1 July 2004. Frequency denotes the number of observations (out of a total of 444) falling into each price range. The cumulative percentage shows the probability of achieving a price equal to or lower than that of a particular histogram category. Data from the Home-Grown Cereals Authority (HGCA), UK and the UK trade journal *Farmers Weekly*.

Lang (2005); these were also adjusted for inflation and deducted from the relevant output combinations to generate crop gross margins. The probability of each gross margin occurring is given by the joint probability of a relevant yield and average class price occurring together.

By using discrete estimates, we lose some information about the extremes of the distribution; however, the variability in the data is retained. For example, for the first winter wheat on Farm A, the average gross margin, exclusive of any support payment, when using prices from the full data set was £397/ha, with a standard deviation of £110/ha. The probability-adjusted average gross margin using the above procedure was £396/ha, with a standard deviation of £108/ha. The gross margin estimates also assume that there is no significant correlation between price and yield. If there was a negative relationship between price and yield (a 'natural hedge'), the variability of the gross margin would be less than as calculated above.

4.3.3 Calculation of gross margins using futures markets

To simulate farmers' use of futures markets, the spot price data were averaged over each marketing year to represent the within-year price stability that would be offered by a futures contract in each year. This procedure gives five average marketing year prices for each crop; it was assumed that each price had an equal chance of occurring. Gross margins were constructed by multiplying each yield by each price, again giving 40 gross margin observations in total (see Section 4.3.2), because each price (P = 0.2) and yield (P = 0.125) has an equal joint chance of occurring (P = 0.025), and the 40 gross margins each have this equal chance of occurring.

Using this procedure, the average gross margin (i.e. the average of the 40 average price gross margins) was £390/ha, with a standard deviation of £91/ha. The figure is lower than the average spot price gross margin figure (£396) because of the uneven number of observations in each marketing year – in some years, there are fewer days with spot price data for all three crops (feed wheat, feed barley and oilseed rape) reported in the HGCA data set. As a result, the assumed probability of P = 0.2 does not quite reflect the number of observations in each of the five marketing years. Therefore, the five marketing year prices were adjusted so that the probability-weighted average gross margin was equal to the probability-weighted average gross margin as calculated using the full spot price data set. For example, for winter wheat, the value required was £0.72/t; for Farm A, this results in a probability-weighted average gross margin of £396/ha for both sets of gross margins.

In constructing the spot price gross margins, we assume a relatively naïve approach to marketing: we simulate a farmer who chooses not to manage any of the risk associated with variable output prices, either through market instruments or through the planned storage and sale of grain. This farmer faces the full variability of prices inherent in the HGCA data, to the extent that this is captured by the approximation technique. The average price gross margins reflect the situation for a farmer who faces a known price at the time of planting (typically September and October), but which varies from year to year. Thus, a farmer growing combinable crops will not be able to remove this year-to-year instability through the use of futures trading. The results from the UEP model show the optimal farm plan that the farmer should implement each year to manage this instability.

4.4 Application

The UEP model was run four times using the 'generalized reduced gradient method' for solving non-linear problems, part of the 'Premium Solver Platform' from Frontline Systems. The method incorporates both 'multistart' and topographic search features to increase the probability of finding a global optimum. In Run 1, we assumed that the farm receives support in the form of fixed area aid payments (AAPs) that are coupled to the production of individual crops. In Run 2, an equal amount of support was assumed, but in the form of a Single Farm Payment (SFP), decoupled from production. This assumption ensures that any differences in utility are due to changes in variability rather than changes in income. In Run 3, in order to simulate the effect of completely removing government support, the SFP was removed and the model run with the spot price distribution as calculated above.

Run 4 used the same assumptions, but with the average price distributions. In all runs, it was assumed that the level of fixed costs did not change and that farmers continued production (or rented out land) even if continued production was not profitable in a long-run sense. Anecdotal evidence suggests that most arable farmers in the UK have continued to crop their land under the SPS and more recently, the BPS.

4.4.1 Run 1: coupled area payments

Figure 4.2 shows the certainty equivalents and utility-maximizing crop areas for the gross margins with area aid payments for six CARA (constant absolute risk aversion) values – ranging from the least (CARA = 1.50×10^{-5}) to the most (CARA = 1.69×10^{-4}) risk-averse case (the upper and lower values originally reported by Oglethorpe, 1995). The maxima and minima correspond to relative risk aversion values of 0.4 to 4.0 and a Net Worth of £23,500. Certainty equivalent for the most risk averse case was £4357, 31% less than the certainty equivalent for the least risk-averse case. The optimal crop mix for the most risk-averse case was more diversified, with less winter wheat and oilseed rape, and more winter barley and winter beans. Under no level of risk aversion was it optimal to let out land at a (risk-free) return of £130/ha.

4.4.2 Run 2: the decoupled Single Farm Payment (SFP)

The area aid payments were deducted from the crop gross margins and an SFP figure calculated using the total level of area aid payments generated under Run 1 for each degree of risk aversion. Table 4.2 shows the impact of the SFP on the certainty equivalent values generated in Run 1. For the most risk-averse case, the certainty equivalent increases by 13%, i.e. the value of the optimal farm plan to the risk-averse farmer increases because of the guaranteed payment. The risk premium to this farmer – the difference between the profit-maximizing farm plan and the utility-maximizing farm plan – falls from £1984 to £1435. The value of the SFP declines with decreasing risk aversion – for the least risk-averse farmer, certainty equivalents are the same with area aid payments and SFP. For all levels of risk aversion, the optimal crop mix remains the same as in Fig. 4.2. Thus, the SFP does not appear to lead to a different mix of crops being selected by the model, but it would allow a risk-averse farmer to maintain utility (for example, maintain

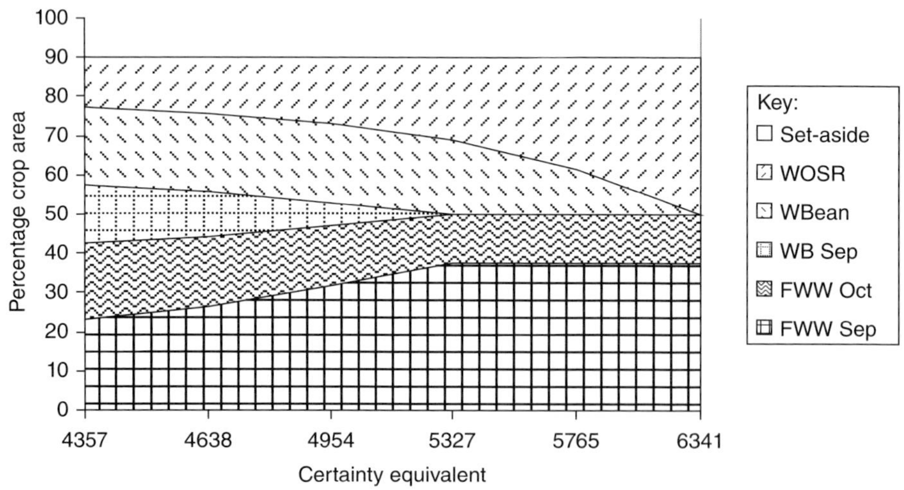

Fig. 4.2. Certainty equivalents and percentage crop areas for different levels of risk aversion, where £4357 is the most risk-averse certainty equivalent and £6341 is the least risk averse. Key: FWW, first winter wheat (sown in September or October); WB, winter barley (sown in September); WBean, winter beans; WOSR, winter oilseed rape. Data derived from the application of the utility efficient programming (UEP) model.

Table 4.2. Certainty equivalent (CE) values in £ with coupled area aid payments (AAPs) and decoupled Single Farm Payments (SFPs) at different CARA (constant absolute risk aversion) values (coefficients) derived from Runs 1 and 2 of the UEP (utility efficient programming) model developed for crop (mainly cereal) farms in the East Midlands of England, UK.

CARA	CE with AAPs	CE with SFP	Difference	SFP
0.000169	4,357	4,906	13%	18,971
0.000138	4,638	5,186	12%	18,972
0.000107	4,954	5,502	11%	18,973
0.000077	5,327	5,846	10%	18,941
0.000046	5,765	6,080	5%	18,731
0.000015	6,341	6,341	0%	18,408

a certainty equivalent of £4357 for the most risk-averse farmer in Table 4.2) and grow crops with a greater gross margin variability.

4.4.3 Runs 3 and 4: spot prices and average prices

In Run 3, the SFP was removed from the model and gross margins constructed using the spot price distributions for each crop. These results were compared with results of the model run with average prices (Run 4), again with no SFP, to simulate the effect of futures markets. The main result is that without the SFP, the farm does not cover fixed costs and certainty equivalents are, therefore, negative. The effect of using futures is to increase welfare by a small margin, 1–4%, in the less and more risk-averse cases respectively. Crop plans are quite different though (see Fig. 4.3), with a more diverse mix of enterprises under the spot price run, suggesting a possible trade off between external financial stabilization measures and internal farm-system measures, such as diversification. The negligible improvement for the risk-averse farmer at first seems surprising: the spot price gross margins all have greater standard deviations; the probability weighting in the model ensures that the chance of low prices is relatively high; and the utility function ensures that negative returns detract from welfare by a greater amount than positive returns add to welfare.[2]

However, from the perspective of a farmer wishing to reduce the variability of income resulting from planting decisions for the case study farm, removal of within-year price variability has little value. This point is further emphasized by forcing the model to accept relatively high-risk crop plans: for example, the optimal position for a risk-neutral farmer would be to accept profit-maximizing levels of September winter wheat and winter oilseed rape of 31.5 and 33.8 ha, respectively. Forcing the risk-averse farmer to accept crop areas greater than or equal to these values reduces utility under both the average and spot price gross margins, though the difference between certainty equivalents in both cases is less than 1%. These results are intended only as an illustration of an application of the UEP technique. None the less, we can draw some conclusions, as the findings are similar to those found by other authors, even though the inclusion of market instruments as a means of dealing with risk is relatively novel.

First, the evidence here suggests that the welfare-improving effects of using futures to stabilize prices are small, even for the risk-averse case. The quintessential paper on why futures markets are not as attractive to farmers as might initially be expected in an unstable market environment is that of Pannell *et al.* (2008). The authors show that reasons for low uptake can be attributed to a wide range of factors, including transaction costs, farmers' price expectations (i.e. they are more optimistic relative to futures market prices), the risks associated with trading itself (i.e. not fulfilling futures contracts due to lower than anticipated crop yields) and the lack of risk aversion among some farmers. To this list can be added crop diversification and other management factors within the farmer's control, such as the use of borrowing (Simmons, 2002) or simply maintaining a healthy (higher Net Worth) balance sheet.

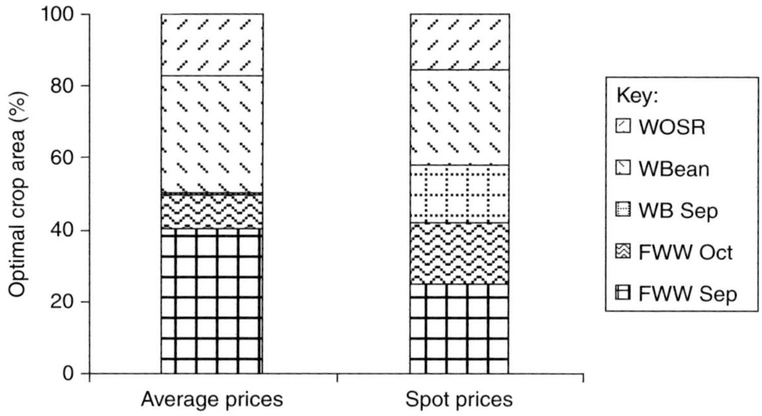

Fig. 4.3. Optimal crop areas (as %) under average prices (to simulate the effect of futures markets) and spot prices in the most risk-averse case (constant absolute risk aversion, CARA = 0.000169; relative risk value, $r_r = 4$). Key: FWW, first winter wheat (sown in September or October); WB, winter barley; WBean, winter beans; WOSR, winter oilseed rape. Data derived from the application of the utility efficient programming (UEP) model.

Secondly, decoupled payments, of the type introduced in the 2005 reforms of the CAP, improve welfare through their stabilizing effect and so it would be possible to maintain pre-reform levels of welfare with a more risky mix of crops. Thirdly, without decoupled payments, there is some indication that the use of market instruments substitutes for on-farm diversification, although the effect of forcing suboptimal plans on utility is small: the more diverse mix of crop choices under the spot price run as shown in Fig. 4.3 is only a marginal improvement on the average price run that simulated futures.

Thus, both government (decoupled payments) and private (use of futures) interventions to manage variability have the potential to reduce the diversity of the enterprise portfolio. Nevertheless, if the opportunity cost of a suboptimal, more diverse plan really is relatively small, the extent of the ecological response to this increased diversity becomes an important research question: if the benefits are there, they may be achievable at a relatively low cost. The simple model presented here is not sufficiently rich to capture the full cost; in particular, the benefits (lower unit costs of production) of specialization are not modelled, but the farm-level approach does suggest that we need to pay much more attention to farm diversity and its links to ecological diversity. There are also implications for the growing debate on agricultural resilience (see Gardner, Ramsden and Hails, 2016) and the ability of farm systems to withstand shocks. The question that arises from the simple modelling approach used here, is: what are the consequences of using policy and market instruments as methods of managing financial risk if they lead to a reduction in the biological resilience of the farm system itself? This is of course an argument about the extent to which financial capital can substitute for natural capital.

4.5 Limitations of the Model and Conclusions

The model described here is relatively simple and only captures some of the elements of the individual farm situation (e.g. yields); it also relies on standard figures (e.g. field work rates) that will differ from farm to farm. Furthermore, because of the approximation technique used to generate gross margins, we lose some information about the distribution of gross margins faced by each of the farms. This matters less for 'high' values, as the utility function used implies diminishing marginal utility of income, particularly in the more risk-averse cases. When using a mid-price figure though, the negative impact on welfare of very low prices is underestimated for the first 'class' of the distribution. The analysis used actual probabilities, determined from historical price and yield distributions, rather than subjectively determined probabilities obtained from farmers. Thus, if farmers expect lower prices to occur with greater frequency, the use of market instruments such as futures would be more appropriate than has been estimated here (it would improve farmer welfare more). There is a dearth of information in the literature relating to farmer risk aversion; the values used here are from a study from as far back as 1995, by Oglethorpe. A study on Dutch pig farmers by Pennings and Leuthold (2000) also stressed the importance of perceived risk as a determinant of farmers' use of futures markets. However, access to information on market prices in the UK, through the websites of organizations such as the AHDB (Agricultural and Horticultural Development Board, successor to the HGCA), has been much improved in recent years.

No attempt has been made here to determine whether yields and prices are independent; we assume that prices are determined by world demand and supply, and that these are not affected by changes in demand and supply in the UK (i.e. we assume a 'small country' case). To test whether this assumption is reasonable, it would be necessary to model the relationship between UK and international supplies of cereals, protein crops and oilseeds. For example, if there was a strong positive correlation between UK wheat supplies and international wheat supplies, we would expect high levels of production in the UK to be associated with lower output prices. Even so, it is not clear that this would increase variability – for example, a negative relationship between price and yield would provide a 'natural hedge' that would reduce the variability of gross margin rather than increase it.

From the limited analysis presented here, changing crop mix was relatively ineffective as a risk-management technique. Forcing a risk-averse farmer to adopt a profit maximizing – and therefore more variable – mix of combinable crops reduced utility by only a small amount; this was for the risk-averse case, without the stabilizing benefit of direct payments. The model retains covariance between crop prices and covariance between crop yields (and hence covariance between crop gross margins), and as both prices and yields are positively covariate (combinable crop yields tend to be 'high' or 'low' in the same year and the same is true for prices), the diversification options for limiting risk with these crops is limited. Moreover, as we have not taken account of the possibility of a 'natural hedge', as discussed above, the model underestimates the scope of crop diversification as a method for reducing risk. It should be noted that the rotation in the model is reasonably flexible and does not restrict the diversification options that would realistically be available on farms if they were selecting from the crops considered here. Clearly, risk-reducing diversification options would be widened with a greater choice of crops, but including more options would have implications for costs, both in terms of additional and often specialized capital requirements (in the UK, potatoes are a good example), and potential cost increases resulting from a less specialized enterprise mix.

The UEP framework is most limited – as are many other studies on risk management in agriculture – by its emphasis on risk and profit, particularly when modelling futures markets. As Pennings and Leuthold (2000) have shown, different farmers have different objectives and the use of futures will be linked to objectives other than reducing risk, as shown by Pannell *et al.* (2008). Within the risk/profit framework, the wider results show that there are incentives for risk-averse farmers to use price stabilization measures to remove within-year price instability within existing price and yield distributions. With the introduction of the BPS in England in 2015, the amount paid directly to farmers has fallen and, other things being equal, the incentive to manage risk will increase. More importantly, relatively high fixed cost levels, if maintained, suggest that production will not be profitable on many combinable crop farms. If farmers and members of the farm family are able to realize the opportunity cost of their own labour, and other non-land fixed costs can be eliminated, ceasing production would be attractive on both profit and risk-management grounds. In reality, as argued by Nix (2013), not all fixed costs are avoidable and keeping land in good agricultural and environmental condition will involve at least some contracting charges. Ceasing production will also have tax implications that are not considered here and would involve a complete change in lifestyle for most farmers. However, the comparison illustrates the dominating effect of the CAP reforms on farm decision making, i.e. that there is little point in managing the variability of farm losses, particularly where a relatively risk-free alternative exists.

Where are the opportunities for farm-level risk modelling? In Europe, there is certainly more scope for work on embedded risk and the extent to which we need more resilient or more adaptable (in the tactical sense) farm systems. We have two responses to climate change and a more uncertain world: one is to adapt and achieve new optimal 'adapted' solutions; the other is to build up resilience – financially, this means building a stronger capital position on farms, somewhat akin to attempts by the banking sector after the financial crisis of the late 2000s. Nevertheless, the potential trade-offs identified in this chapter suggest that we need to understand the implications of substituting financial for environmental capital if this leads to a reduction in the biological resilience of farm systems. This, in turn, suggests that optimization models need to put more emphasis on capturing biological tipping points and thresholds: at what point do natural systems fail and is the chance of failure exacerbated by the uptake of external adaptations, such as the use of futures and options trading? This is not to argue that more 'ecological' recommendations – often based on arguments around greater self-sufficiency – should not also be tested; for instance, are farmers with a self-imposed 'reduced set' of adaptation options more vulnerable to increased uncertainty? As other authors have also noted, there is also a need for more models of non-embedded risk and the consequences of getting adaptations wrong. The real risks to farmers operating in competitive, undifferentiated commodity markets come not from the variable commodity prices, but from the effect of this variability when combined with misplaced large-scale investment.

Notes

[1] The requirement for set-aside land (land not in food production) has since been set to zero; in the model here it can be viewed as a risk-free payment for environmental services.
[2] For example, at a CARA coefficient of 1.69×10^{-4}, a total gross margin of £5000 adds 0.57 to utility; a total gross margin of –£5000 subtracts 1.33 from utility.

References

Benton, T.G., Vickery, J.A. and Wilson, J.D. (2003) Farmland biodiversity: is habitat heterogeneity the key? *Trends in Ecology and Evolution* 18, 182–188.

Chavas, J.P., Chambers, R.G. and Pope, R.D. (2010) Production economics and farm management: a century of contributions. *American Journal of Agricultural Economics* 92, 356–375.

Chavez, M.D., Berentsen, P.B.M. and Oude Lansink, A.G.J.M. (2014) Analyzing diversification possibilities on specialized tobacco farms in Argentina using a bio-economic farm model. *Agricultural Systems* 128, 35–43.

Davis, J., Caskie, P. and Wallace, M. (2013) Promoting structural adjustment in agriculture: the economics of New Entrant Schemes for farmers. *Food Policy* 40, 90–96.

Dono, G. and Mazzapicchio, G. (2010) Uncertain water supply in an irrigated Mediterranean area: an analysis of the possible economic impact of climate change on the farm sector. *Agricultural Systems* 103, 361–370.

Dorward, P. (1999) Modelling embedded risk in peasant agriculture: methodological insights from northern Malawi. *Agricultural Economics* 21, 191–203.

Flaten, O. and Lien, G. (2007) Stochastic utility-efficient programming of organic dairy farms. *European Journal of Operational Research* 181, 1574–1583.

Frank, S., Fürst, C., Koschke, L., Witt, A. and Makeschin, F. (2013) Assessment of landscape aesthetics – validation of a landscape metrics-based assessment by visual estimation of the scenic beauty. *Ecological Indicators* 32, 222–231.

Freund, R.J. (1956) The introduction of risk into a programming model. *Econometrica* 24, 253–263.

Gardner, S., Hails, R. and Ramsden, S. (eds) (2016) *Agricultural Resilience: Perspectives from Ecology and Economics*. British Ecological Society, London and Cambridge University Press, Cambridge, UK (forthcoming).

Gibbons, J.M. and Ramsden, S.J. (2008) Integrated modelling of farm adaptation to climate change in East Anglia, UK: scaling and farmer decision making. *Agriculture, Ecosystems and Environment* 127, 126–134.

Hardaker, J.B., Patten, L.H. and Pannell, D.J. (1988) Utility-efficient programming for whole-farm planning, *Australian Journal of Agricultural Economics* 32, 88–97.

Hardaker, J.B., Huirne, R.B.M., Anderson, J.R. and Lien, G. (eds) (2004) *Coping with Risk in Agriculture,* 2nd edn. CAB International, Wallingford, UK.

Hazell, P.B.R. (1971) A linear alternative to quadratic and semivariance programming for farm planning under uncertainty. *American Journal of Agricultural Economics* 53, 53–62.

Hoag, D.L. (2010) *Applied Risk Management in Agriculture.* CRC Press, Taylor & Francis, Boca Raton, Florida.

Hope, J. and Lingard, J. (1992) The influence of risk aversion on the uptake of set-aside: a MOTAD and CRP approach. *Journal of Agricultural Economics* 43, 401–411.

Janssen, S. and van Ittersum, M.K. (2007) Assessing farm innovations and responses to policies: a review of bio-economic farm models. *Agricultural Systems* 94, 622–636.

John, M., Pannell, D. and Kingwell, R. (2005) Climate change and the economics of farm management in the face of land degradation: dryland salinity in Western Australia. *Canadian Journal of Agricultural Economics* 53, 443–459.

Kingwell, R.S., Pannell, D.J. and Robinson, S.D. (1993) Tactical responses to seasonal conditions in whole-farm planning in Western Australia. *Agricultural Economics* 8, 211–226.

Lambert, D.K. and McCarl, B. (1985) Risk modeling using direct solution of nonlinear approximations of the utility function. *American Journal of Agricultural Economics* 67, 846–852.

Lang, B. (2005) *Report on Farming in the Eastern Counties of England, 2003/2004*. Department of Land Economy, University of Cambridge, Cambridge, UK.

Lien, G. and Hardaker, J.B. (2001) Whole-farm planning under uncertainty: impacts of subsidy scheme and utility function on portfolio choice in Norwegian agriculture. *European Review of Agricultural Economics* 28, 17–36.

Lien, G., Hardaker, J.B., van Asseldonk, M.A.P.M. and Richardson, J.W. (2009) Risk programming and sparse data: how to get more reliable results. *Agricultural Systems* 101, 42–48.

Markowitz, H. (1952) Portfolio selection. *The Journal of Finance* 7, 77–91.

Mosnier, C., Agabriel, J., Lherm, M. and Reynaud, A. (2009) A dynamic bio-economic model to simulate optimal adjustments of suckler cow farm management to production and market shocks in France. *Agricultural Systems* 102, 77–88.

Nix, J. (2013) *Farm Management Pocketbook*. Agro Business Consultants, Melton Mowbray, UK.

Oglethorpe, D.R. (1995) Sensitivity of farm plans under risk-averse behaviour: a note on the environmental implications. *Journal of Agricultural Economics* 46, 227–232.

Oglethorpe, D.R. and Sanderson, R.A. (1999) An ecological–economic model for agri-environmental policy analysis. *Ecological Economics* 28, 245–266.

Ogurtsov, V.A., van Asseldonk, M.A.P.M. and Huirne, R.B.M. (2015) Modelling of catastrophic farm risks using sparse data. In: Plà-Aragonés, L.M. (ed.) *Handbook of Operations Research in Agriculture and the Agri-food Industry*. International Series in Operations Research and Management Science, Springer, New York, pp. 259–275.

Pannell, D.J. (2006) Flat earth economics: the far-reaching consequences of flat payoff functions in economic decision making. *Review of Agricultural Economics* 28, 553–566.

Pannell, D.J., Malcolm, B. and Kingwell, R.S. (2000) Are we risking too much? Perspectives on risk in farm modeling. *Agricultural Economics* 23, 69–78.

Pannell, D.J., Hailu, G., Weersink, A. and Burt, A. (2008) More reasons why farmers have so little interest in futures markets. *Agricultural Economics* 39, 41–50.

Pennings, J.M.E. and Leuthold, R.M. (2000) The role of farmers' behavioural attitudes and heterogeneity in futures contracts usage. *American Journal of Agricultural Economics* 82, 908–919.

Simmons, P. (2002) Why do farmers have so little interest in futures markets? *Agricultural Economics* 27, 1–6.

Spink, J.H., Semere, T., Sparkes, D.L., Whaley, J.M., Foulkes, M.J., Clare, R.W. and Scott, R.K. (2000) Effect of sowing date on the optimum plant density in winter wheat. *Annals of Applied Biology* 137, 179–188.

Stott, A.W., Lloyd, J., Humphry, R.W. and Gunn, G.J. (2003) A linear programming approach to estimate the economic impact of bovine viral diarrhoea (BVD) at the whole-farm level in Scotland. *Preventive Veterinary Medicine* 59, 51–66.

Tauer, L.W. (1983) Target MOTAD. *American Journal of Agricultural Economics* 65, 606–610.

Tomek, W.G. and Peterson, H.H. (2001) Risk management in agricultural markets: a review. *The Journal of Futures Markets* 21, 953–985.

Wallace, M.T. and Moss, J.E. (2002) Farmer decision making with conflicting goals: a recursive strategic programming analysis. *Journal of Agricultural Economics* 53, 82–100.

5 Modelling Farm-level Biosecurity Management

Arnaud Rault[1]* and David A. Hennessy[2]
[1]*BIOEPAR, INRA - Oniris, 44307 Nantes, France;*
[2]*Michigan State University, East Lansing, Michigan, USA*

5.1 Introduction

Heterogeneity is a prominent feature of farmed animal health management. It is manifest in many ways, including disease control strategies and the incidence of costs. The control of a herd's health status concerns all sector participants as animal diseases can cause a wide variety of economic losses. Animal diseases may be categorized as: exotic diseases, which are usually absent in a territory and can have large consequences when they occur; endemic transmissible diseases, which are prevalent in a region and are often costly, but might be successfully managed at the farm level; and, finally, non-infectious diseases, which can be acute or chronic but do not threaten herd health. Diseases within these categories have various biological and economic consequences, and rely on disparate control actions. Indeed, herd health management can include many practices and diverse strategies can coexist to control a particular disease.

At higher levels of aggregation (e.g. the market or region), much attention has been devoted to preventing the entry and spread of animal diseases in order to avoid potentially serious consequences after any entry. At the farm level, specific *ex post* curative actions may be undertaken when a disease has been detected. *Ex ante* preventive practices can be specific to a disease (e.g. vaccination), but such practices are typically generic – monitoring, record keeping and investments to promote hygiene are examples – and most biosecurity decisions are not disease specific.

As revealed by Mugnier *et al.* (2012), the importance attributed to health and breeding management is heterogeneous across farmers. However, depending on the disease, animal health management may be afforded high priority at the public level in light of the public goods provided. In the case of disease prevention at the farm level, i.e. biosecurity management, individual actions may have cross-effects on various diseases. Their potential total effect may be large and broad, but this is difficult to measure because reductions in probability and the extent of avoided damage are hard to estimate when little evidence is available at an appropriate counterfactual level for comparison. As a consequence, the primary research issue for biosecurity actions is less about measuring or modelling the consequences of implementing such actions than about studying action levels and impediments to adoption.

The objective of this chapter is to give insights into current veterinary, agricultural and economic research into biosecurity management at the farm level, where particular attention

*E-mail: arnaud.rault@nantes.inra.fr

is paid to characterizing heterogeneity in choice. Section 5.2 outlines the pertinent economic and veterinary literature on biosecurity management practices. Section 5.3 presents some methods for farm-level modelling in the biosecurity management arena. Section 5.4 provides a data set and applies the methods described in the previous section to reveal the diversity among US Midwest beef farmers in approaches to biosecurity management. The final main section is a discussion of the limitations of the methods that have been previously described.

5.2 Review of Existing Methods

Understanding biosecurity management practices in livestock production is important for public sector officials who seek to guarantee animal health as a public good and to ensure food safety. In both cases, the potential is large for adverse externalities that may be difficult to address in the absence of regulatory intervention. In addition, as governments are often involved in insuring animal agriculture against income downturns, governments may also seek to reduce potential future liabilities by taking policy actions that alleviate the extent of or consequences arising from the production risks that livestock farmers are exposed to. As farmers are the primary health managers of their herds, the adoption of biosecurity practices by the herd owner is crucial for managing farm-level disease. This adoption is difficult to incentivize because it may be impractical to monitor voluntary actions, while efforts to mandate actions may face legal and political difficulties. For example, an effort to require the identification of premises among cattle owners in the USA was formally abandoned in 2013 (Federal Register, 2013). Although the motives for studying biosecurity actions are largely economic, technical and administrative complexities surround the identification of optimal biosecurity choices, so it is not surprising that most research in the area has been led by veterinary scientists.

Biosecurity management decisions can be assumed to depend on the profit-motivated trade-offs facing farmers. However, privately optimal and socially optimal actions diverge in the presence of infectious disease externalities. This issue is the concern of several lines of inquiry into disease management at regional and country scales of analysis. Strategic behaviours have been modelled by various authors (e.g. Hennessy, 2007; and Kobayashi and Melkonyan, 2011), because the spread of transmissible diseases is partly a consequence of individual preventive actions. Heterogeneity in economic behaviour and suboptimality of efforts have been addressed when considering the design of collective animal health management policies (e.g. Hennessy, 2008; Ceddia *et al.*, 2009; Wang and Hennessy, 2014). Nevertheless, the economic literature is sparse on how such issues translate to the individual farm level. This is in part because, with some notable exceptions, including imperfect competition and auctions, econometric estimation of strategic interactions is in its infancy (Bajari *et al.*, 2013). Such estimations are often based on carefully thought through structural models of behaviour, but considerable knowledge about outcomes and context are often required.

Most of the empirical literature has ignored spillover effects from biosecurity decisions, focusing instead on within-farm modelling issues. Stott and Gunn (2008), and also Stott (2009), have emphasized how biosecurity practices, and more generally animal disease control, comprise one part of general livestock management, and that these actions compete with other investment opportunities for the limited resources of the farm operation. By comparing biosecurity expenditures and their returns, Stott and Gunn (2008) argued that alternative expenditure allocations to act against disease entry compete for resources. In their paper, they relied on epidemiological modelling of bovine viral diarrhoea (BVD) when assessing the effects of alternative biosecurity choices on farm income. Their study demonstrated the complexity of the relationships between public health objectives on the one hand and individual incentives and constraints on the other. When identifying possible trade-offs in biosecurity management, Stott and Gunn (2008) compared the net benefits from various biosecurity practices to reveal health net benefit isoclines between biosecurity alternatives that compete for limited resources.

Stott and Gunn's technique is restrictive in practice because of the information constraints and difficulties encountered when seeking to collect data about specific benefits (e.g. economic, epidemiological, technical) afforded by one particular biosecurity practice in comparison with an alternative. As a consequence, most studies of biosecurity management at the farm level rely on data obtained from one-time surveys. Using a survey of livestock farmers that collected information on management priorities, farm structure and farm performance data, Mugnier *et al.* (2012) found that farmers consider health management to be a major decision domain in livestock farming. None the less, a principal component analysis, combined with a clustering of farmers according to their priorities, showed that farmers' livestock health management choices are heterogeneous along various dimensions of farm characteristics, including farmer's age and farm production intensity.

The case of bovine leukosis virus (BLV) has been studied by Gramig *et al.* (2010), who developed a simple theoretical model of decision making at the farm level that reflects trade-offs when deciding whether to adopt alternative biosecurity measures. Empirical application of this model was based on logistic regression with the intent to estimate the relationship between farm practices and disease prevalence, and also to model the binary decision on each biosecurity practice. As already stated, biosecurity practices comprise one part of the farmers' resource allocation choices, and these practices are expected to generate spillovers for other diseases. Gramig and Wolf (2007) investigated the resource allocation question using econometric methods. Their approach was based on combinations of numerous specific disease-control functions for the estimation of various disease prevalence levels. Unlike most other studies, in which control functions are estimated for one specific disease at a time, Gramig and Wolf's approach permits simultaneous estimations of several disease control functions at the farm level. The right-hand variables comprise biosecurity and other management practices that are common to several diseases. Econometric identification is a matter of concern in this line of research because biosecurity practices are discretionary choices. Such choices reflect management attributes as well as the farm's situation, and so biosecurity choices made cannot be assumed to be exogenously assigned explanatory variables.

As biosecurity practices are given variable priority at the farm level, operator characteristics shared by those who adopt, and the sorts of technology packages that are adopted, have been subjected to many empirical studies. Hoe and Ruegg (2006) observed high correlations between farm size and management practices, with the implication that larger production operations face either lower costs or higher benefits when making biosecurity choices. For Danish swine, Boklund *et al.* (2004) showed that herd size influences both biosecurity practices and the herd's declared health status. Using factor analysis, they partitioned biosecurity indicators according to correlation structure, and so revealed jointness in biosecurity management decisions.

More generally, by using survey responses and farm structure data, farmers can be partitioned into relatively homogeneous categories that reveal the extent of heterogeneity in their biosecurity management practices and so potentially shed light on their drivers. Costard *et al.* (2009) used multiple factor analysis (MFA), while Ribbens *et al.* (2008) used multiple correspondence analysis (MCA), and both studies used hierarchical cluster analyses (HCA) to scrutinize surveys on biosecurity practices. The methods used permitted the variables to be condensed into factor sets of biosecurity practices. In the case of biosecurity, management choices are often represented as categorical data, so MCA is an appropriate method for statistical analysis. MFA is more appropriate when categorical and quantitative variables are prominent in the same data set because this type of analysis can be interpreted as an extension of other factor analysis techniques. Just as factor analyses allow the identification of correlated variables, cluster analytical methods permit the analyst to create groups of statistical subjects. Effectively, HCA aims at defining groups of farmers with respect to their health management practices.

Other lines of research have focused on finding the drivers for biosecurity actions. Statistical regressions are then the main tools, and logistic regression is most prominent when the choices are discrete. Such regressions identify correlations between biosecurity practices (represented as categorical variables) and farm characteristics. Sayers *et al.* (2013) applied logistic regression to reveal regional differences in biosecurity practices and opinions among Irish dairy farms. An inference drawn was that heterogeneities in the type and size of livestock operations across different regions had consequences for the level of technical support available and therefore for the biosecurity advice and incentives that growers countenanced.

Heffernan *et al.* (2008) used cluster analysis and logistic regression analysis to highlight the drivers of the relative lack of interest of UK farmers' in taking biosecurity measures. In this analysis, a dependent variable is membership status in a biosecurity scheme, thus emphasizing the fact that individual decisions on biosecurity management are also partly driven by the socio-economic context. In order to analyse the impacts of the main a priori determinants of biosecurity actions among British farmers, Toma *et al.* (2013) implemented a structural equation modelling (SEM) framework. SEM is a combination of statistical methods that accommodates the modelling of latent variables and their causal relationships. The authors suggested that (among other determinants) the biosecurity behaviour of farmers may be a priori driven by their perceived effects of past diseases, perceptions about and knowledge of the effects of biosecurity measures, and whether a farm is certified organic. A separate line of inquiry in veterinary science research has focused on psychological and behavioural approaches to decision analysis in order to describe farmers' perceptions about biosecurity (Kristensen and Jakobsen, 2011) and to compare farmers' *ex ante* attitudes to biosecurity practices with their *ex post* behaviours (Gunn *et al.*, 2008).

Hence, understanding farm management choices in general and biosecurity practices in particular is dependent on many technical, institutional, and socio-economic factors. In the next section, we briefly explain three techniques used in the area: MCA, HCA and logistic regression.

5.3 The Core Model: Adoption Decisions in Biosecurity Management

The analysis of farm-level biosecurity management aims to identify effective preventive health practices as well as understand the determinants of their adoption. Preventive strategies in animal health are many and varied. Individual biosecurity strategies will be chosen in view of how they complement or substitute for other strategies as well as for any scale and scope economies that they provide. Practice choices are usually represented by discrete choice models, which are useful for describing and explaining decisions such as whether to test for a disease upon (livestock) purchase. These models allow the identification of statistical relationships between the choices made and attributes of the decision maker. For example farm management choices can be statistically related to some farm (e.g. location, climate) or farmer (e.g. age, education) attributes. Unlike models aiming to optimize over continuous choice variables, discrete choice models analyse alternative decisions under given modelling assumptions that might rationalize the use of methods such as logistic or probit regression. Basically, this kind of model aims to compare the utility (or net benefit) derived from alternative actions in order to determine whether an action is likely to be chosen.

In the specific case of agricultural health management choices, economic modelling can be founded on the presumption that expected utility of profit is maximized, which, in turn, depends on the own-herd health benefits and costs, which are themselves incident upon making alternative choices of biosecurity practice (an example is given in Gramig *et al.*, 2010). However, rigorous empirical validation of this kind of theoretical model is challenging as a large integrated data set (i.e. accounting information, and data on farm structure and management practices) is required. Such data sets can rarely be assembled from publicly available sources.

The key point in biosecurity management at the farm level is that biosecurity measures cannot be considered in isolation. They might best be viewed as a security plan, with practices chosen to work as an ensemble in reducing the probability of entry and/or spread of any health disorder in the herd. Thus, economic and statistical research stresses the multiplicity of choices that are made by farmers and, more specifically, their types of actions and the interconnections of these. From both economic and epidemiological points of view, the question is to understand how some combinations of biosecurity actions can complement one another in terms of reducing health hazards and improving livestock sector competitiveness. With an emphasis on the totality of practices, three of the most widely used tools in the farm-level analysis of biosecurity management are explained in this section: MCA, HCA and logistic regression. These statistical methods generally rely on surveys of management practices and farm structures.

5.3.1 Multiple correspondence analysis (MCA)

MCA is a statistical method adapted for the analysis of categorical variables. As a consequence, it is commonly used for farm survey analysis, especially when adoption decisions are binary, as is often the case for biosecurity practices. MCA methods aim to analyse relationship patterns among several categorical dependent variables, in a way similar to that in which principal component analysis can be used to analyse quantitative data.

In a survey, the *i*th farmer, among *I* individuals, is represented a set of variables that constitute their response profile. In the case of biosecurity actions, these variables are question responses and are mostly categorical. There are *V* response variables in all and we refer to a representative element as the *v*th variable. For instance, the testing of animals for specific diseases before entry into a herd (yes vs no) is a nominal variable with two levels.

In MCA, the data table is organized as a complete disjunctive table (or indicator matrix) in which each individual is represented by a row and each category – or level – of variable is coded into a column (taking the value 1 or 0), so that columns comprise *K* dummy variables whenever *K* discrete response levels are available. This representation allows for a separated consideration of each variable category in the MCA. The use of MCA on the indicator matrix allows for a reduction in the number of dimensions to a few synthetic dimensions. In order to obtain this synthetic data representation, two-by-two distances are calculated between individuals on the one hand, and variable categories on the other.

Following Le Roux and Rouanet (2010), the distance, *d*, between individuals is computed as follows. First, for question *v*, the part of the distance between two individuals *i* and *i′* due to question *v* is null, or $d_v^2(i,i')=0$, whenever *i* and *i′* choose the same answer. On the contrary, whenever *i* and *i′* choose different answers to the question, a non-zero distance exists between the two individuals. Assume now that two different categories *k* and *k′* can exist for question *v* when *i* chooses *k* and *i′* chooses *k′*. Then the part of the squared distance between *i* and *i′* due to question *v* is:

$$d_v^2(i,i')=\frac{1}{f_k}+\frac{1}{f_{k'}}$$

where $f_k \equiv I_k/I$ is the relative frequency of individuals who have chosen category *k*. More generally, the overall distance between *i* and *i′* is:

$$d^2(i,i')=\frac{1}{V}\sum_{v\in V} d_v^2(i,i')$$

As a consequence, a cloud of the *I* individuals is created in which each individual is represented as a point M^i. The distance (i.e. the difference) between *two* individuals grows as the correlations among the responses given decline. A weighted cloud of categories is also built, in which each of the *K* categories is represented as a point M^k and has a weight I_k, so that for each question *v*, the sum of the category point weights is *I* and the sum for all the *K* categories is *IK*.[1] Each point M^k has the weight $p_k \equiv I_k/(IV)$. Assuming that $I_{k,k'}$ is the number of individuals choosing both categories *k* and *k′*, then the squared distance between two category points M^k and $M^{k'}$ is:

$$\left(M^k M^{k'}\right)^2=\frac{I_k+I_{k'}-2I_{k,k'}}{I_{k,k'}/I}$$

where the numerator is the number of individuals who chose either *k* or *k′* but not both. The denominator represents the theoretical frequency for the (*k*, *k′*) answer combination.[2]

Similar to the distance between individuals, the distance between two variable categories grows with the heterogeneity of the individuals they comprise. Two individuals are measured as being closer whenever they share more variable levels, and are further apart whenever their variable levels are distinct. This is also how one should interpret distances between two variables in an *I*-dimensional space.

Distance metrics for variable categories and for individuals are then projected on to synthetic axes that aim to maximize the sum of their correlations. An advanced explanation of the geometrical implementation of these statistical data analysis techniques can be found in Greenacre and Blasius (2006) and Abdi and Valentin (2007). These synthetic axes – or synthetic dimensions – are the quantitative variables F_n, $n \in \{1,\ldots,N\}$, and are referred to as the factors of the MCA. The factors F_n are obtained by maximizing the sum of their correlations with the variable and with the individuals. In others words, when considering a $V \times I$-dimensional space for variable levels and individuals, the factors F_n are the orthogonal (i.e. uncorrelated) dimensions most correlated to them, where the F_n axes cross at the cloud's centroid.

The correlation ratios between the qualitative terms and quantitative factors F_n are obtained so that all levels of each variable have a degree of correlation with each MCA factor. More precisely, the contributions of the levels of the variables and of the individuals to the factors are calculated and the correlations help to locate individuals with reference to a given factor. The individuals are projected orthogonally on to the axes (factors) at the centre of gravity of the levels of variables they comprise while the levels of variables are projected orthogonally on to the axes at the centroid of the individuals having them.

Each factor F_n can be interpreted as a combination of levels of variables. More precisely, the projection and location of some M^k category points on specific F_n axes and the interpretation of their location on both sides of these axes provide a definition of the meaning of the F_n latent quantitative variables of the analysis. Note that both sides of the F_n axes are characterized by a certain degree of correlation with some of the M^k category points, meaning that for a given question v with two categories k_1 and k_2, the association of a side of the F_n factor with k_1 does not necessarily imply the association of the other side of F_n with k_2. Indeed, the category point cloud has a centroid which is the intersection of all of the F_n factors. Thus, the less frequently some category k_1 is observed, the more it contributes to variance in the category point cloud and the further the point M^{k_1} is from the cloud's centroid. In contrast, a frequent category k_2 is closer to the cloud's centroid and tends to have a lower degree of association with the F_n factor sides. As with other factor analysis methods, the importance of the axes F_n in explaining variance in the data decrease as n grows, i.e. factor F_1 explains more data variation than does F_2, and so on.

The combination of two orthogonal factors allows for a graphic representation of the data as scatter plots in the Cartesian plane. This plane representation of the data is called a factor map. In MCA, the number of dimensions that are analysed depends on their eigenvalues as well as their interpretability. Typically, only the first factor map (representing factors F_1 and F_2) is used for analysis because it integrates the most representative dimensions of the data set. In the factor map, the M^i individual points are depicted and located with respect to their association with the F_n axes. Levels of variables shared by the highest number of individuals are proximate, while levels of variables that are not correlated are as separated as possible. As a consequence, individuals with many levels of variables in common are near each other, but the most dissimilar individuals are separated. Groups of individuals can be revealed by the use of clustering methods.

5.3.2 Hierarchical clustering analysis (HCA)

Cluster analysis consists of grouping sets of individuals or objects according to the similarities observed among them. The approach seeks to sort into groups, or clusters, composed of individuals that are more similar to each other than to those in other groups. Many clustering techniques exist, but the most relevant for our purpose is the hierarchical clustering method.

As in MCA, consider a data set comprising I individuals and V variables. Each individual is characterized by its own answers (or categories) to the V questions, and differences are observed in the K categories among the I individuals. The dissimilarities among individuals can be measured as distances.

HCA is a stepwise clustering method in which the I individuals are initially in different clusters. At each step, the closest pair of clusters (initially the least dissimilar pair of individuals) is merged into a single cluster. The within-cluster dissimilarity value is called the intra-class inertia, and the between-cluster dissimilarity value is referred to as the inter-class inertia. So, by construction, in HCA, the intra-class inertia of the clusters obtained increases at each step. Many different methods exist to measure the distance between observations and the distance between clusters (sets of observations). Consequently, measurements of intra- and inter-class inertia can be sensitive to modelling choices.

Outcomes from HCA can be presented as a dendrogram (or binary tree, so called because at each step two objects – or clusters of objects – are merged; see an example in Section 5.4, Empirical Application), where all the clusters appear as branches. Intra- and inter-class inertias are computed as branch lengths such that the levels of dissimilarity within a cluster and between clusters are identifiable. In this framework, one way to identify clusters is to choose the partition set that maximizes the inter-class to intra-class inertia ratio. This criterion is, however, inappropriate in some cases, as it may identify too many or too few clusters in comparison with alternative approaches. The optimal choice of clusters can also be determined by visual inspection of the dendrogram.

5.3.3 Logistic regression

Logistic (or logit) regression is a statistical method for modelling and measuring relationships between a qualitative dependent variable and independent categorical or continuous variables. It is used to estimate the parameters of qualitative response models. This method is particularly useful when analysing production and management choices as a function of farm types, which are all categorical variables.

Consider an endeavour to estimate the probability that a farmer will adopt a biosecurity practice. The variable is categorical, having the value $y=1$ whenever the farmer adopts the biosecurity practice and the value $y=0$ otherwise. The two alternatives – adoption ($y=1$) vs non-adoption ($y=0$) – support different utility functions for the farmers (U_0 if $y=0$ and U_1 if $y=1$). These utilities cannot be directly observed. They are heterogeneous among farmers and depend on a set of farm characteristics (e.g. farm operator's age, disease prevalence, management costs, incentives) represented by x_i. Expected utilities from the alternative choices are then represented as:

$$U_0 = a_0 + \sum_i a_{0i} x_i + \varepsilon_0;$$

$$U_1 = a_1 + \sum_i a_{1i} x_i + \varepsilon_1$$

where a_0, a_{0i}, a_1 and a_{1i} are regression coefficients, and ε_0 and ε_1 are residuals or error terms.

The difference between these two functions is not observable as it is latent, or hidden, but the decision on whether or not to adopt the biosecurity practice depends on this difference. Specifically, the farmer adopts the biosecurity practice whenever $U_1 \geq U_0$. The probability, Pr, of adoption, P_{adopt}, is:

$$P_{adopt} = \Pr(y=1) = \Pr(U_1 \geq U_0)$$

$$= \Pr\left(a_1 - a_0 + \sum_i (a_{1i} - a_{0a}) x_i \geq \varepsilon_0 - \varepsilon_1\right)$$

$$\Pr(\beta x \geq \varepsilon_0 - \varepsilon_1) = F(\beta x)$$

The adoption probability P_{adopt} is a cumulative distribution function $F(\cdot)$ in which x is a vector of the explanatory variables x_i and β is a vector of parameters representing differences in the underlying coefficients with $\beta_0 = a_1 - a_0$ and $\beta_i = a_{1i} - a_{0i}$. The parameter vector β is to be estimated. In a logit model distribution function, $F(\cdot)$ is a logistic function, whereas in a probit model, the distribution is cumulative normal. In general, results are not sensitive to the model used. An estimate of the coefficient vector β is then generally obtained by maximum likelihood methods. Further explanation of qualitative variables in econometrics and the logistic regression can be found in, for example, Maddala (1983) and Hosmer *et al.* (2013).

The statistical and econometric methods presented in this section are widely used in both applied economics and veterinary epidemiology research. They allow for the analysis of multiple choices, which are a common feature of biosecurity choice sets available at the farm level. However, extant research has seldom inquired into epidemiological and economic issues simultaneously. Progress will be made if the study of biological and socio-economic issues can be integrated into a unified modelling framework of the many decisions that comprise animal health management at the farm level. Such research should assist in the design and evaluation of agricultural and health policies with the objectives of improving animal health and welfare, farm income and human health.

5.4 Empirical Application

In this section, we apply the methods described in the previous section to a survey of farmer biosecurity choices. The findings highlight how these techniques can bring out the underlying structure of a data set by revealing complementary practices, alternative strategies and heterogeneous management patterns that underlie it.

5.4.1 Description of data

In 2007, Iowa State University's College of Veterinary Medicine, the University of Nebraska's Great Plains Veterinary Educational Center and Kansas State University's College of Veterinary Medicine administered a joint survey that focused on cattle producers and their biosecurity practices in Iowa, Nebraska and Kansas. Data were generously provided by Dr Annette O'Connor at Iowa State University. The surveys were designed to better understand how farmers integrate biosecurity practices into routine management activities.

The survey questions were focused on biosecurity management practices. A first set of questions addressed regular management procedures, such as the testing of cattle upon introduction into a herd, testing for specific purposes (e.g. for specific diseases, or diagnosing cause of death), quarantine activities and cleaning procedures. A second set focused on monitoring potential sources of disease entry, and on contacts with others herds, contacts with wildlife or contacts with professionals working at more than one farm. A last set of questions queried the maintenance of herd health and performance records. The whole survey comprised 26 questions for cow-calf producers (seed stock producers, who produce purebred or registered cattle for breeding; and commercial producers, who keep a permanent herd of cows to produce calves for later sale) and 18 for feedlot producers (who finish cattle in feedlots for sale or slaughter). Responses were categorical, typically Yes/No or Never/Seldom/Sometimes/Often/Always, and many of the questions were asked of both feedlot and cow-calf producers. A preliminary examination of the answers to these questions provides a sense of the variety and complexity of the biosecurity decisions that cattle farmers make.

Table 5.1 confirms that biosecurity practices can differ significantly among farmers, depending on enterprise type. Although some practices, such as cleaning medical equipment, were common across enterprise type, others (such as routine post-mortem examinations) were seldom practised on any farm. In addition, uptake of some practices varied greatly across enterprise type. For example, isolation pens for sick animals were used by 24% of cow-calf seed stock producers, 18% of cow-calf commercial producers and 55% of feedlot producers.

5.4.2 Analysis of data by MCA and hierarchical clustering

In order to represent the data set's underlying structure, MCA was performed separately for both cow-calf producers and feedlot producers. Hierarchical clustering was also carried out so as to group farmers with respect to their practices. The two methods were combined so as to extract information on farm profiles and farmer biosecurity management strategies.

In addition to collecting information on farmers' biosecurity practices, the surveys collected three other variables that characterized the nature of the production operation and the producer's views about biosecurity. These were: (i) attendance at meetings or presentations about livestock biosecurity (for both cow-calf producers and feedlot producers); (ii) type of production (either seed stock or commercial); and (iii) bringing some herd animals to shows or sales. Items (ii) and (iii) relate only to cow-calf producers. For our analysis, these variables were selected as additional explanatory variables and were not taken into account when building factor maps. These additional explanatory variables are analysed in Section 5.4.4.

Table 5.1. Descriptive statistics from a survey of biosecurity practices among US Midwest farmers: number of farmer responses by state and examples of heterogeneity in practices among farmers

Farmer responses/Farmer types	Cow-calf seed stock producers	Cow-calf commercial producers	Feedlot producers
Number of responses by state:			
Iowa	8	44	25
Kansas	13	45	30
Nebraska	33	33	28
Farmers who:			
always isolate cattle on entry	14.8%	16.4%	22.9%
always test cattle for specific diseases before they join the herd	11.1%	9.8%	2.4%
always keep treated sick cattle in isolated pens	24.1%	18.0%	55.4%
always have dead animals examined	5.6%	6.6%	10.8%
routinely clean feeding areas and equipment	41.7%	23.8%	34.3%
routinely clean medicating equipment	85.2%	79.9%	79.5%
have control measures to limit the access to cattle areas by wildlife animals	51.9%	41.0%	53.6%
maintain a record of visitors to their operation	27.8%	6.6%	15.7%

5.4.3 MCA and interpretation of the dimensions it generates

An MCA was performed on cow-calf producer data, the principal outcome of which was a design of factor maps arising from correlations among the variable categories. The results show a large diversity among answers. When programmed to perform the analysis along 20 dimensions, the eigenvalues show that the first ten axes of the MCA explain 55.1% of sample variation, revealing high heterogeneity in practice choices. The first two axes are the most representative and interpretable, and explain, respectively, 12.4 and 7.1% of the sample variation. These axes represent synthetic or aggregated variables. Categories of the initial variables are associated with the different axes in the MCA decomposition. The categories associated with the first dimensions are summarised in Table 5.2.

The analysis of variable categories associated with a dimension of the factor map allows us to encapsulate what each axis seeks to characterize. Axis 1 represents rather opposite behaviours, i.e. diverse categories of variables. Side A can be interpreted as an aggregate of 'good' biosecurity practices, including the use of animal examinations, scheduled cleaning practices, control of wildlife contacts and record keeping for farm visits and herd production. The opposite side of the first axis (side B) reveals the variable categories characterizing individuals that are distant from those who implement the 'good' biosecurity practices that are aggregated on side A. Quite different practices define this side of the axis, as characterized by the absence of: (i) scheduled cleaning practices; (ii) control measures for wildlife; and (iii) record keeping for farm operations. To sum up, axis 1 helps to separate cow-calf producers who implement regular biosecurity practices from cow-calf producers who place much less emphasis on biosecurity practices.

Axis 2 gathers heterogeneous practices, and characterizes individuals who have very different approaches to biosecurity management. On side A, biosecurity control occurs through monitoring practices (pre-breeding examinations on bulls, recording pregnancy rates) although this side does not include control of food, feeding areas, contacts with other herds or contact with wildlife. Axis side B collects quite opposite behaviours. On the one hand it is characterized by high levels of certain practices (scheduled cleaning, control measures for wildlife) and equipment categories, but on the other hand, less attention is paid to the animals (e.g. there are no tests for specific diseases and no herd production records).

Table 5.2. Categories of variables associated with the two sides (A and B) of the first dimensions (axes 1 and 2) of a multiple correspondence analysis (MCA) of data from US Midwest cow-calf producers.

Axis 1			
Side A		Side B	
Variable	Level	Variable	Level
Isolation of cattle on entry	[yes]	Scheduled cleaning practices	[no]
Sick animals treated separately	[yes]	Control measures for wildlife access	[no]
Pre-breeding examination of heifers	[yes]	Records of farm operations	[no]
Scheduled cleaning practices	[yes]		
Control measures for wildlife access	[yes]		
Records of visits	[yes]		
Records of herd production	[yes]		

Axis 2			
Side A		Side B	
Variable	Level	Variable	Level
Pre-breeding examination of bulls	[yes]	Scheduled cleaning of equipment and buildings	[yes]
Records of pregnancy rate	[yes]	Control measures for wildlife access	[yes]
Control of food	[no]	Test for specific diseases	[no]
Control of feeding areas	[no]	Examination of bulls	[no]
Control of contacts with other herds	[no]	Immunization schedules	[no]
Control measures for wildlife access	[no]	Records of herd production	[no]

An MCA was also performed (over 20 dimensions) for feedlot producers. The most interpretable and significant explanatory axes were again the first two. The first axis accounted for 18.0% of the data variation while the second axis accounted for 10.3%. The variable categories characterizing the two first dimensions are as given in Table 5.3.

Axis 1 gathers quite opposite health management strategies. To sum up, side A is associated with only a few category points, including the absence of both regular cleaning practices and control measures for wildlife. Side B is associated with a large set of category points. It comprises many biosecurity practices, including the regular cleaning of pens, the use of different equipment to clean pens and handle feed, the isolation of cattle upon entry and the examination of dead animals, the implementation of control measures for wildlife animals and keeping records of visitors to the operation. We can conclude that, in the factor map, on one side of the MCA's first dimension/axis are feedlot owners who implement multiple biosecurity practices, while on the other side are feedlot owners who may have diverse practices (revealed by the relatively low amount of significantly associated category points), the common denominator being that cleaning practices and wildlife control are not implemented.

The results for axis 2 reveal heterogeneity in the acceptance of biosecurity practices on some operations compared with others. Side B pools routine cleaning practices, tests for specific diseases and requires an animal's history before admission into a herd, but sick animals are not treated separately and dead animals are not examined. On the other side, A, distant individuals are characterized by higher use of facilities for biosecurity purposes; specific facilities are used to treat sick animals and new arrivals, and different equipment is used to clean pens and handle feed. However, cleaning practices are not regularly scheduled.

To sum up, in this step of the analysis, and for both feedlot producers and cow-calf producers, MCA identifies some trends in individual biosecurity practices. For both feedlot and cow-calf producers, the MCA's first dimension, which is the most representative of each data set, reveals

Table 5.3. Categories of variables associated with the two sides (A and B) of the first dimensions (axes 1 and 2) a multiple correspondence analysis (MCA) of data from US Midwest feedlot producers

Axis 1			
Side A		Side B	
Variable	Level	Variable	Level
Regular cleaning practices	[no]	Isolation of cattle on entry	[yes]
Control measures for wildlife access	[no]	Different equipment to clean pens and handle feed	[yes]
		Scheduled examination of dead animals	[yes]
		Regular cleaning practices	[yes]
		Control measures for wildlife access	[yes]
		Records of visitors	[yes]

Axis 2			
Side A		Side B	
Variable	Level	Variable	Level
Different equipment to clean pens and handle feed	[yes]	Testing for specific diseases	[yes]
Sick animals treated separately	[yes]	Request of health and history of cattle on entry	[yes]
Different facilities for new arrivals and sick animals	[yes]	Sick animals treated separately	[no]
Scheduled cleaning practices	[no]	Examination of dead animals	[no]
		Scheduled cleaning practices	[yes]

the combined implementation of many biosecurity practices on one side and instances of poor biosecurity management practices on the other. For both data sets, the second dimension of analysis reveals heterogeneities among operations in chosen biosecurity practices.

5.4.4 Additional explanatory variables

In order to better understand cluster formation for both cow-calf and feedlot operations, some additional explanatory variables were also selected, as noted in Section 5.4.2. For both cow-calf and feedlot operations, we considered whether the respondent attended meetings about livestock biosecurity. For cow-calf operations we considered type of production, be it seed stock or commercial production, and also whether herd animals were brought to shows or sales. These additional variables can be projected on to the factors (dimensions) of the corresponding MCA to serve as illustrative elements in the analysis. We calculate correlations between the levels of the additional explanatory variables and the factors. Factor scores for the additional variables allow us to locate the individuals represented by these variable categories on the factor maps, and thus to enrich the interpretation of dimensions emerging from the MCA. The results are given in Table 5.4, in which factor scores reveal the degree of association between these levels of variables and both sides of the MCA factors, and a null factor score reveals no association.

Intuition would suggest that the additional explanatory variables are positively correlated with biosecurity actions. Indeed, when compared with animals in commercial production, animals in seed stock production have potentially high individual value added which may justify increased willingness to protect seed stock herds from disease entry. Moreover, attendance at meetings about livestock biosecurity is expected to be correlated with greater owner attention to livestock biosecurity.

Table 5.4. Association between levels of the explanatory variables and the first factors of the multiple correspondence analysis (MCA) for US Midwest cow-calf producers and feedlot producers (factor scores in brackets).

	Axis 1		Axis 2	
	Side A	Side B	Side A	Side B
Cow-calf producers				
Type of production	Seed stock (0.33)	Commercial (0.17)	Seed stock (0.09)	Commercial (0.04)
Participation in meetings about biosecurity	Yes (0.15)	No (0.06)	Yes (0.24)	No (0.10)
Participation in exhibits or sales	Yes (0.40)	No (0.13)	Yes (0.11)	No (0.04)
Feedlot producers				
Participation in meetings about biosecurity	No (0.28)	Yes (0.26)	Yes (0.17)	No (0.28)

The results for cow-calf producers are not surprising per se. First, the factor scores for exhibit participants and seed stock producers are uniformly of similar magnitude across the table. This is intuitive, as seed stock producers are expected to be more inclined to show their animals; they participate in exhibits as well as attending biosecurity meetings (to a lesser extent, on the basis of its factor score), all of which are calculated to be correlated with side A of both the first and second dimensions of the MCA. Recall that side A of the MCA's first axis is interpreted as 'good' biosecurity practices, while side A of the second axis characterizes individuals paying relatively high attention to their animals' health status but making comparatively little use of equipment and infrastructure for biosecurity purposes. The additional explanatory variables confirm the presumption that seed stock producers tend to display high levels of biosecurity management and that attendance at biosecurity information meetings is a good indicator of the importance that a farmer attaches to his herd's health status and to taking multiple biosecurity actions. Second, the factor scores for commercial producers, for non-participants in biosecurity meetings and for farmers not exhibiting animals at shows are relatively close to each other and are correlated with side B for each of the first and second MCA axes. These results indicate that the profile of commercial producers is close to that of non-participants in biosecurity meetings, and that these types of farmers tend to implement biosecurity practices at relatively low levels (corresponding to side B in axis 1 in Table 5.2). In addition, their biosecurity management tends to rely more on cleaning practices and control of wildlife than on monitoring and control of herd health status (side B in axis 2 in Table 5.2).

For attendance at biosecurity meetings, evidence from feedlot producer responses is consistent with that from cow-calf producer responses. Attendance at biosecurity meetings is associated with a relatively high level of biosecurity practice adoption (side B of axis 1 in Table 5.3) and with a biosecurity strategy based more on monitoring herd health status (side B of axis 2 in Table 5.3) than on the use of specific infrastructure (side A of axis 2 in Table 5.3). Of course, we cannot draw causal inference from the above as attendance may be 'caused' by the biosecurity risks faced and we have no information available to support a claim that attending such meetings changed biosecurity behaviour.

5.4.5 Clustering and characterization of the biosecurity practices

We conducted hierarchical cluster analysis in order to identify groups among farmers according to their biosecurity actions. This technique permits us to distinguish behavioural types among cow-calf and feedlot producers. As a consequence, the interviewed farmers can be categorized into clusters. The result is the formation of groups of farmers who are similar to each other but dissimilar from other groups, so that the scatter plot of individuals in the MCA can be disaggregated according to these clusters.

In this analysis, the Euclidian metric was chosen to measure inter-individual distance while inter-cluster distance (also called the linkage criterion) was obtained by the use of Ward's criterion. For cow-calf producers, a total of three classes (clusters) emerged from the HCA. As previously explained, hierarchical clustering can be represented by a dendrogram. Figure 5.1 illustrates the dendrogram for cow-calf producers, and shows these three clusters.

Recall that MCA can be scrutinized by a factor map in which the axes (factors) are synthetic variable projections of the original response variables and where individuals are represented by points on the map. Hierarchical clustering allows us to create significantly different groups of individuals in the factor map's scatter plot so that each cluster can be interpreted independently. Figure 5.2 represents the MCA's factor map for cow-calf producers according to its two first dimensions; the three clusters obtained by the HCA (see Fig. 5.1) are represented by different symbols.

Considering our interpretation of the MCA, the three behavioural profiles of cow-calf producers, which correspond to the three HCA clusters, are as follows. Around 18% of farmers

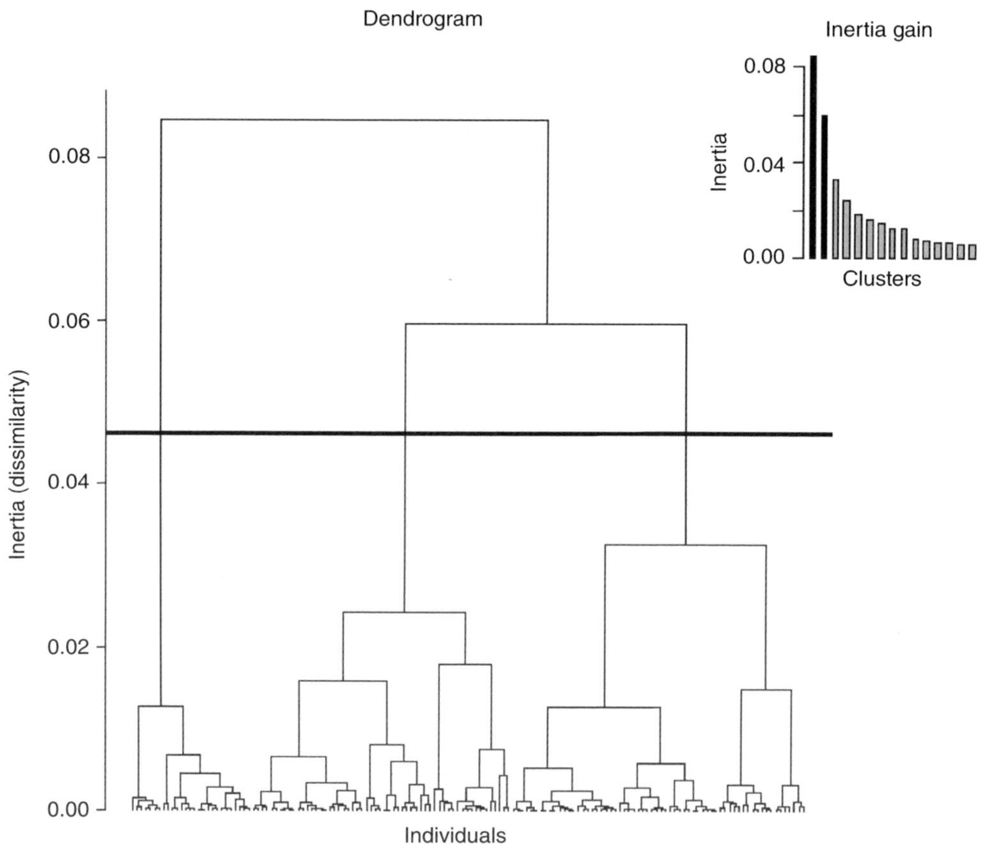

Fig. 5.1. Large diagram left: dendrogram (or binary tree) resulting from the hierarchical clustering of US Midwest cow-calf producers according to their biosecurity actions. The suggested number of different clusters is obtained from the maximization of the intra-cluster inertia. As represented in the figure at top right, the maximal gain of intra-cluster inertia is observed between the second and the third cluster (see the cluster cut-off shown on the dendrogram). As a result, three different clusters of individuals are identified.

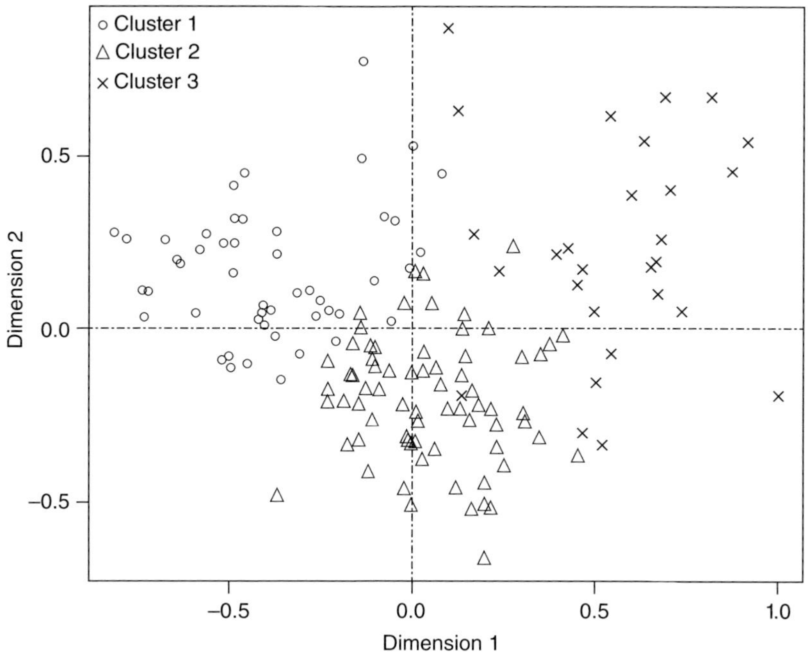

Fig. 5.2. Factor map and hierarchical clustering of the biosecurity practices of US Midwest cow-calf producers. Sides A and B of dimension 1 (axis 1) are, respectively, on the left and right, and sides A and B of dimension 2 (axis 2) are, respectively, down and up (see text for further details).

interviewed (cluster 3) are found to pay little attention to and/or adopt few biosecurity practices (right side of axis/dimension 1). They are more inclined to manage animal health by scheduled cleaning practices and the control of wildlife entry than by monitoring herd health status (right side of axis/dimension 2). In light of our analysis of additional explanatory variables, we can assume that these farmers have the profile of commercial producers. In contrast, 39% of farmers (cluster 1) tend to implement many biosecurity practices (side A of axis/dimension 1). This cluster may include seed stock producers (by correlation with the corresponding additional explanatory variable), but a large fraction of the cluster is also characterized by the minimal adoption of disease tests in the producers' approach to biosecurity (side B of axis/dimension 2). The most numerous cluster of cow-calf producers (43% of the sample, cluster 2) display intermediate behaviours towards biosecurity, which are between the two extremes represented by the other clusters. The analysis reveals that: (i) producers in this cluster take a wide range of approaches to managing external sources of disease; (ii) most herd owners in this cluster pay attention to their herd's disease status; and (iii) recording herd performance is important for these growers. In addition, this cluster is also correlated with the seed stock producer profile.

Concerning feedlot producers, two clusters, which are interpretable and significant (in size), emerge from hierarchical clustering. These are represented in the MCA factor map provided in Fig. 5.3.

In light of the MCA's first two dimensions, we can now draw some profile characteristics for the two clusters. The 26 farmers who comprise cluster 2 differ from the other cluster according to the MCA's first dimension/axis. Most (cluster 1) do not implement many biosecurity measures on their farm (dimension/axis 1), although some attention is paid to cleaning practices (dimension/axis 2). A smaller group of farmers (cluster 2), though heterogeneous in terms of practices (dimension/axis 2), place more emphasis on biosecurity actions in general and tend to combine various complementary actions in order to maintain high biosecurity levels for their herds.

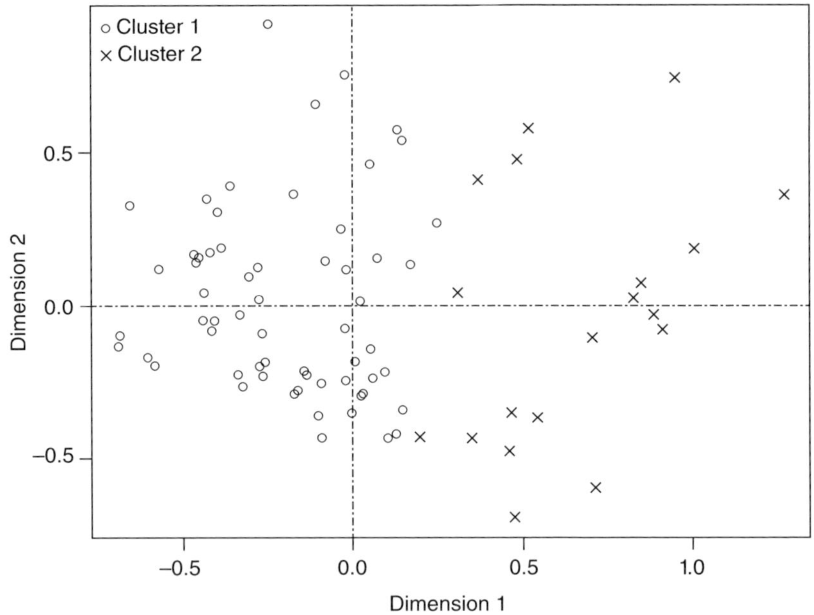

Fig. 5.3. Factor map and hierarchical clustering of the biosecurity practices of US Midwest feedlot producers. Sides A and B of dimension 1 (axis 1) are, respectively, on the left and right, and sides A and B of dimension 2 (axis 2) are, respectively, down and up (see text for further details).

5.4.6 Logistic regression

Perspective on the main differences in the drivers of biosecurity management among seed stock and commercial cow-calf producers is provided by a logistic (logit) regression. One of the few variables in the sample that can be correlated with biosecurity practices is the type of production carried out by the interviewed farmers. More specifically, it is expected that commercial producers and seed stock producers differ in the incentives they face for biosecurity practice adoption because seed stock herd animals are likely to be more valuable. For this logistic regression, the dependent variable is the type of production (seed stock or commercial production).

First, a univariate analysis permits the selection of variables that associate with 'type of production'. A multivariate logistic regression is then undertaken with intent to identify any link between the type of production and various biosecurity practices. After the selection of variables, maximum likelihood estimation gives the results provided in Table 5.5.

Overall, the regression results clearly highlight some different views on health management among commercial and seed stock producers. When compared with commercial producers, seed stock producers pay more attention to control measures against rodents and other wildlife animals in order to reduce the likelihood that disease enters an operation. In addition, they report that they routinely clean their equipment as well as their feeding and holding areas.

When compared with commercial herds, seed stock production herds show a lower propensity to share grassland with other cattle. Nevertheless, as these herd producers often exhibit animals, they are more exposed to other animals by this channel. Finally, the regression points out that, when compared with commercial breeders, seed stock producers may pay less attention to the examination of pre-breeding bulls – although it may be that most bulls in seed stock production are raised rather than purchased, so that such examinations are unnecessary.

5.5 Limitations and Discussion

Health management is a central feature of animal production and herd owners take many different

Table 5.5. Cow-calf seed stock production status (seed stock/commercial) regressed on management and biosecurity practices using data from US Midwest cow-calf producers. Maximum likelihood estimated values of parameters from the logistic regression.

Type of production/variable	Estimated value	Standard error	Pr > χ^{2}[a]
Regression intercept	0.28	0.34	0.406
Cattle grazed in common with other cattle	−0.54	0.25	0.028
Pre-breeding examinations on bulls	−0.71	0.23	0.002
Cleaning areas and equipment	0.55	0.27	0.041
Cattle exposed to other animals at exhibits or sales	1.06	0.23	<0.0001
Control measures that limit access to cattle pens by wild animals	0.47	0.22	0.031

[a]Probability that estimated maximum likelihood is greater than χ^2, *P* value.

approaches to disease control. For veterinary and economic research, as well as for policy analysis, modelling health and biosecurity management at the farm level is crucial to understanding agricultural producer behaviour and to appropriately designing health management policies. In this chapter, we have presented some statistical and econometric tools that can be used to characterize biosecurity management practices at the farm level. Our review of the literature has revealed that, except for rare examples, most studies on animal sector biosecurity have been disease specific and/or based on descriptive statistics obtained from producer surveys. In the absence of alternatives, one-time farm surveys have been the primary information source for studies of farm-level biosecurity management decisions. These studies provide little information on either animal health or farm performance outcomes, and have tended to focus on specific diseases in order to highlight possibilities for targeted actions. Longitudinal studies are more expensive and are not often carried out, even for human health issues, but would allow for deeper inquiries into the connections between actions and outcomes. The design of specifically adapted one-time surveys on individual biosecurity practices and a particular disease is an inexpensive approach to obtaining original information on specific issues, but ignores the fact that actual farm approaches to biosecurity are designed to address many animal disease and food safety issues.

A few studies have recognized that disease management activities compete for limited resources, that different producers have different objectives and that any given biosecurity choice can address multiple disease risks. Insights on heterogeneity among farmers' biosecurity strategies have been obtained primarily from statistical and econometric tools in order to determine practice adoption. Of these tools, factor analysis approaches (including multiple correspondence analysis) can indicate complementarities among biosecurity practices and identify management profiles among farmers in their preventive health practices. In addition, econometric tools such as logit regression (for categorical variables) are useful for estimating determinants of biosecurity actions among operations.

The application of these methods to cattle producers in the US Midwest has illustrated the strengths and weaknesses of such research methods. As shown in the empirical application, a statistical treatment of biosecurity surveys highlights patterns in biosecurity strategies among the farmers surveyed. Various health management profiles are identified among these producers, and operation types are correlated with particular biosecurity actions. The main strength of this approach is its ability to reveal the extent of heterogeneity in biosecurity practices among the producers. However, heterogeneity in approaches to biosecurity among farmers points to heterogeneity in the economic incentives faced by these farmers. The heterogeneity in incentives that growers face when making biosecurity decisions has received little, if any, attention, but it is an issue that deserves attention.

If progress at the interface between veterinary research and economic research is to be made, then the complex interplays between disease entry and propagation mechanisms, biological responses, and management decisions at various levels will need to be addressed. For evaluation or estimation purposes, one challenge facing multidisciplinary animal health projects is that the associated biological and economic analyses are likely to be data intense, so that further efforts to link the data may overburden respondents. An integrated analysis would involve linking multiple sources of information, including herd health status, accounting data and management information, in order to estimate the actual effect of various control measures on both farm profitability and disease prevalence.

Common agricultural databases (at least in the USA and the European Union) provide detailed farm-level financial accounting information, but very limited information on management practices, and almost no information on animal health management. Most research in the field of food, animal disease and health is performed by veterinary scientists *sensu lato* (e.g. epidemiologists, parasitologists), and the collected data seldom permit linkage between health practices and outcomes with farm-level economic data. The dearth of integrated animal health, financial, grower choice and grower socio-demographic data leaves the study of animal health economics at a disadvantage relative to other topic areas in agricultural economics, such as agri-environmental economics, agricultural finance, industrial organization and food consumption, where broad data sets have been assembled by the use of such sources as remote sensing, financial market information and supermarket scanner data. In the absence of data, the incentive to develop conceptual and modelling frameworks is lower. Endeavours to co-construct conceptual and data modelling frameworks at the interface between biological and economic approaches to animal health management have been limited in scope and depth (but see Cho *et al.*, 2013, for example). This is unfortunate, because efforts at building closer correspondences between economic incentives and disease outcomes may unearth innovative insights for the design of health policies in agriculture.

A further problem with existing research is that one cannot necessarily infer from survey data or observed behaviour that correlations between practices and outcomes imply causality, as the effects may not be adequately identified in the econometric sense. Elsewhere (e.g. Cohen and Dupas, 2010), economists have worked with healthcare personnel in the field to provide robust insights on the benefits and costs of biosecurity practices. Their primary method has been randomized controlled trials. Such careful analysis is costly, even when conducted in less developed countries and applied to discrete stand-alone technologies. Further, the approach may be difficult to replicate in production agriculture where technologies often come as packages.

Other problems also confound inquiry into biosecurity choices. One of particular relevance to empirical modelling is behaviourally grounded critiques of the standard microeconomic paradigm. Biosecurity management at the farm level amounts to preventive behaviour aimed at protecting assets from the potential occurrence of animal diseases. Preventive animal health practices generally address multiple disease risks and, in many cases, these risks occur with low probability and the disease may not be well understood among experts. Our comprehension of how people manage low and ambiguous probability events is limited (see, for instance, Schade *et al.*, 2012), and how given choices are framed and presented can determine the choice made (Tversky and Kahneman, 1981). Studies of farm animal health decisions that emphasize what farmers know and find to be salient may enable recent progress in behavioural economics to inform our understanding of and public policy towards biosecurity management choices.

Notes

[1] In the categorical data analysis literature, 'cloud' refers to data as represented by a large set of discrete points displayed on the Cartesian plane.

[2] In a K_v by $K_{v'}$ two-way table, the theoretical frequency represents the theoretical proportion of answers in the (k,k') cell, under the hypothesis that the K categories follow a discrete uniform distribution.

Acknowledgements

This work was written while David Hennessy was at Iowa State University. We are very grateful to Annette O'Connor for sharing the data and for insights that she provided to us.

References

Abdi, H. and Valentin, D. (2007) Multiple correspondence analysis. In: Salkind, N.J. (ed.) *Encyclopedia of Measurement and Statistics*. Sage, Thousand Oaks, California, pp. 651–657.

Bajari, P., Hong, H. and Nekipelov, D. (2013) Game theory and econometrics: a survey of some recent research. In: Acemoglu, D., Arellano, M. and Dekel, E. (eds) *Advances in Economics and Econometrics, Tenth World Congress, Volume III, Econometrics*. Econometric Society Monograph No. 51. Cambridge University Press, Cambridge, UK, pp. 3–52.

Boklund, A., Alban, L., Mortensen, S. and Houe, H. (2004) Biosecurity in 116 Danish fattening swineherds: descriptive results and factor analysis. *Preventive Veterinary Medicine* 66, 49–62.

Ceddia, M.G., Heikkilä, J. and Peltola, J. (2009) Managing invasive alien species with professional and hobby farmers: insights from ecological-economic modelling. *Ecological Economics* 68, 1366–1374.

Cho, J., Tauer, L.W., Schukken, Y.H., Smith, R.L., Lu, Z. and Grohn, Y.T. (2013) Cost-effective control strategies for Johne's disease in dairy herds. *Canadian Journal of Agricultural Economics* 61, 583–608.

Cohen, J. and Dupas, P. (2010) Free distribution or cost-sharing? Evidence from a randomized malaria prevention experiment. *Quarterly Journal of Economics* 125, 1–45.

Costard, S., Porphyre, V., Messad, S., Rakotondrahanta, S., Vidon, H., Roger, F. and Pfeiffer, D.U. (2009) Multivariate analysis of management and biosecurity practices in smallholder pig farms in Madagascar. *Preventive Veterinary Medicine* 92, 199–209.

Federal Register (2013) Traceability for livestock movement interstate; final rule. *Federal Register* 78 FR 2039 (6, January 9), 2039–2075.

Gramig, B.M. and Wolf, C.A. (2007) Estimating within-herd preventive spillovers in livestock disease management. *American Journal of Agricultural Economics* 89, 1219–1225.

Gramig, B.M., Wolf, C.A. and Lupi, F. (2010) Understanding adoption of livestock health management practices: the case of ovine leukosis virus. *Canadian Journal of Agricultural Economics* 58, 343–360.

Greenacre, M. and Blasius, J. (eds) (2006) *Multiple Correspondence Analysis and Related Methods*. Chapman and Hall/CRC, Taylor & Francis, Boca Raton, Florida.

Gunn, G., Heffernan, C., Hall, M., McLeod, A. and Hovi, M. (2008) Measuring and comparing constraints to improved biosecurity amongst GB farmers, veterinarians and the auxiliary industries. *Preventive Veterinary Medicine* 84, 310–323.

Heffernan, C., Nielsen, L., Thomson, K. and Gunn, G. (2008) An exploration of the drivers to bio-security collective action among a sample of UK cattle and sheep farmers. *Preventive Veterinary Medicine* 87, 358–372.

Hennessy, D.A. (2007) Biosecurity and spread of an infectious animal disease. *American Journal of Agricultural Economics* 89, 1226–1231.

Hennessy, D.A. (2008) Biosecurity incentives, network effects, and entry of a rapidly spreading pest. *Ecological Economics* 68, 230–239.

Hoe, F.G.H. and Ruegg, P.L. (2006) Opinions and practices of Wisconsin dairy producers about biosecurity and animal well being. *Journal of Dairy Science* 89, 2297–2308.

Hosmer, D.W. Jr, Lemeshow, S. and Sturdivant, R.X. (2013) *Applied Logistic Regression*. Wiley, Hoboken, New Jersey.

Kobayashi, M. and Melkonyan, T. (2011) Strategic incentives in biosecurity actions: theoretical and empirical analyses. *Journal of Agricultural and Resource Economics* 36, 242–262.

Kristensen, E. and Jakobsen, E. (2011). Danish dairy farmers' perception of biosecurity. *Preventive Veterinary Medicine* 99, 122–129.

Le Roux B. and Rouanet H. (2010) *Multiple Correspondence Analysis*. Quantitative Applications in the Social Sciences 63, Sage, Thousand Oaks, California.

Maddala, G.S. (1983) *Limited-dependent and Qualitative Variables in Econometrics*. Econometric Society Monograph No. 3. Cambridge University Press, Cambridge, UK.

Mugnier, S., Magne, M.A., Pailleux, J.Y., Poupart, S. and Ingrand, S. (2012). Management priorities of livestock farmers: a ranking system to support advice. *Livestock Science* 144, 181–189.

Ribbens, S., Dewulf, J., Koenen, F., Mintiens, K., De Sadeleer, L., de Kruif, A. and Maes, A. (2008) A survey on biosecurity and management practices in Belgian pig herds. *Preventive Veterinary Medicine* 83, 228–241.

Sayers, R.G., Sayers, G.P., Mee, J.F., Good, M., Bermingham, M.L., Grant, J. and Dillon, P.G. (2013) Implementing biosecurity measures on dairy farms in Ireland. *The Veterinary Journal* 197, 259–267.

Schade, C., Kunreuther, H. and Koellinger, P. (2012). Protecting against low-probability disasters: the role of worry. *Journal of Behavioral Decision Making* 25, 534–543.

Stott, A.W. (2009) Optimization methods for assisting policy decisions on endemic diseases. In: Rushton, J. (ed.) *The Economics of Animal Health and Production*. CAB International, Wallingford, UK.

Stott, A.W. and Gunn, G.J. (2008) Use of a benefit function to assess the relative investment potential of alternative farm animal disease prevention strategies. *Preventive Veterinary Medicine* 84, 179–193.

Toma, L., Stott, A.W., Heffernan, C., Ringrose, S. and Gunn, G. (2013). Determinants of biosecurity behaviour of British cattle and sheep farmers – a behavioural economics analysis. *Preventive Veterinary Medicine* 108, 321–333.

Tversky, A. and Kahneman, D. (1981) The framing of decisions and the psychology of choice. *Science* 211, 453–458.

Wang, T. and Hennessy, D.A. (2014) Modelling interdependent participation incentives: dynamics of a voluntary livestock disease control programme. *European Review of Agricultural Economics* 41, 681–706.

6 Modelling Farm Efficiency

Patrick Gillespie,[1]* Fiona Thorne,[1] Thia Hennessy,[1] Stephen Hynes[2] and Cathal O'Donoghue[1]
[1]*Rural Economy and Development Programme (REDP), Teagasc, Athenry, Ireland;* [2]*National University of Ireland, Galway (NUIG), Ireland*

6.1 Introduction

Productivity and efficiency studies are manifold in agricultural economics, partly due to the flexibility that such a framework provides in answering a wide range of research questions. These methods have addressed topics as diverse as the interaction between agricultural production and the environment (Färe *et al.*, 2012a,b), intercountry comparisons and economic development (Trueblood and Coggins, 2003), production under risk (Orea and Wall, 2012) and agricultural policy reform (Matthews *et al.*, 2007; Walden *et al.*, 2012). Furthermore, sectoral competitiveness has emerged as a focal point among policy makers and industrial stakeholders as global trade policy trends towards liberalization. For European agriculture, this interest has tended to focus in force on periodic reviews of the Common Agricultural Policy (CAP), hence productivity and efficiency analyses of various policy changes are in high demand; an understanding of production costs and farm efficiency provides a safeguard route to many farms in achieving a range of objectives from sustainability to maximizing farm profits.

This chapter will provide the context for why it is important for a farm to be efficient in terms of its use of inputs and how these translate into outputs. In particular, the global macroeconomic and policy conditions that provide an impetus to examine efficiency at the farm level are outlined (Section 6.1.1). A brief introduction to the definition of efficiency and the theory relating to efficiency is then presented in Section 6.1.2. Following this brief introduction to the rationale for examining efficiency and the theory relating to efficiency, the remainder of the chapter is divided into the following sections: a review of existing methodologies and models (Section 6.2), an examination of the efficiency of Irish dairy farms using the true random effects model, including a model overview and results (Section 6.3) and a concluding section that looks at the relevance of the research to policy makers (Section 6.4).

6.1.1 Policy context

Agricultural policy, in the European Union (EU) in particular, has traditionally focused on improving farm income and maintaining the numbers of people working and living in rural areas. While the improvement of efficiency and productivity has been a long-standing goal of

*E-mail: patrick.gillespie@teagasc.ie

European agricultural policy – see Article 33 of the Treaty of Rome (European Union, 2006) and Article 39 of The Treaty on the Functioning of the European Union (European Union, 2008) – Matthews *et al.* (2007) commented that it was traditionally not a priority objective. Through successive CAP reforms, starting with the MacSharry reforms, the income objectives of agricultural policy have been pursued through direct income support measures. As greater proportions of farm income have been derived from direct income support schemes, the market orientation of farming has diminished and, as a result, there has been less emphasis on productivity and efficiency. However, the policy and market environment within which European agriculture operates may now be changing. The budgetary resources for agricultural income support will, for the foreseeable future, be (at best) fixed in nominal terms, and increasing policy emphasis will be placed on the environmental impact of agriculture. The sustainable intensification of agriculture, therefore, is one of the great challenges facing European agriculture.

At a global level, population growth and increasing income pose a challenge and opportunity for the global agricultural sector and an opportunity for food-exporting countries. Furthermore, the removal of the milk quota regime from European dairy farming after 30 years represents the abolition of a major constraint on output potential. Expanding production is likely to be a key feature of the dairy sector of many European countries and, as such, there may be a renewed focus on farm efficiency. Efficiency (and the ensuing productivity-based growth) – i.e. getting more output for a given volume of inputs – will, as ever, be the key to sustainable growth in agricultural incomes and will be the basis for the environmental sustainability of global agriculture.

Against this policy backdrop, research aimed at understanding and measuring farm efficiency is both timely and relevant, and is likely to be of particular use to policy makers in monitoring the achievement of policy objectives relating to agricultural efficiency and productivity.

6.1.2 Defining efficiency

In this section, we provide some definitions of the terms relating to efficiency and a short overview of the evolution of the theory of efficiency analysis. The roots of efficiency analysis can be traced back to at least as far as Debreu (1951) with his seminal contribution proving the existence of a 'coefficient of resource utilization'. Farrell (1957) applied Debreu's result to the specific question of measuring productive efficiency. Farrell's primary focus was on quantifying efficiency. Bravo-Ureta *et al.* (2007) provided a useful definition of the term efficiency as the ability of a firm to transform a given set of inputs into maximum achievable output given the available technology.

Technical efficiency

For the purpose of this chapter, we define input- and output-orientated Debreu–Farrell efficiency, as outlined by Kumbhakar and Lovell (2000, p. 43), as follows:

Definition 1: An input-oriented measure of technical efficiency (TE_I) is represented by the function:

$$TE_I(y,x) = \min\{\theta : \theta_x \in L(y)\}$$

Definition 2: An output-oriented measure of technical efficiency (TE_O) is represented by the function:

$$TE_O(x,y) = \left[\max\{\phi : \phi y \in P(x)\}\right]^{-1}$$

where $L(y)$ and $P(x)$ are feasible input and output sets, respectively. In the case of output-oriented efficiency as defined above (Definition 2), this is simply the proportion of observed output relative to maximum attainable output from a given input vector.

The efficiency function ranges over the interval [0,1], and its difference from unity can be interpreted as the fraction of output which is lost to inefficiency. This is summarized graphically in Fig. 6.1, in which the fraction of the distance to the production frontier, *pf*, achieved by the farm at point *A* is OY'/OY^* (where *O* represents the origin), which is just $Y'/Y^* = TE_O$.

Cost efficiency

The real contribution of Farrell's 1957 paper was the decomposition of efficiency into technical efficiency (TE) and allocative efficiency (AE) using a cost frontier.[1] He started by sketching an isoquant in a two-input commodity space, which is reproduced in Fig. 6.2. Farrell's nomenclature uses y as a separate *input* here, with the isoquant sketched representing a level set of input bundles producing the same output as in the definition of isoquants in the standard theory. He reasoned that if the so-called efficient production function that generated this isoquant were known in practice, it could provide a benchmark for the observed input–output sets of other firms. The isoquant SS' represents all minimal combinations of inputs that could produce a given level of output. Now suppose that at point Q an efficient firm were able to produce the same

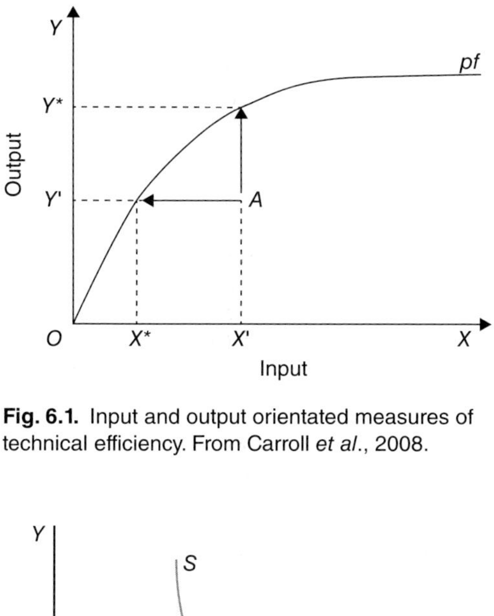

Fig. 6.1. Input and output orientated measures of technical efficiency. From Carroll *et al*., 2008.

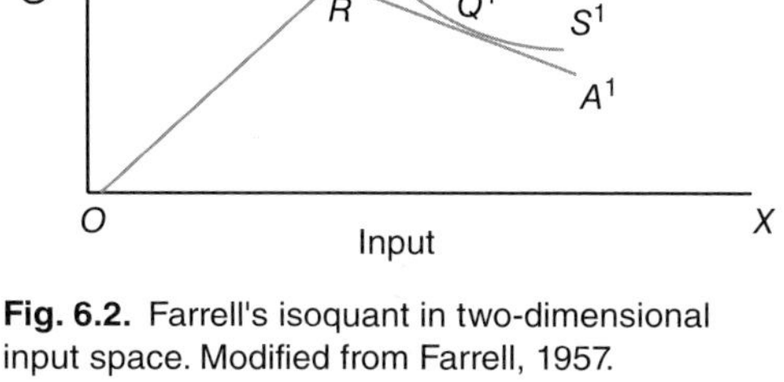

Fig. 6.2. Farrell's isoquant in two-dimensional input space. Modified from Farrell, 1957.

as the firm located at point P. The measure of TE that Farrell proposed would then be equal to the ratio $OQ = OP$. Furthermore, the curve AA' represents the trade-off in terms of costs of the two inputs; it is the isocost line, which has a slope that is equivalent to the ratio of the two input costs. The new efficient point would then be Q' instead of Q, and AE is then given by the ratio $OR = OQ$, which is the fraction of costs faced by a firm producing Q' relative to those faced by one at Q (by the nature of isocost lines). Finally, Farrell defines a measure of overall efficiency – understood to be CE (cost efficiency) – as the product of AE and TE, a quantity that is equivalent to the ratio $OR = OP$. Graphically, it can readily be seen that the improvement in overall efficiency (and thus the cost of production) is composed of the improvement in TE – a movement from P to Q – and AE – a movement from to Q to Q'.

Farrell's 'efficient production function' is known as a production frontier (*pf*) in modern terms, and the isoquant that results from the production frontier is a cost frontier that is known to exist by virtue of duality theory. When one considers 'overall' efficiency in the context of a cost frontier (i.e. cost efficiency), one refers to CE. Hence, CE is composed of both TE and AE, and it follows from this that for all firms, TE ≥ CE, because AE will also be bounded [0,1].

Total factor productivity

While the focus in this chapter is mainly on efficiency measurement, the terms efficiency and productivity are often used interchangeably, and it is important to explain how the two differ. Total factor productivity (TFP) finds its beginnings in growth theory and is fundamentally an aggregate measure. It relates growth in output to growth in inputs for a sector, industry or economy as a whole. It has historically been seen as a measure of technical change but, strictly speaking, technical change only accounts for part of TFP. TFP can be decomposed into technical change, change in TE and scale efficiency change (Coelli *et al*., 2005). So, although the two terms efficiency and productivity are often used interchangeably, TE is, in fact, only one component of TFP.

6.2 Review of Alternative Methodologies to Examine Efficiency

This section provides a review of the alternative methodologies available in the literature for examining farm efficiency. The empirical estimation of TE has evolved in the literature under two very different methodologies – data envelopment analysis (DEA) and stochastic frontier analysis (SFA). Both methods estimate production frontiers representing the maximum level of output for a group of firms, with individual firm inefficiency represented as the potential proportional increase in output to the frontier previously specified. The approaches differ in that DEA is a deterministic process and, by definition, does not account for any statistical noise in the data. Any deviation in observed output from the maximum feasible frontier output is fully attributed to inefficiency. In farming data, such noise could be due to many factors, such as the weather.

SFA analysis explicitly takes account of statistical noise when calculating technical inefficiency. The approach defines the production technology for a particular industry using a stochastic production frontier, whereby output is expressed as a function of inputs, a random error component and a one-sided technical inefficiency component which captures deviations below the frontier output level. However, the SFA approach is not without its shortcomings. One of the main issues associated with the specification of an SFA model is that it requires the specification of a functional form, the alternatives of which are often debated in the literature. Nevertheless, given that statistical noise is very prevalent in farming data sets, the SFA method is considered to be a more favourable approach than DEA, and it is discussed in further detail below.

6.2.1 Stochastic frontier analysis (SFA)[2]

The estimation of stochastic production frontiers using cross-sectional data was simultaneously proposed by Aigner *et al.* (1977) and Meeusen and van den Broeck (1977). Assuming a Cobb–Douglas production technology, the stochastic frontier (SF) is written as:

$$\ln y_i = \beta_0 + \sum_{k=1}^{K} \beta_k \ln x_{ki} + e_i \quad \text{where } e_i = v_i - u_i \quad (1)$$

and in which y_i is the farm's output level and x_{ki} is a vector of k production inputs. The composite error term e_i is made up of a statistical noise component v_i and a non-negative technical inefficiency component u_i. In order to extract the inefficiency component from the composite error, distributional assumptions must be made about both components. The model can then be estimated using maximum likelihood.

In standard practice in production economics, the noise term of any econometric model is often assumed to follow a standard distribution, but a number of alternative distributional assumptions have been suggested in the literature for the inefficiency component; these include the half-normal or exponential distribution (both of which were suggested in the original SFA papers), the truncated normal distribution (Stevenson, 1980) and the gamma distribution (Greene (1980) and Stevenson (1980)). Despite the large range of assumptions evident in the literature, Kumbhakar and Lovell (2000) found that efficiency estimates were largely unaffected by the distributional assumptions and that estimates were generally highly correlated between alternative distributions.

Pitt and Lee (1981) extended the cross-section model presented in Equation (1) above to a general panel data specification by adding the subscript t to output, inputs and the error term, and assuming a time-invariant inefficiency term (Equation 2).

$$\ln y_{it} = \beta_0 + \sum_{k=1}^{K} \beta_k \ln x_{kit} + v_{it} - u_i \quad (2)$$

In this specification, it is assumed that technical inefficiency is time invariant. This becomes less plausible when the number of time periods increases. For example, inefficient farms may become more inefficient over time. Similarly, in unbalanced panels of data, it is likely that some farms become less efficient through time before exiting and leaving the sample. The time-invariant assumption that is imposed will depend upon the length of the panel, the nature of the sample (balanced or unbalanced) and, also, the competitive structure of the sector in question. To account for time-varying inefficiency specifications, the following models have been frequently employed in the literature:

$$u_{it} = u_i / \left[1 + \exp\left(\alpha t + \gamma t^2\right)\right]$$

Kumbhakar (1990) (3)

$$u_{it} = u_i \times \exp\left[-\eta\left(t - T\right)\right]$$

Battese and Coelli (1995) (4)

where, $t = 1, 2, ..., T$ is time and α, γ and η are parameters to be estimated. Both models assume a time-invariant inefficiency term, u_i, and allow this to follow a temporal trend. The drawback of such specifications is that they impose the same temporal pattern of inefficiency on all farms and, as such, the efficiency ranking of a farm would not change through time. Again, this is a somewhat restrictive assumption.

The model proposed by Cuesta (2000) allows each farm to follow its own temporal inefficiency pattern:

$$u_{it} = u_i \times \exp\left[-\xi_i\left(t - T\right)\right] \quad (5)$$

where, ξ_i are farm-specific parameters responsive to different patterns of temporal variation. While this model is theoretically appealing, it has proved difficult to estimate in practice.

Another problem identified in the literature in relation to earlier studies examining efficiency is in relation to estimates of inefficiency also capturing farm heterogeneity that is unrelated to inefficiency. A wide range of models have attempted to overcome this problem, such as those of Kumbhakar and Hjalmarsson (1993), Kumbhakar and Heshmati (1995) and Kumbhakar and Lovell (2000). However, the most recent models specified in the literature addressing this issue of farm heterogeneity captured in inefficiency estimates are the 'true' fixed and random effects models outlined in Greene (2004) and Greene (2005), and employed by Farsi and Filippini (2004) and Farsi *et al.* (2006). These models are similar to that of Kumbhakar and Hjalmarsson (1993) but differ in the method of estimation. It is assumed that the inefficiency component estimated in conventional approaches also captures the time-invariant heterogeneity that is inappropriately being labelled as inefficiency. In the true fixed effects (TFE) model, unobserved heterogeneity is modelled directly in the production function using farm-specific dummy variables and estimated in a one-step maximum likelihood approach:

$$\ln y_{it} = a_i + \sum_{k=1}^{K} \beta_k \ln x_{kit} + v_{it} - u_{it} \quad (6)$$

where a_i are farm-specific, time-invariant dummy variables and the inefficiency term is a time-varying random variable. Greene's true random effects (TRE) model is similar in its theoretical foundation but has a random farm-specific time-invariant constant term (which is assumed to be normally distributed). This model may be estimated using maximum simulated likelihood estimation:

$$\ln y_{it} = \left(\beta_0 + w_i\right) + \sum_{k=1}^{K} \beta_k \ln x_{kit} + v_{it} - u_{it} \quad (7)$$

where w_i is a time-invariant, farm-specific random term, again intended to capture cross-farm time-invariant heterogeneity. Both the 'true' fixed and the 'true' random effects models assume that the error term and the inefficiency term are independently and identically distributed normally and half-normally, respectively.

6.3 Examining the Efficiency of Irish Dairy Farms

6.3.1 Background to the research question

The removal of the milk quota system at the end of March 2015, along with the introduction of the Basic Payment Scheme (BPS), has presented the European dairy sector with the first growth opportunity in over 30 years and hence there is much interest in the effects that it will have from both an academic and an industrial perspective. Much of the research attempting to forecast the EU response to quota abolition predicts a large expansion in Irish production (Lips and Rieder, 2005; Binfield *et al.*, 2008; Bouamra-Mechemache *et al.*, 2008). It would appear, therefore, that Ireland is uniquely situated to take advantage of a post-quota era in milk production. Donnellan *et al.* (2009) found that Irish dairying enjoys a high degree of competitiveness owing to its low-cost production system. However, there are concerns about the impact of the prolonged quota period on Ireland's competitive position. The Donnellan study itself notes the deterioration in

Ireland's competitive position when the economic value of unpaid labour and land input is considered. Furthermore, the study also notes that competitor countries that adopted more liberalized quota trading systems over the last 30 years have tended to do better in total economic cost terms, although causality has not been established.

Hennessy *et al.* (2009) constructed a linear programming model to quantify the inefficiencies associated with the 'ring-fenced' design of Irish quota trade that was in operation in Ireland at that time. The study concluded that ring fencing in the form of regional restrictions on quota trade had a negative impact on the efficiency of the sector. Other examples of mathematical programming exercises of this sort employed on a wider European level include those of Kirke and Moss (1987), and later Colman (2000), specifically on dairy production in Northern Ireland in the former and the UK more broadly in the latter. Colman *et al.* (2002) employed an econometric approach using the specification of ad hoc cost functions to simulate various changes to quota implementation, and to draw conclusions in relation to structural change and efficiency. All of these studies concluded that the milk quota was associated with decreased efficiency, or that the distribution of quota was far from optimal with respect to an efficient allocation.

Clearly then, the milk quota system has had an impact on the productivity and efficiency of Irish dairy farming. Many studies have attempted to measure this impact, for example, Matthews (2000), Boyle (2004) and Donnellan *et al.* (2009) invoked aggregate-level approaches, while O'Neill and Matthews (2001), O'Neill *et al.* (2002), Newman and Matthews (2006, 2007) and Carroll *et al.* (2008), used micro-level data to examine the productivity and efficiency of Irish dairy farming. Newman and Matthews (2006, 2007) used SF methods to construct TFP indices while analysing Irish dairy farm data. They found higher rates of TFP growth among specialist dairy farms, but also some evidence of a slowing down of this growth. O'Neill and Matthews (2001) also completed an earlier SF study in which TE was measured, and they too found that it tended to be higher on dairy farms.

Although there have been a number of studies of changing efficiency levels in Irish dairy farming, due to limited data availability, none of these studies have examined efficiency change from the pre-quota period. Examining efficiency change both before and after the introduction of the quota would facilitate a greater understanding of the impact of the production constraint on efficiency levels. The recent digitalization of farm data from the early 1980s has now made such a study possible. In Section 6.3.1 below, we describe an SFA model of TE specified for the period both preceding and following the introduction of the milk quota.

Section 6.3.2 outlines the theoretical specification of the model, Section 6.3.3 outlines the data used in the model and Section 6.3.4 outlines the results from the model and the validation of the results.

6.3.2 The theoretical model

In this chapter, the TRE model (as detailed in Section 6.2 above) was the model of choice for the purpose of examining TE of the Irish dairy sector from 1979 to 2012. The model is based on a panel of specialist dairy farms and includes observations for 5 years before quota introduction in 1984. Hence, it is possible to observe the development of TE (and other components of TFP, not reported here) in the years before and initial years following quota implementation and, therefore, to provide an indication of the impact of the policy on the productive capacity of the Irish dairy sector.

A translog production technology was assumed with annual time dummy variables to capture neutral technical change. The full specification is given by Equation (8).

$$\ln y_{it} = \alpha + \sum_{k=1}^{K} \beta_k \ln x_{kit} + 0.5 \sum_{k=1}^{K} \sum_{j=1}^{K} \beta_{kj} \ln x_{kit} \ln x_{nit} + \sum_{t=1}^{T} \delta_t D_t + v_{it} - \mu \quad (8)$$

in which TRE gives

$$\alpha = \beta_0 + w_i \quad \text{and} \quad \mu = u_{it}$$

and where D_t are annual dummy variables.

6.3.3 Data sources

The main source of data used in this analysis was the Teagasc (Agriculture and Food Development Authority of Ireland) National Farm Survey (NFS). The NFS is the Irish component of the

EU-wide Farm Accountancy Data Network (FADN). The NFS is an unbalanced panel of farms with observations recorded from 1973 to as recently as 2013. The survey is conducted annually by Teagasc. Participation in the survey is voluntary. The survey yields a high-quality micro-level data set with measures of economic, sociographic and technical variables. The responses are collected during an interview between professional farm recorders and a member of the farm household. Teagasc attempts to collect data from the same farms over multiple years. However, the voluntary nature of the survey means that some farms will inevitably end their participation, either due to unwillingness or inability to provide data (e.g. a farm may cease operation in a particular year), and farms typically provide data for only a few years before exiting the sample. These facts lead to panel data that are unbalanced.

6.3.4 Model outputs

Technical efficiency of Irish dairy farms: 1979–2012

The panel data employed for this analysis was constructed from the Teagasc, NFS. The sample includes data from 1844 dairy farms which participated at any point during 1979–2012, yielding a total of 9434 observations. Although the panel spans 34 years, most respondents contributed for a far shorter period than this, with average duration at 5.1 years; the interquartile range of participation is 2–7 years. Table 6.1 presents summary statistics for the variables of interest in this analysis. These consist of the portion of farm gross output that is directly attributable to the dairy enterprise as the dependent variable, and five regressors that capture the various types of inputs of this business activity. These were: the 12 month average number of dairy cows; variable costs; capital expenditure; labour input in standard man days; and dairy forage area in hectares.

The inclusion of monetized variables was both a concession due to a lack of quantity variables in the cases of certain inputs and an allowance for some aggregation of otherwise unreconcilable measures, for instance a tractor or a farm building. A similar argument could be made for different classes of outputs; for example, cull cows and litres of milk cannot be meaningfully combined except in monetary terms. The need for some variable aggregation stems from the nature of the translog production function used to estimate the model, as it specifies several interactions and squared terms for each input in the primary model. As Newman and Matthews (2006, p. 196) have pointed out, the aggregation process also alleviates problems concerning the inability of the translog functional form to manage zero-valued observations of modelled variables.

Variable costs include concentrate feeds, costs associated with pasture production (mainly fertilizers) and expenditure on lime. Capital expenditure consists of upkeep of land, machinery and buildings, all of which were not specific to the dairy enterprise, but have been allocated proportionally to the ratio of dairy gross output to total farm gross output per farm annually. These measures, being nominal, have all been deflated using appropriate commodity-specific price indices obtained from the Irish Central Statistics Office (CSO), hence allowing them to be used as an aggregated volume measure. The same procedure has been carried out for the other monetized variable in the model, the dairy gross output, which was the dependent variable.

The parameters for the SF model are reported in Table 6.2. Herd size and direct costs

Table 6.1. Panel data on the technical efficiency of 1844 Irish dairy farms, 1979–2012. Summary statistics are authors' calculations using National Farm Survey micro-level data from Teagasc.

	Specialist dairy farms (n = 9434 sets of observations)			
	Mean	SD	Min.	Max.
Farm size (ha)	22	12	2	136
Dairy gross output (€, base year = 2000)	56,122	45,915	279	466,257
Herd size (dairy cows)	41	26	9	283
Direct costs (€, base year = 2000)	15,432	12,857	366	155,898
Capital (€, base year = 2000)	46,528	45,070	110	491,999

Table 6.2. Parameters for the stochastic frontier (SF) model based on data from 1844 Irish dairy farms, 1979–2012.

Parameter	Value	Degree of confidence[a]	Standard error
Log(L)	6,538		
χ^2	13,076		
Significance	0.00000		
Production function			
Herd (head)	0.70	***	(0.007)
DC (real €)	0.20	***	(0.003)
Capital (real €)	0.05	***	(0.002)
Labour (days)	0.04	***	(0.005)
Forage (ha)	0.03	*	(0.005)
Herd[b]	−0.08	***	(0.02)
Herd: DC	−0.10	***	(0.02)
Herd: Capital	0.03	*	(0.009)
Herd: Labour	0.10		(0.02)
Herd: Forage	0.03	***	(0.02)
DC[b]	0.03	***	(0.005)
DC: Capital	−0.01	**	(0.005)
DC: Labour	0.01	***	(0.01)
DC: Forage	0.02	***	(0.01)
Capital[b]	0.00		(0.002)
Capital: Labour	0.00	***	(0.006)
Capital: Forage	−0.01	***	(0.006)
Labour[b]	−0.03	***	(0.006)
Labour: Forage	−0.03	*	(0.02)
Forage[b]	−0.01	***	(0.009)
Sampling method controls			
Size class 2 control	−0.16	***	(0.012)
Size class 3 control	−0.19	***	(0.013)
Size class 4 control	−0.21	***	(0.013)
Size class 5 control	−0.22	***	(0.015)
λ	1.95	***	(0.052)
σ	0.13	***	(0.001)
σ_u	0.11822		
σ_v	0.06053		
α_i (random effect)	0.07	***	(0.014)
σ_{wi}	0.03	***	(0.002)

Technological change (time dummies, base year = 1979)

Year	Change	Degree of confidence[a]	Standard error
1980	0.02	*	(0.009)
1981	0.02	**	(0.008)
1982	0.07	***	(0.009)
1983	0.09	***	(0.008)
1984	−0.09	***	(0.015)
1985	−0.04	***	(0.014)
1986	−0.06	***	(0.014)
1987	−0.05	***	(0.015)
1988	−0.06	***	(0.015)
1989	−0.08	***	(0.015)
1990	−0.04	***	(0.015)
1991	−0.04	**	(0.016)

Continued

Table 6.2. Continued.

Year	Change	Degree of confidence[a]	Standard error
1992	−0.03	*	(0.016)
1993	−0.03	**	(0.015)
1994	−0.03	*	(0.015)
1995	−0.03	**	(0.016)
1996	−0.01		(0.015)
1997	0.01		(0.015)
1998	−0.03	*	(0.015)
1999	−0.02		(0.015)
2000	0.02		(0.015)
2001	0.05	***	(0.015)
2002	0.03	**	(0.015)
2003	0.06	***	(0.015)
2004	0.09	***	(0.015)
2005	0.07	***	(0.016)
2006	0.06	***	(0.015)
2007	0.06	***	(0.016)
2008	0.05	***	(0.015)
2009	0.03	*	(0.015)
2010	0.05	***	(0.016)
2011	0.08	***	(0.015)
2012	0.04	**	(0.016)

[a]*, **, ***, 90, 95, and 99% confidence.
[b]Herd, 12 months average no. dairy cows; DC, Direct costs, i.e. variable costs directly attributable to the dairy enterprise on the farm; Capital = investment in buildings and machinery; Labour = standard man days; Forage = farm forage area allocated to dairy herd.

(mainly feed and fertilisers) were the dominant inputs, and four of the five primary terms were statistically significant at the 99 per cent level. Most of the quadratic terms are also statistically significant. The time dummy variables included in the model are all statistically significant at the 99 per cent level as well, barring 4 individual years in the mid and late 1990's. The sampling method controls are included to account for the fact that the NFS moved to a weighting methodology in 1984. The controls were dummy variables indicating that a particular observation belonged to one of the NFS size-classes which determine the sample weight, and this was interacted with a dummy variable marking the switch to the new methodology in 1984. All were statistically significant, and their sign indicates that the overall effect of the methodology change was to lower the predicted production output across relative to the largest size-class, which was taken as the base level.

Table 6.3 reports (weighted) mean output elasticities and TE scores resulting from the TRE version of the model for the years 1979–2012. The function generally increased with inputs, as theory dictates. Slightly increasing returns to scale (RTS) were consistently present (the RTS parameter indicates increasing RTS when >1), and within a few years there were negative signs on the labour and land inputs. This result has been observed before (Carroll *et al.*, 2008) and may be indicative of some variability in soil quality (farms with poorer land also tend to be larger in physical terms).

Moreover, Table 6.4 shows that the model's input variables are correlated, although not so much as to prevent the model from estimating with statistically significant parameters. Multicollinearity can result in some parameters having incorrect signs, so this may have contributed somewhat to the theoretical violations observed. The estimated returns to scale associated with this production function were 1.075 with a standard deviation of 0.007 over the 34 yearly observations.

The overall level of TE was generally high, with minimal movement over time. This may in part be due to the increasing homogeneity of the group of farms classified as specialist dairy producers. The TRE specification also has the effect

Table 6.3. Technical efficiency (TE) scores, (weighted) mean output elasticities ($\in$) and returns to scale (RTS) from the true random effects (TRE) version of the model based on data from 1844 Irish dairy farms, 1979–2012.

Year	TE	$\in_1$	$\in_2$	$\in_3$	$\in_4$	$\in_5$	RTS
1979	0.91	0.86	0.17	0.05	−0.01	0.01	1.07
1980	0.91	0.85	0.17	0.05	−0.01	0.01	1.07
1981	0.91	0.88	0.18	0.04	−0.03	0.01	1.08
1982	0.92	0.84	0.18	0.05	−0.01	0.02	1.07
1983	0.91	0.83	0.18	0.04	−0.01	0.02	1.06
1984	0.91	0.78	0.19	0.05	0.01	0.03	1.06
1985	0.91	0.79	0.18	0.05	0.01	0.03	1.06
1986	0.90	0.80	0.19	0.05	0.00	0.03	1.07
1987	0.91	0.79	0.19	0.05	0.00	0.03	1.06
1988	0.92	0.79	0.20	0.05	0.00	0.03	1.07
1989	0.91	0.78	0.20	0.04	0.01	0.03	1.06
1990	0.92	0.77	0.20	0.05	0.01	0.03	1.06
1991	0.92	0.77	0.19	0.05	0.02	0.03	1.06
1992	0.92	0.77	0.20	0.05	0.01	0.03	1.06
1993	0.92	0.75	0.20	0.04	0.02	0.04	1.05
1994	0.91	0.75	0.21	0.04	0.02	0.04	1.05
1995	0.92	0.74	0.21	0.04	0.02	0.04	1.05
1996	0.91	0.74	0.20	0.05	0.02	0.04	1.05
1997	0.91	0.74	0.19	0.05	0.03	0.04	1.04
1998	0.91	0.73	0.20	0.04	0.03	0.04	1.04
1999	0.91	0.72	0.21	0.04	0.03	0.04	1.04
2000	0.91	0.72	0.20	0.04	0.03	0.04	1.04
2001	0.92	0.71	0.20	0.04	0.04	0.04	1.03
2002	0.91	0.70	0.21	0.04	0.04	0.04	1.03
2003	0.91	0.71	0.20	0.04	0.04	0.04	1.03
2004	0.91	0.70	0.20	0.05	0.04	0.04	1.03
2005	0.91	0.70	0.20	0.05	0.04	0.04	1.02
2006	0.91	0.70	0.21	0.04	0.04	0.03	1.02
2007	0.91	0.71	0.21	0.04	0.04	0.03	1.02
2008	0.91	0.71	0.19	0.05	0.05	0.03	1.02
2009	0.90	0.69	0.19	0.05	0.05	0.04	1.02
2010	0.91	0.69	0.20	0.05	0.05	0.04	1.02
2011	0.92	0.67	0.19	0.05	0.06	0.04	1.00
2012	0.91	0.65	0.20	0.05	0.06	0.04	1.00

Table 6.4. Correlations among input variables (x) for the true random effects (TRE) version of the model based on panel data on the technical efficiency (TE) of 1844 Irish dairy farms, 1979 to 2012.

Variable	x_1	x_2	x_3	x_4	x_5
x_1	1.00	0.89	0.77	0.85	0.75
x_2	0.89	1.00	0.77	0.72	0.68
x_3	0.77	0.77	1.00	0.60	0.58
x_4	0.85	0.72	0.60	1.00	0.66
x_5	0.75	0.68	0.58	0.66	1

of increasing TE, as the random farm effect (meant to capture farm heterogeneity) absorbs some of the variation in output that would otherwise be attributed to inefficiency. In relation to the hypotheses set out earlier in the chapter, this analysis found that TE was very slightly negatively affected in the early years of the milk quota, but aside from this minor adjustment, it remained remarkably stable for the duration of the milk quota era, with average TE scores quite high throughout.

The results presented by this model appear to contradict the theory put forward by Hennessy and Shrestha (2007) that the milk quota has been associated with a loss of efficiency, but that this has become less egregious as more liberalized trading of the quota has been allowed, thus giving more efficient producers the ability to expand their market share. However, it is important to remember the specific definition of efficiency that this model's measure of TE uses. The model estimates TE on the basis of separation of the most efficient producers from the rest of the data. Furthermore, this measure is cast in terms of physical output and inputs, whereas the Hennessy and Shrestha study examined costs. Table 6.3 shows that the milk quota has not resulted in any appreciable separation of producers as TE (a measure of relative efficiency) is 0.9 or above (i.e. ≤90% efficiency), and that TE remains high and little changed throughout. A valid interpretation is that the milk quota has had little effect on each farm's level of efficiency relative to the rest of the dairy farms, but that changes to the other components of TFP may still mean that the entire sector's productivity (TFP) could have been affected by the policy. Additional results reported in Gillespie *et al.* (2015) suggested that movements of the productivity frontier have been far more influential than the average distance from it. Irish dairy farms are, and have been, highly technically efficient throughout the sample when a full account was given for the technology that has hitherto been available in Ireland.

Model validation

The statistical assumptions underlying the variant of panel SF models employed above have already been discussed, but the researcher must also check that the estimated models are consistent with microeconomic theory. The details of the post-estimation tests which must be carried out are given below.

APPROPRIATENESS OF VARIABLE INCLUSION IN MODEL. Table 6.5 details the likelihood ratio

Table 6.5. Specification tests employed for analysis of the stochastic frontier (SF) models used to analyse dairy enterprise production technology.

Test	
H_0	Cobb–Douglas production function
H_A	Translog production function
	Statistic: 165.40
	Critical value: $\chi_{15,0.01}$ = 30.58
	Reject null at 1% significance
H_0	Translog, no technical change
H_A	Translog, neutral technical change
	Statistic: 1116.00
	Critical value: $\chi_{33,0.01}$ = 54.78
	Reject null at 1% significance
H_0	Translog, neutral technical change, no sampling controls
H_A	Translog, neutral technical change, with sampling controls
	Statistic: 18.35
	Critical value: $\chi_{5,0.01}$ = 15.09
	Reject null at 1% significance
Hausman test for random effects (RE) vs fixed effects (FE) using estimates from OLS (ordinary least squares) panel regression:	
H_0	Farm effect is uncorrelated with error term (RE is favoured)
H_A	Farm effect is correlated with error term (FE is favoured)
	Statistic: 685.81
	P value: $P_{57,0.05}$ = 15.09
	Reject null at 5% significance

tests used to determine the appropriateness of the variables included. The tests start from the core specification using a Cobb–Douglas functional form only. The Cobb–Douglas form is nested inside the translog functional form, so adding in the quadratic terms to the Cobb–Douglas form allows a test statistic that follows a χ^2 distribution with 15 degrees of freedom (parameter restrictions). The test is statistically significant at a 99% confidence level, and this supports the use of a translog functional form.

The next test (the Translog test with no technical change) listed in Table 6.5 examines the joint significance of the 33 time-dummy variables that were specified in the model to represent Hick's neutral technical change. This test is statistically significant with 33 degrees of freedom at the 99% confidence level, so the test supports the existence of technical change in the model.

The third test (the Translog test with neutral technical change) verifies the need for the control variables that were specified to remove the effects of changes in the sampling methodology from the other parameter estimates. The controls are dummy variables which indicate the size strata that the farms belong to and these are interacted with a dummy variable indicating the first 5 years of the sample. The test has five degrees of freedom and is statistically significant at the 99% confidence level. This supports the inclusion of the sampling controls in the model.

The fourth test listed in Table 6.5 is not a likelihood ratio test, but the standard Hausman test to determine whether fixed effects (FE) or a random effects (RE) type of panel model is more appropriate. The RE model (i.e. the TRE model) is more efficient, but its parameters will be biased if the Hausman rejects the null hypothesis of no correlation between the individual effect and the model's error term. The Hausman test is an F test. In this case, there are 57 degrees of freedom, and the null hypothesis is rejected at the 95% confidence level. This suggests that an FE type of model (i.e. the TFE model) should be used. However, the theoretical tests described below strongly support the use of the TRE model instead.

THEORETICAL CONSISTENCY TESTING. The importance of the underlying theoretical consistency of stochastic frontiers was highlighted by Sauer *et al.* (2006). The assessment of theoretical consistency involves evaluating the modelled production function using the estimated parameters and, in the case of the translog functional form, also the individual values of each variable in the model. The first order (monotonicity) condition for the translog production function is:

$$\frac{\partial y}{\partial x_i} = \left(\frac{y}{x_i}\right) \times \left(a_i + \sum_{j=1}^{n} a_{ij} \ln x_j\right) > 0 \quad (9)$$

The second order (diminishing marginal product) condition requires the second derivatives with respect to each input of the model to be strictly negative. The inequality can be written for the translog as:

$$\frac{\partial^2 y}{\partial x_i^2} = \left[\begin{array}{c} a_{ii} + \left(a_i - 1 + \sum_{j=1}^{n} a_{ij} \ln x_j\right) \\ \times \left(a_i + \sum_{j=1}^{n} a_{ij} \ln x_j\right) \end{array}\right] \times \left(y / x_i^2\right) < 0 \quad (10)$$

It can be seen from Equations (9) and (10) that these two conditions depend upon the model parameters, as well as depending on the data points themselves, so it is incumbent on the researcher to check the extent to which any individual data set violates theory. Table 6.6 reveals that the more standard Pitt–Lee and Battese–Coelli variants of the SF model generally outperform the more recent 'true effects' models. In particular, the TFE model almost always violates theory for this model specification and data set. So despite the possibility of some bias in an RE type of model as suggested by the Hausman test, the TFE's inconsistency strongly dissuades the researcher from implementing it over the TRE model. Furthermore, Table 6.7 shows that there is a high correlation between the inefficiency parameter estimates from the TRE and TFE models, suggesting that while some bias may exist in the parameter estimates from the TRE version, the severity of this bias may not be too great.

REGULARITY CONDITION TESTING. The regularity conditions are simplified at the point of approximation if the data have been transformed such that they are equal to unity at that point. Researchers typically choose the means of the variables as the point of approximation, so the procedure is to simply divide each variable by its own mean. Recalling Equation (9), it can be

Table 6.6. Percentages of samples meeting theoretical regularity conditions for the four types of stochastic frontier (SF) models discussed.

Model	Monotonicity	Diminishing marginal product	Monotonicity and diminishing marginal product
Pitt–Lee	84	96	84
Battese–Coelli	84	96	84
True random effects (TRE)	76	86	76
True fixed effects (TFE)	2	0	0

Table 6.7. Correlations among estimates of inefficiency parameters of conditions for the four types of stochastic frontier (SF) models discussed.

Model	Pitt–Lee (PL)[a]	Battese–Coelli (BC)[a]	True random effects (TRE)	True fixed effects (TFE)
PL	1.00	0.99	0.42	0.28
BC	0.99	1.00	0.43	0.28
TRE	0.42	0.43	1.00	0.90
TFE	0.28	0.28	0.90	1.00

seen that the second term of the inequality involves ln x_i, but since x_i has been transformed in this way, it will become the logarithm of unity, which is always equal to zero. Hence, the second term simplifies to a_i, which is the main effect. This can now be directly interpreted as the elasticity, just as in the Cobb–Douglas functional form, albeit only when evaluating the function *at this point*. In essence, the regularity conditions become functions of the parameters alone for this special case. This is convenient, and it facilitates the checking of the quasi-concavity condition.

Table 6.8 below summarizes the results of the regularity checks. The TFE model does not fulfil any of the theoretical conditions, whereas the other three versions of the SF model meet the first and second order conditions, but fail to satisfy the concavity condition.

RETURNS TO SCALE. Table 6.9 gives the results of the test of the hypothesis that the technology exhibits constant returns to scale. The test is a Wald test which follows a χ^2 distribution. There are six degrees of freedom to the test, and each restriction of the model adds a degree of freedom. The restrictions do two things: first, the main effects are forced to sum to unity; secondly, each input's quadratic terms are forced to sum to zero. Failure to reject these restrictions lends support to the existence of constant returns to scale, while a rejection supports variable returns to scale. The table shows that the null hypothesis of constant returns to scale (CRS) is clearly rejected for all variants of the SF model.

The Pitt–Lee and Battese–Coelli models tend to be the most consistent with theory in this data set, but these variants of the SF modelling framework are also more rigid in terms of their treatment of inefficiency. As discussed above, the Pitt–Lee model assumes that inefficiency is completely time invariant; the inefficiency component is determined in the first period and assumed to be the same for the rest of the time horizon. Over long panels of data, this assumption becomes more restrictive, as it seems plausible that many farms may experience changes in their efficiency levels.

The Battese–Coelli model does allow inefficiency to develop over time, but the path of this development is prescribed by the model to be identical for all farms; they differ only in their starting points. This also seems restrictive in a 34 year long panel.

The 'true effects' models are far less restrictive in this regard, as they allow each farm's efficiency to follow its own development path over time. As noted above, the TFE model performs quite poorly in terms of theoretical consistency, so the TRE variant of the SF model is preferred for this analysis.

Table 6.8. Regularity conditions for the translog production function of the four types of stochastic frontier (SF) models (evaluated at the sample means of the data set).[3]

Model	Monotonicity[a]	Diminishing marginal product	Quasi-concavity
Pitt–Lee	F	F	NF
Battese–Coelli	F	F	NF
True random effects (TRE)	F	F	NF
True fixed effects (TFE)	NF	NF	NF

[a]F, condition fulfilled; NF, condition not fulfilled.

Table 6.9. Hypothesis tests of constant returns to scale using the Wald test on the four types of stochastic frontier (SF) models discussed.

Model	Wald statistic	*P* value	Conclusions (CRS or VRS)[a]
Pitt–Lee	157.35	0.00	VRS
Battese–Coelli	154.67	0.00	VRS
True random effects (TRE)	210.42	0.00	VRS
True fixed effects (TFE)	147.03	0.00	VRS

[a]CRS, constant returns to scale; VRS, variable returns to scale.

6.4 Conclusions and Relevance of Research Findings to Policy Makers

This chapter has provided the reader with an overview of the current state of the art in productivity and efficiency measurement. The discussion has covered a wide range of topics, beginning with the theoretical foundations and then moving on to a practical application of SFA. It now concludes with discussion of the real-world implications of the results.

There are three key results from Section 6.3. The most important is that there was no apparent effect on TE of the implementation of the milk quota as it remained quite high through time. The second result relates to the time dummy parameters; output initially increases each year, before a negative effect takes over in 1984, and output eventually becomes positive again in 1997. The last key result is that increasing the number of dairy cows became less strongly associated with increased output from 1984 onwards; the average output elasticity of the herd input dropped from 0.83 to 0.78, and continued decreasing throughout the panel. Taken together, these results suggest that farms adapted well to the milk quota policy, and that this may have come in the form of changing from an expansion strategy to a new focus on increasing yield per cow. Rather than a reduction in TE, it was likely that other components of total factor productivity were the drivers of any changes in sectoral productivity.

Several studies have examined how technical efficiency has evolved in other EU Member States, and these make useful comparisons with the results from this chapter. Hadley (2006) used a different variant of the SFA model to measure TE in England and Wales. He reported a value of 0.95 for TE in 1982, but this fell to 0.85 by 2002. Akande (2012) used DEA to calculate average TE levels, and in aggregate these ranged from 0.57 to 0.97 for Irish agriculture. The average of Irish TE across all years was 0.66, which was significantly lower than the EU-15 average, but this was due to unusually low scores in the 3 years from 2000 to 2002. The range of yearly mean TE scores is 0.85–0.97 when these years are excluded, and this is much closer to the scores reported in this chapter.

Newman and Matthews (2006, p. 203) found evidence of a slowdown in productivity growth immediately post the milk quota introduction in 1984, and they suggest that this may reflect the 'cost of the dislocation and adjustments required on dairy farms immediately following the introduction of the quota regime'. They also say that a later recovery in productivity growth rates could represent 'the re-establishment of more "normal" growth rates as farmers learned to live with the quota regime and as the regime in recent years has become less restrictive'. The signs of the time dummy variables in the analysis given in this chapter agree with these assertions; this helps to explain the apparent disconnect between the unaffected level of TE found here and the hypothesis that the milk quota adversely affected TE, which is often claimed in the literature. If the milk quota primarily shifted the production frontier then this would result in lower sectoral productivity, but it would not result in lower TE scores.

Outside the EU, New Zealand is often compared with Ireland owing to similarities in climate and a focus on grass input. Evidence from New Zealand suggests that the growth of TFP averaged only 1.5% annually during its high subsidy period (1972–1984), but that since the elimination of subsidies it has improved to 2.5% p. a. (Lattimore, 2006). While there are various factors which, when combined, mean that the experience in New Zealand may not be entirely representative of the Irish situation post quota removal (Donnellan *et al.*, 2015), there appears to be a certain acceptance that at least some of the New Zealand experience will be replicated in Ireland. For example, Donnellan *et al.* (2014) reported that 20% of dairy farmers intended to reduce production post the milk quota abolition and that a further 13% intended to quit production completely. The impact that this will have on future TE performance will depend on which farms these constitute. There is some evidence suggesting that smaller less efficient farms will leave the system, in which case we would expect an improvement in overall levels of TE in the system as a whole.

Technical efficiency has ramifications for the Irish dairy sector's competitiveness in the future, particularly in the face of a stagnant or declining price of output and inflationary increases in costs. The level of TE reported in this chapter was relatively high, but it did imply that Irish farms have room to improve further, i.e. a level of 0.91 implies that approximately 9% of output is lost due to inefficiency on average. Continued investment in farm extension services are a potential way to improve the TE of dairy farms. However, Newman and Matthews (2006) found that it is technical change, rather than change in TE, that has driven productivity change in Ireland in recent history. Therefore, investments in research and development may be even more important than improving access to extension services, because research and development can expand the technological possibilities.

Finally, the future direction of farm-level efficiency research looks promising. From the EU perspective, following the 2013 CAP reform, there is much greater emphasis now placed on the need for monitoring and evaluating the effects of agricultural policy, including the new European Innovation Partnership for Agricultural Productivity and Sustainability (EIP-Ag). One of the indicators to be used to measure the impact of EIP-Ag is the growth in agricultural productivity, of which efficiency is one component. Hence, it is clear from a policy perspective that a greater understanding of productivity and its drivers is of paramount importance.

Notcc

[1] Farrell called AE price efficiency, a measure of the efficiency of the firm in choosing an optimal set of inputs according to their prices.

[2] The description of the SFA approach presented here is based on a review by Carroll *et al.* (2008).

[3] Quasi-concavity evaluated on the basis of eigenvalues. Negative values for all eigenvalues indicate a negative semi-definite matrix, which is a necessary and sufficient condition for local concavity.

References

Aigner, D., Lovell, C.A.K. and Schmidt, P. (1977) Formulation and estimation of stochastic frontier production function models. *Journal of Econometrics* 6, 21–37.

Akande, O.P. (2012) *An Evaluation of Technical Efficiency and Agricultural Productivity Growth in EU Regions*. Wageningen University, Wageningen, The Netherlands.

Battese, G.E. and Coelli, T.J. (1995) A model for technical inefficiency effects in a stochastic frontier production function for panel data. *Empirical Economics* 20, 325–332.

Binfield, J., Donnellan, T. and Hanrahan, K. (2008) An examination of milk quota expansion at EU member state level with specific emphasis on Ireland. *Paper prepared for presentation at the 107th EAAE Seminar "Modelling of Agricultural and Rural Development Policies", Sevilla, Spain, January 29th–February 1st, 2008.* Available at: http://ageconsearch.umn.edu/bitstream/6467/2/cp08bi16.pdf (accessed 29 February 2016).

Bouamra-Mechemache, Z., Jongneel, R. and Réquillart, V. (2008) Impact of a gradual increase in milk quotas on the EU dairy sector. *European Review of Agricultural Economics* 35, 461–491.

Boyle, G.E. (2004) Competitiveness concerns at the production and processing level: the example of the dairy sector. In: *Workshop on Enhancing Competitiveness in the Agro-food Sector: Making Policies Work, 16–17 June (Vol. 203), Vilnius, Lithuania*. Available at: http://www.oecd.org/tad/agricultural-policies/32562861.pdf (accessed 1 March 2016).

Bravo-Ureta, B., Solís, D., Moreira, V., Maripani, J., Thiam, A. and Rivas, T. (2007) Technical efficiency in farming: a meta-regression analysis. *Journal of Productivity Analysis* 27, 57–72.

Carroll, J., Newman, C. and Thorne, F.S. (2008) *An Examination of the Productivity of Irish Agriculture in a Decoupled Policy Environment*. Trinity College Dublin. Available at: http://t-stor.teagasc.ie/bitstream/11019/832/1/eopr-5507.pdf (accessed 1 March 2016).

Coelli, T.J., Rao, D.S.P., O'Donnell, C.J. and Battese, G.E. (2005) *An Introduction to Efficiency and Productivity Analysis*. Springer, New York.

Colman, D. (2000) Inefficiencies in the UK milk quota system. *Food Policy* 25, 1–16.

Colman, D., Burton, M., Rigby, D. and Franks, J. (2002) Structural change and policy reform in the UK dairy sector. *Journal of Agricultural Economics* 53, 645–663.

Cuesta, R.A. (2000) A production model with firm-specific temporal variation in technical inefficiency: with application to Spanish dairy farms. *Journal of Productivity Analysis* 13, 139–158.

Debreu, G. (1951) The coefficient of resource utilization. *Econometrica* 19, 273–292. Available at: http://www.jstor.org/stable/1906814 (accessed 1 March 2016).

Donnellan, T., Hennessy, T.C. and Thorne, F.S. (2009) Perspectives on the competitiveness of EU dairy farming. *EuroChoices* 8(1), 23–29.

Donnellan, T., Hennessy, T.C., Fenelon, M. and O'Callaghan, D.J. (2014) The potential for economies of scale in the dairy processing sector. *International Journal of Dairy Technology* 67, 129–134.

Donnellan, T., Hennessy, T. and Thorne, F.S. (eds) (2015) *The End of the Quota Era: A History of the Irish Dairy Sector and Its Future Prospects*. Teagasc, Athenry, Co. Galway, Ireland. Available at: http://www.teagasc.ie/publications/2015/3541/End_of_the_Quota_Era_final.pdf (accessed 29 February 2016).

European Union (2006) Article 33, Consolidated Version of the Treaty Establishing the European Union. In: *Consolidated Versions of the Treaty on European Union and of the Treaty Establishing the European Community. Official Journal of the European Union* 49, C321 E/54. Available at: http://eur-lex.europa.eu/legal-content/EN/TXT/PDF/?uri=CELEX:C2006/321E/01&from=EN (accessed 29 February 2016).

European Union (2008) Article 39,, Consolidated Version of the Treaty on the Functioning of the European Union. In: *Consolidated Versions of the Treaty on European Union and the Treaty on the Functioning of the European Union. Official Journal of the European Union* 51, C 115/62–63. Available at: http://eur-lex.europa.eu/legal-content/EN/TXT/PDF/?uri=OJ:C:2008:115:FULL&from=EN (accessed 29 February 2016).

Färe, R., Grosskopf, S., Margaritis, D. and Weber, W. (2012a) Technological change and timing reductions in greenhouse gas emissions. *Journal of Productivity Analysis* 37, 205–216.

Färe, R., Grosskopf, S., Pasurka, C.A. and Weber, W.L. (2012b) Substitutability among undesirable outputs. *Applied Economics* 44, 39–47.

Farrell, M.J. (1957) The measurement of productive efficiency. *Journal of the Royal Statistical Society, Series A (General)* 120, 253–290. Available at: http://www.jstor.org/stable/2343100 (accessed 23 June 2016).

Farsi, M. and Filippini, M. (2004) Regulation and measuring cost efficiency with panel data models: application to electricity distribution utilities. *Review of Industrial Organization* 25, 1–19.

Farsi, M., Filippini, M. and Greene, W.H. (2006) Application of panel data models in benchmarking analysis of the electricity distribution sector. *Annals of Public and Cooperative Economics* 77, 271–290.

Gillespie, P.R., Hynes, S., O'Donoghue, C., Thorne, F.S. and Hennessy, T.C. (2015) Dairy and grassland economics in an era of possible expansion. PhD thesis, National University of Ireland, Galway, Ireland.

7 Quantifying Agricultural Greenhouse Gas Emissions and Identifying Cost-effective Mitigation Measures

Michael MacLeod* and Vera Eory
Land Economy, Environment and Society (LEES) Research Group, Scotland's Rural College (SRUC), Edinburgh, UK

7.1 Introduction

During the last few decades, there has been growing concern about the risk of climate change arising from the emission of anthropogenic greenhouse gases (GHGs). Policies such as the Kyoto Protocol have set national emission reduction targets and one of the challenges for government is to meet these targets while also achieving other important policy goals in areas such as economic development and food security. Governments therefore need to know: (i) what scale of reduction (or mitigation) in GHG could be achieved in different sectors of the economy (including agriculture); and (ii) what the most cost-effective ways are of achieving the mitigation. These are complex questions, particularly for agriculture, in which emissions arise as a result of complex biological processes. However, techniques have been developed to assist in the quantification of agricultural GHG emissions and the appraisal of mitigation measures, and it is these that are the subject of this chapter.

As shown in Fig. 7.1, the major GHGs associated with agricultural production are:

- CH_4 (methane) arising mainly from the anaerobic decomposition of organic matter during enteric fermentation and manure management, but also from flooded rice cultivation;
- N_2O (nitrous oxide) arising from the microbial transformation of N in soils and manures – during the application of manure, composts and synthetic fertilizers to land, via urine and dung deposited by grazing animals and from crop residues; and
- CO_2 (carbon dioxide) arising from: (i) energy use pre farm, on farm and post farm; and (ii) changes in above and below ground carbon stocks arising from land use and land use change.

Globally, emissions from the agriculture, forestry and land use (AFOLU) sector (see Fig. 7.1) were estimated to account for about 25% of total anthropogenic emissions in 2010 (Edenhofer *et al.*, 2014, p. 44). Figure 7.2 shows the change in AFOLU emissions over time and provides a breakdown of emissions by source. The decrease in 2000–2009 is primarily due to reductions in FOLU emissions (deforestation is thought to have reduced and afforestation increased), but agricultural emissions have increased by 25% since the 1970s, leading Smith *et al.* (2014, p. 816) to conclude that 'leveraging the mitigation potential in the sector is extremely important in meeting emission reduction targets'.

*E-mail: michael.macleod@sruc.ac.uk

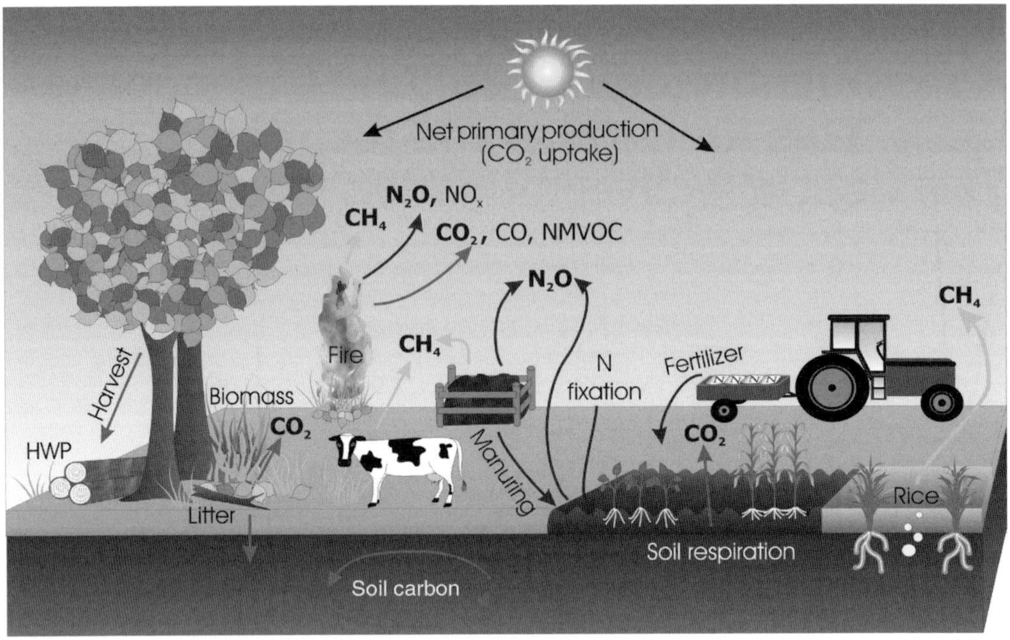

Fig. 7.1. The main agricultural greenhouse gas (GHG) emission sources, removals and processes in managed ecosystems. Note: carbon sequestration is not explicitly represented but also plays a role in the GHG balance. Key: HWP, harvested wood products; NMVOCs, non-methane volatile organic compounds. Modified from IPCC, 2006b, p. 1.6.

The chapter is arranged as follows. The next section (7.2) provides a brief explanation of how and why GHG emissions arise in agricultural supply chains, describes the methods available for estimating emissions – both on the farm and beyond the farm gate – and summarizes existing studies of GHG emissions in food supply chains. It then explains the role of cost-effectiveness in identifying appropriate mitigation measures and discusses some of the challenges and limitations that remain. Section 7.3 describes the application of some models, tools and schemes that have been developed for assessing the impact of agricultural practices and policies on the environment. Section 7.4 offers some concluding remarks.

7.2 Quantifying On-farm GHG Emissions

In order to solve a problem, one must first understand the nature of the problem. In this context, if we wish to find ways of reducing the GHG emissions from food supply chains, we first need to understand where and why emissions are arising. In recent years, a growing number of tools have been developed to support the quantification of agricultural GHG emissions. The World Resources Institute (2014, p. 88) lists over 40 such tools, ranging from fairly simple and user-friendly carbon calculators to more complex process-models, such as DNDC (DeNitrification DeComposition) and DayCent (which simulates fluxes of C and N among the atmosphere, vegetation and soil). The range of tools is expanding to meet the needs of specific users; for example, AgRE Calc (Agricultural Resource Efficiency Calculator; http://www.agrecalc.com/) has recently been developed to enable UK farmers to benchmark their performance and identify practical steps that can be taken on farm to reduce emissions. The European Commission Joint Research Centre (2015) has developed a carbon calculator designed for assessing the life cycle of GHG emissions from different types of farming systems across the whole of the European Union (EU). There are also numerous whole-farm models to assess GHG emissions (and in some cases farm

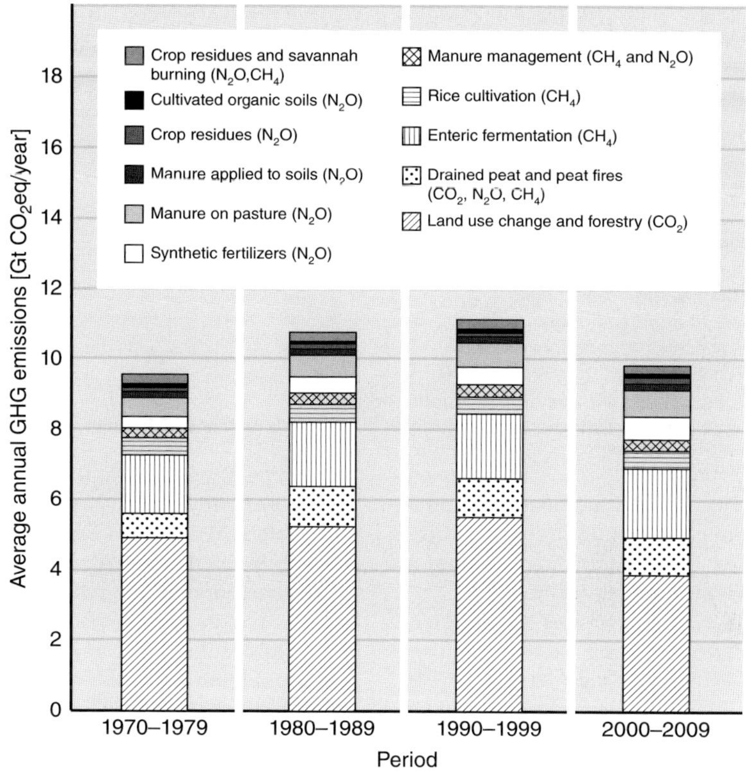

Fig. 7.2. Agriculture, forestry, and other land use (AFOLU) greenhouse gas (GHG) emissions by source. Modified from Smith *et al.*, 2014, p. 820.

economics), mainly developed for the ruminant sector (Schils *et al.*, 2007; Crosson *et al.*, 2011).

Many carbon calculators and models use the methods set out by the Intergovernmental Panel on Climate Change (IPCC) in *Volume 4, Agriculture, Forestry and Other Land Use*, of the *IPCC Guidelines for National Greenhouse Gas Inventories* (IPCC, 2006b). These guidelines were developed to support the United Nations Framework Convention on Climate Change (UNFCC), which requires signatory nations to establish and regularly update a national GHG inventory. These inventories adopt a territorial approach to reporting emissions, i.e. they estimate the emissions that arise from the economic activity that occurs within a defined territory. As mentioned above, *Volume 4* of the *IPCC Guidelines for National Greenhouse Gas Inventories* (IPCC, 2006b) provides guidance for quantifying emissions from AFOLU, see Table 7.1.

Key features of the IPCC guidance include: (i) an introduction to the emissions source; (ii) an overview of the alternative methods for quantifying the emissions (Tier 1, Tier 2 or Tier 3); (iii) flow diagrams to aid method selection; (iv) formulae required to undertake the Tier 1 and Tier 2 calculations; (v) default values for some key parameters (such as a methane conversion factor and nitrogen excretion rates); and (vi) uncertainty ranges for key parameters.

The differences between the Tier 1, 2 and 3 approaches depend on the emission category. In general, the Tier 1 approach involves multiplying activity data by a default emission factor (e.g. multiplying the number of dairy cattle in the UK by the Western Europe Tier 1 enteric fermentation emission factors (EF) for dairy cattle (IPCC, 2006b, p. 10.29), Tier 2 involves determination of the EF through a series of calculations (e.g. calculation of the dairy cow's net and gross energy requirement and feed dry matter intake based on the animal's size, milk yield, ration etc.), while Tier 3 involves the use of more complex process models. In *Volume 1, General Guidance and Reporting*, of the *IPCC Guidelines for National Greenhouse*

Table 7.1. Emissions sources as given in *Volume 4, Intergovernmental Panel on Climate Change (IPCC) Guidelines for National Greenhouse Gas Inventories* (IPCC, 2006b).

Source	Gas
Emissions and removals resulting from C stock changes in biomass, dead organic matter and mineral soils for all managed lands	CO_2
Emissions from fire on all managed land	CO_2 and non-CO_2
Emissions from all managed soils	N_2O
Emissions associated with liming and urea application to managed soils	CO_2
Emissions from rice cultivation	CH_4
Emissions from cultivated organic soils	CO_2 and N_2O
Emissions from managed wetlands	CO_2 and N_2O
Emissions from livestock enteric fermentation	CH_4
Emissions from manure management systems	CH_4 and N_2O
C stock change associated with harvested wood products	CO_2

Gas Inventories (IPCC, 2006a, p. 4.5) it is recommended that Tier 2 and 3 methods are used for key categories of emissions, though where this is not possible, the Tier 1 method can be used, as long as its use is clearly documented.

One of the reasons why the simpler Tier 1 approach should be seen as a last resort for some key emission categories, such as enteric CH_4, manure CH_4 and manure N_2O, is that it employs regional default values for key parameters. These default values can obscure important differences within regions and between farm types. The Tier 2 approach (illustrated in Fig. 7.3) means that the calculated EFs reflect differences in many parameters and so they enable variation in these parameters to be captured. They also enable estimates to be revised in light of new evidence; for example, the assumptions about how manure is managed in the UK were revised in 2014, leading to significant changes in the manure management EFs (see Webb *et al.*, 2014, p. 503).

Emissions estimates based on Tier 2 or Tier 3 methods can, therefore, provide useful insights into the drivers of GHG emissions. However, even when these more complex methods are used, the national GHG inventories still have certain limitations:

- Agricultural activities are modelled as separated rather than integrated processes.
- Temporal links are not fully captured, which can make it difficult to represent multi-annual processes such as crop rotations.
- The inventories estimate total emissions rather than emission intensities (i.e. emissions per unit of output).

The last point means that the inventories reflect the scale of activities, rather than the efficiency with which the activities are undertaken. While they undoubtedly provide useful information to governments on the relative importance of different sectors within their policy ambit, they can be misleading, as changes in emissions can arise from changes in the scale of agricultural activity rather than from changes in emissions intensity. Given the forecast increases in demand for food, reducing agricultural output in one country is likely to lead to a rise in prices and a consequent increase in output in another country, thus displacing rather than reducing emissions. The challenge is to improve the efficiency of production so that emissions can be reduced as output is increased. Consequently, we need to supplement the estimates of total emissions with estimates of emissions intensity. Potential measures of emissions intensity (EI) and their functional units (FUs) for agricultural products are summarized in Table 7.2.

Ultimately, a FU reflects the function that the product or activity performs in a particular context, as the same product can have different functions in different contexts, e.g. where diets are protein deficient it may be appropriate to use kg protein as the FU for food products, rather than total mass or energy content.

7.2.1 Moving beyond the farm gate: life cycle analysis (LCA)

The agricultural emissions quantified by many models are essentially those arising on farm, but

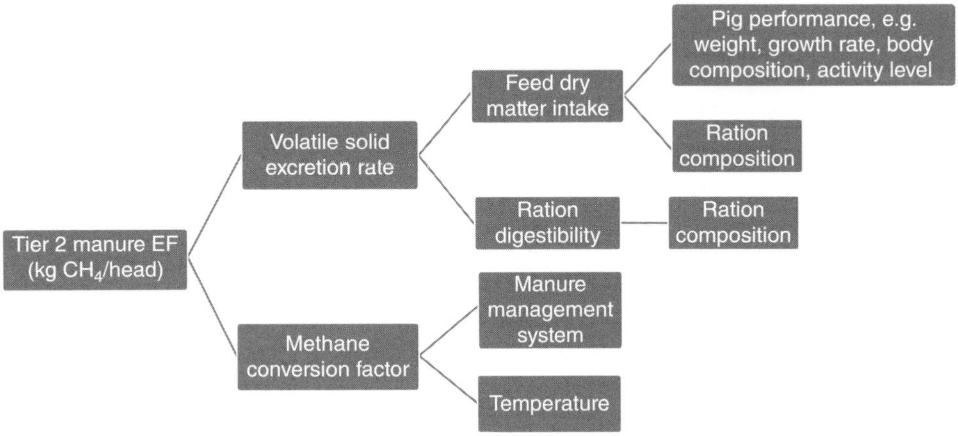

Fig. 7.3. A simplified schematic representation of the IPCC (2006a) Tier 2 approach to determining a manure management emission factor (EF) for methane emissions.

Table 7.2. Measures of emissions intensity (EI) and functional units for food and drink products.

Measure of EI[a]	Functional unit (FU)	Summary
kg CO_2e/country p.a.	Territory	The metric used in current national greenhouse gas (GHG) inventories. This is useful for identifying the contribution of different sectors, but provides no information on the efficiency of production.
kg CO_2e/ha p.a. or kg CO_2e/head p.a.	Land area or number of livestock	This is useful for identifying the contribution of different locations or land types, but provides no information on the efficiency of production.
kg CO_2e/kg product	Total edible mass	This is a commonly used FU that enables the emissions intensity of the same product to be studied in different locations and systems. It is of limited use in comparing products with different nutritional properties.
kg CO_2e/kg protein	Mass of edible protein	These FUs are similar to total mass, with the added advantage that they express results for products with different nutritional properties in ways that enable like-for-like comparison.
kg CO_2e/MJ DE	Digestible energy content	
kg CO_2e/£ GVA	Economic value	A useful FU when the primary function of production is economic growth. It can be difficult to interpret its meaning in practice, e.g. a decrease in EI over time could be due to a change in: (i) efficiency of production; (ii) increase in value; or (iii) both.

[a]CO_2e, carbon dioxide equivalent; DE, digestible energy; GVA, gross value added

farms do not exist in isolation. The production of food, fuel and fibre involves complex supply chains that deliver inputs to the farm (such as energy, fertilizer and machinery), and transport and transform farm outputs to meet consumer demand (see Table 7.3). Limiting the analysis to on-farm emissions can, therefore, omit important processes, providing a partial view of their full impact and ignoring opportunities for mitigation. In order to avoid this, we can use life-cycle analysis (LCA).

Table 7.3. Examples of pre- and post-farm processes.

Pre-farm process	Post-farm process
Production of crop feed materials	Transportation and processing of agricultural commodities
Production of non-crop feeds (e.g. synthetic additives, fishmeal)	Manufacture of packaging
CO_2 and N_2O emitted during fertilizer manufacture	Retail
Feed blending	Consumption
Production of fuel and electricity	Waste management
Manufacture of capital goods	
Transportation of inputs	

What is LCA?

LCA is a technique for assessing the environmental aspects of and potential impacts associated with a product, process or service, by:

- compiling an inventory of relevant energy and material inputs and environmental releases;
- evaluating the potential environmental impacts associated with the inputs and releases identified; and
- interpreting the results.

The defining characteristic of LCA is that (in theory) it includes all stages of the product's life cycle, from the manufacture of inputs, through the use of the product to its final disposal at the end of its life. For instance, a life cycle analysis of a plastic bag would include the impacts arising from:

- oil extraction and transportation;
- oil refining;
- ethylene polymerization;
- bag manufacture;
- distribution;
- use; and
- disposal.

Why use LCA?

LCA helps to avoid simply shifting the burden to another part of the supply chain. For example, swapping a petrol-driven car for an electric car reduces the exhaust emissions dramatically. However, it also leads to additional impacts upstream and downstream arising from the manufacture, charging and disposal of the battery, and these need to be included to make a valid comparison. In theory, LCA provides an assessment of the total burden required for a product and so enables a clearer picture of where burdens are arising along supply chains, thereby improving the efficiency of production, informing consumer choices and enabling more efficient policy making. How might it help us better understand food supply chains? LCA is commonly divided into two types: (i) attributional LCA, which describes the situation that currently exists; and (ii) consequential LCA, which is predictive and tries to answer 'what if ...?' questions. Table 7.4 provides examples of the sorts of questions that can be answered with attributional and consequential LCA.

What are the main steps in LCA?

Life-cycle analysis can be broken down into four main steps (Curran, 2006):

- *Goal definition and scoping* – define and describe the product, process or activity. Establish the context in which the assessment is to be made and identify the boundaries and environmental effects to be reviewed for the assessment.
- *Inventory analysis* – identify and quantify inputs (e.g. energy, water, materials) and outputs, i.e. products and environmental releases (e.g. air emissions, solid waste disposal, waste water discharges).
- *Impact assessment* – assess the potential human and ecological effects of energy, water and material usage, and the environmental releases identified in the inventory analysis.
- *Interpretation* – evaluate the results of the inventory analysis and impact assessment to select the preferred product, process or

Table 7.4. Examples of the sorts of questions that can be answered with attributional and consequential life cycle analysis (LCA).

Attributional	Consequential
What is the total life cycle impact of a commodity, e.g. UK or global sugar production, and what contribution do different processes make? How do two different commodities compare, e.g. UK beef and UK lamb? How does the same commodity, produced in different locations and/or by different systems compare, e.g. UK lamb vs New Zealand lamb, free range eggs vs cage eggs, fresh fish vs frozen fish?	What happens if: A farm converts to organic? Nuclear power stations are replaced with coal? Recyclable packaging is introduced? Genetically modified (GM) feed soy is banned? New welfare regulations for laying hens are introduced? A disease eradication scheme is introduced?

service, with a clear understanding of the uncertainty and the assumptions used to generate the results.

Guidance for undertaking LCA of food supply chains

Undertaking an LCA can be a complex and data-intensive task requiring significant amounts of time, money and expertise. However, there are increasing amounts of methodological guidance (Table 7.5), as well as data and tools (Table 7.6), now available to assist.

7.2.2 Existing studies of GHG emissions in food supply chains

There is a growing body of literature on the GHG emissions arising from food supply chains. Table 7.7 provides a (non-exhaustive) list of references on livestock systems that have been published over the last 10 years or so to illustrate the range of studies. The focus tends to be on: (i) cattle, pigs and chickens; (ii) developed countries; and (iii) emissions arising up to the farm gate. The majority of these studies do not include post-farm processes. Most of the studies focus on one species in a particular country. While such studies are valuable, differences in approach makes it difficult to investigate variations between species and countries. However, some studies are available that enable the comparison of different species across a range of countries – see Lesschen *et al.* (2011), Weiss and Leip (2012), MacLeod *et al.* (2013) and Opio *et al.* (2013).

7.2.3 Identifying ways of reducing GHG emissions

To address the dual challenges of climate change and food security, we need to find ways of reducing the emissions intensity (i.e. the emissions per unit output, e.g. kg CO_2e (CO_2 equivalents)/kg milk) of food supply chains. There are many ways in which the emissions intensity can be reduced on farm. For example, MacLeod *et al.* (2010) identified 97 mitigation measures that could be used to reduce emissions from crops and soils in the UK. These represent a subset of the mitigation measures, which could also include, depending on the supply chain in question, mitigation via livestock management, housing and manure, energy efficiency and land use change. In addition, mitigation may also be achieved pre farm (e.g. by using lower EI inputs) and post farm (e.g. by improving logistics and reducing food waste).

Given the range of potential mitigation measures, the challenge is to identify the 'best' mitigation measures for a particular situation. This is a complex task, although applying selection criteria can help to simplify matters. For instance, we may wish to apply the following criteria to a prospective measure:

- Does it work?
- What effect does it have on emissions?
- How does its effect vary (e.g. between countries, soil types, farm types)?
- What is the certainty of the effect?
- What might the unintended consequences be?

Table 7.5. Methodological guidance for life cycle analysis (LCA) and greenhouse gases (GHG) in the agricultural supply chain.

Name of guidance and website	Description
Livestock Environmental Assessment and Performance (LEAP) Partnership (http://www.fao.org/partnerships/leap/en/)	LEAP is a multi-stakeholder partnership which develops comprehensive guidance and methodologies for understanding the environmental performance of livestock supply chains.
Guidelines for National Greenhouse Gas Inventories (http://www.ipcc-nggip.iges.or.jp/public/2006gl/index.html)	Volume 4 of the IPCC guidelines provides methods for quantifying the GHG emissions arising from agriculture, forestry and other land use (IPCC, 2006b).
Greenhouse Gas Protocol (http://www.ghgprotocol.org/)	The Greenhouse Gas Protocol (GHG Protocol) is a widely used international accounting tool for greenhouse gas emissions that was developed by the World Resources Institute (WRI) and the World Business Council for Sustainable Development (WBCSD).
European Food Sustainable Consumption and Production (SCP) Round Table (EFSCPRT) (http://www.food-scp.eu/)	This is an initiative, co-chaired by the European Commission and food supply chain partners, which aims to establish the food chain as a major contributor towards sustainable consumption and production in Europe. It produced the ENVIFOOD (Environmental Assessment of Food and Drink) Protocol in 2012.
European Platform on Life Cycle Assessment (http://eplca.jrc.ec.europa.eu/)	This European Commission platform supports business and government needs for the availability, interoperability and quality of life cycle data and studies.
UNEP/SETAC Life Cycle Initiative (http://www.lifecycleinitiative.org/)	UNEP (United Nations Environment Programme)/SETAC (Society of Environmental Toxicology and Chemistry) launched this initiative in 2002 to enable users around the world to put life cycle thinking into effective practice.

- What is its mitigation potential, i.e. how much could it reduce emissions by in practice?
- What are the barriers to uptake?
- How amenable is it to different policy approaches? (i.e. could incentives be provided and compliance monitored?)
- Is it economically efficient?
- Are the total economic (i.e. monetary and non-monetary) benefits of the measure greater than the total economic costs?
- Is the measure cost-effective, i.e. does it achieve emission reduction at a lower financial cost than the social cost of carbon (which is defined by the US EPA (2016) as 'an estimate of the economic damages associated with a small increase in carbon dioxide')?

These are challenging questions. Nevertheless, there is now a growing body of work on the cost-effectiveness of agricultural GHG mitigation measures, which is reviewed in MacLeod *et al.* (2015).

7.2.4 Challenges and limitations in the quantification of emissions

Quantifying the emissions from food supply chains, and measures to reduce them, have many challenges. Providing an overview of all the challenges is not possible in this chapter, so instead we highlight four.

Improving the comparability of results

LCA is essentially a methodological framework, and provides the user with a degree of latitude. The results of an LCA are sensitive to the input data, assumptions and methodological choices, so it is important that these are clearly reported. Various standards and guidelines for LCA have been developed, such as the British Standards Institution's (BSI's) PAS (Publically Available Specification) document PAS 2050 (BSI, 2011) and the International Standards Organization's (ISO's) TS14067 (ISO, 2013). These guidelines are, by necessity, somewhat generic, but there

Table 7.6. Data and tools for food life cycle analysis (LCA) and greenhouse gases (GHG) in food supply chains.

Name of data source or tool (and website)	Description
ecoinvent (http://www.ecoinvent.org/) GaBi (http://www.gabi-software.com/) SimaPro (http://www.simapro.co.uk/)	These are commercially available LCA databases and tools.
AGRIBALYSE (http://ur-hortsys.cirad.fr/en/research-projects/agribalyse)	AGRIBALYSE is a public Life Cycle Inventory (LCI) database for the main French agricultural products (at the farm gate) that follows a homogeneous methodology shared between agricultural partners.
FAOstat (http://faostat3.fao.org/home/)	The database of the Food and Agriculture Organization of the United Nations (FAO), which contains global data sets on a range of relevant parameters on food production, prices, trade, food balances, inputs, etc.
Eurostat (http://ec.europa.eu/eurostat/web/agriculture/data/main-tables)	Eurostat provides a range of agricultural production data for European Union (EU) Member States.
OECD.Stat (http://stats.oecd.org/)	The Organisation for Economic Co-operation and Development provides data for a range of agricultural and environmental parameters for OECD countries and selected non-member economies.
FertiStat (http://www.fao.org/ag/agp/fertistat/)	FAO's database of fertilizer use statistics.
Aquaculture Feed and Fertilizer Resources Information System (AFFRIS) (http://www.fao.org/fishery/affris/affris-home/en/)	FAO's database for providing information on aquaculture feed and nutrient resources and guidelines to using information on feeding strategy.
FeedPrint (http://webapplicaties.wur.nl/software/feedprint/)	The CFPAN calculation tool and database FeedPrint from Wageningen UR calculate the carbon footprint of feed raw materials during their complete life cycle.
Feedipedia (http://www.feedipedia.org/)	Feedipedia is an open access information system on animal feed resources that provides information on the nature, occurrence, chemical composition, nutritional value and safe use of nearly 1400 livestock feeds worldwide.
Sub-Saharan Africa Feed Composition Database (http://192.156.137.110/ssafeed/)	'SSA Feeds' is web-based software and a database from CGIAR that provide information on the nutritive value of 14,571 samples of 459 livestock feeds from 14 countries in sub-Saharan Africa (SSA).

are also some sector-specific guidelines, e.g. from the International Dairy Federation for dairy (IDF, 2010), and from the BSI for horticulture (BSI, 2012a) and seafood (BSI, 2012b).

Dealing with interactions between measures

When applied in combination, measures can interact, meaning that the 'stand alone' abatement rate and cost-effectiveness can be quite different from the 'combined' cost-effectiveness (see MacLeod *et al.*, 2010; Hristov *et al.*, 2013, p. 121). For example, if a farm plants legumes and decreases the amount of N fertilizer applied, then the extent to which emissions can be further reduced by improving the efficiency of N application will diminish, making N efficiency measures less cost-effective. Conversely, optimizing the soil water level via improved drainage may enhance the mitigation effect of nitrification inhibitors or make the timing of nutrient applications easier. It is, therefore, important that any targeted policy measures take into account such synergies.

Characterizing variability and uncertainty

The components of cost-effectiveness, i.e. abatement rate and costs, often show a large variation

Table 7.7. Example of studies of livestock GHG emissions undertaken since 2004.

		Cattle						Chickens	
Study	Country	Beef	Dairy	Sheep	Goats	Buffalo	Pigs	Meat	Eggs
Basset-Mens *et al.* (2004)	France						x		
Cederberg and Flysjö (2004)	Sweden						x		
Casey and Holden (2005)	Ireland		x						
Eriksson *et al.* (2005)	Sweden						x		
Casey and Holden (2006)	Ireland	x							
Mollenhorst *et al.* (2006)	The Netherlands								x
Williams *et al.* (2006)	United Kingdom	x	x	x			x	x	x
Dalgaard (2007)	Denmark						x		
Ogino *et al.* (2007)	Japan	x							
Dekker *et al.* (2008)	The Netherlands								x
Pelletier (2008)	USA							x	
Thomassen *et al.* (2008)	The Netherlands		x						
Cederberg *et al.* (2009)	Sweden	x					x		x
Kool *et al.* (2009)	Denmark, England, Germany, The Netherlands						x		
Vergé *et al.* (2009a)	Canada						x		
Vergé *et al.* (2009b)	Canada								x
Beauchemin *et al.* (2010)	Canada	x							
Halberg *et al.* (2010)	Denmark						x		
Pelletier *et al.* (2010)	USA						x		
Prudencio da Silva *et al.* (2014)	Brazil, France							x	
Ripoli-Bosch *et al.* (2010)	Spain			x					
Wiedemann *et al.* (2010)	Australia						x		
Christie *et al.* (2011)	Australia		x						
Flysjö *et al.* (2011)	New Zealand, Sweden		x						
Foley *et al.* (2011)	Ireland	x							
Henriksson *et al.* (2011)	Sweden		x						
Kristensen *et al.* (2011)	Denmark		x						
Ledgard *et al.* (2011)	New Zealand			x					
Lesschen *et al.* (2011)	EU-27	x	x				x	x	x
Nielsen *et al.* (2011)	Denmark							x	
Thoma *et al.* (2011)	USA						x		
Wiedemann and McGahan (2011)	Australia								x
Hagemann *et al.* (2012)	Global		x						
Lafontaine *et al.* (2012)	Canada		x						
Leinonen *et al.* (2012a)	United Kingdom							x	
Leinonen *et al.* (2012b)	United Kingdom								x
Weiss and Leip (2012)	EU-27	x	x	x			x	x	x
Wiedemann *et al.* (2012)	Australia							x	
MacLeod *et al.* (2013)	Global						x	x	x
Opio *et al.* (2013)	Global	x	x	x	x	x			
Thoma *et al.* (2013)	USA		x						
Jones *et al.* (2014)	UK			x					
Prudencio da Silva *et al.* (2014)	Brazil, France							x	

spatially, temporally and across farm types. For instance, as fertilizer and crop prices vary across years and between countries, the cost of growing more maize and less grass for animal feed might be very different for different farmers. Such variability is rarely accounted for in most cost-effectiveness studies, which, instead, describe the assumed average situation. Because of this, most cost-effectiveness results should be seen as guidelines rather than prescriptions applicable to individual farms or individual years.

A related issue is the uncertainty of the estimates presented in cost-effectiveness studies, stemming from the uncertainty of the underlying data and assumptions. Some of the most significant uncertainties are those of the abatement rates and the estimated future uptake of the measures (Eory *et al.*, 2014a). These uncertainties are rarely quantified and presented in the analyses. Uncertainty analysis (or at least sensitivity analysis) should ideally accompany cost-effectiveness studies in order to provide guidance on the likely range of the mitigation potential and the costs.

Data quality and availability, particularly in developing countries

Most of the studies of GHG mitigation focus on OECD (Organisation for Economic Co-operation and Development) countries. While there are some global studies, few focus specifically on developing countries, which is where most growth in agricultural emissions is likely to occur. This lack of data is being addressed through a number of initiatives with remits that include developing countries, such as:

- the CGIAR Research Program on Climate Change, Agriculture and Food Security (CCAFS) (https://ccafs.cgiar.org/);
- the Global Research Alliance on Greenhouse Gas Emissions (http://globalresearchalliance.org/); and
- the Global Agenda for Sustainable Livestock, a partnership of livestock sector stakeholders committed to the sustainable development of the sector originally set up by FAO, which provides an Agenda Support Team (http://www.livestockdialogue.org/)

In order to simplify the task, many studies draw the system boundary at the farm gate. This may be reasonable given that post-farm emissions usually represent a small proportion of the overall total, but it means that post-farm losses (which can have a significant effect on the emissions intensity of some products) are rarely considered.

7.3 Application

7.3.1 The Global Livestock Environmental Assessment Model (GLEAM)

A significant recent initiative has been the development of the Global Livestock Environmental Assessment Model (GLEAM). This has been developed by FAO, which describes the model thus (FAO, 2015a):

> GLEAM is a modelling framework that simulates the interaction of activities and processes involved in livestock production and the environment. The model is developed to assess livestock's impacts, adaptation and mitigation options at (sub)national, regional and global scale.
> GLEAM differentiates key stages along livestock supply chains such as feed production, processing and transport; herd dynamics, animal feeding and manure management; and animal products processing and transport. The model captures the specific impacts of each stage, offering a comprehensive and disaggregated picture of livestock production negative side effects and valuable information for intervention.

GLEAM is comprehensive in scope and uses georeferenced information for computation. Geography is highly important to the assessment of agroecological processes, which depend on factors such as soil quality, climate and land use that have contrasting spatial patterns. This methodological development is a major improvement on other global assessments, which typically rely on national averages, and the GIS (geographical information system) platform provides flexibility in combining data sets and aggregating results. GLEAM can also compensate for the shortage of global data sets on animal production and related resource use by enabling livestock statistics to be disaggregated into different systems and animal cohorts, and also enabling the determination of feed rations, when no data sets are available. Furthermore, GLEAM allows a wide range of parameters to be varied, thereby enabling predictive modelling and design of mitigation interventions.

To date, GLEAM has been used to undertake a variety of different tasks (see Gerber *et al.* 2013; FAO, 2015b). These include both descriptive modelling, such as the global LCAs of livestock GHG emissions reported in MacLeod *et al.* (2013) and Opio *et al.* (2013), and predictive modelling to (for example) estimate the effects of using sexed semen in UK dairy herds (Eory *et al.*, 2014b) or treating trypanosomosis in East Africa cattle (MacLeod, M., Eory, V., Wint, G.R.W., Shaw, A.P.M., Robinson, T.P., Gerber, P., Cecchi, G. and Mattioli, R.C. Assessing the greenhouse gas mitigation effect of intervening against bovine trypanosomosis in Eastern Africa, submitted for publication).

7.3.2 The CAPRI model and the MITERRA-Europe assessment tool

As the importance of the additional information that a life-cycle approach can provide is increasingly being recognized, models and tools are being expanded to incorporate elements of LCA emissions. The European agricultural sector model CAPRI (Common Agricultural Policy Regionalised Impact Modelling System; http://www.capri-model.org) and the integrated assessment tool MITERRA-Europe have been further developed to assess the life-cycle impacts of GHG and N control actions and policies at the European scale.

The CAPRI model is a partial equilibrium economic model that was developed to assess agricultural policies in the EU in a global context. It has been equipped with modules for calculating GHG emissions from agricultural activities at the NUTS2 (Level 2 Nomenclature of Units for Territorial Statistics) regional level. In addition to calculating on-farm emissions, CAPRI is also capable of accounting for cradle-to-gate emissions (Weiss and Leip, 2012). The off-farm emission sources include feed production, transport and processing, fertilizer and pesticide production, electricity production and land use change.

The MITERRA-Europe integrated assessment tool quantifies the effects and interactions of agricultural policies and measures in EU-27 countries on the N and P balances of farms (including atmospheric and water pollution from reactive N) and GHG emissions from farms. It is partly based on the existing models CAPRI and GAINS (Greenhouse Gas and Air Pollution Interactions and Synergies; developed by the International Institute for Applied Systems Analysis, IIASA), supplemented with an N leaching module and a module with sets of measures (Velthof *et al.*, 2009). The development of MITERRA-Europe was led by Wageningen University in a service contract with the European Commission. The LCA elements recently added to this tool include feed production, fertilizer production and electricity production.

Both CAPRI and MITERRA-Europe have been used to compare the GHG emission intensity of agricultural products, including regional differences in the EU (Lesschen *et al.*, 2011) and the assessment of the nitrogen footprint of products (Leip *et al.*, 2014); they also allow for the assessment of mitigation measures (Leip *et al.*, 2010).

7.3.3 The Livestock Environmental Assessment and Performance (LEAP) partnership

The LEAP Partnership (see Table 7.5) was founded in 2012 and involves stakeholders across the livestock sectors with a shared interest in improving the environmental performance of livestock supply chains. The objective is to develop comprehensive guidance and methodology for understanding the environmental performance of livestock supply chains. To date, the Partnership has produced a database of GHG emissions for the five main feed crops and guidelines for the following (FAO, 2015c):

- environmental performance of animal feed supply chains;
- greenhouse gas emissions and fossil energy demand from poultry supply chains;
- greenhouse gas emissions and fossil energy demand from small ruminant supply chains;
- a review of indicators and methods to assess biodiversity;

- principles for the assessment of livestock impacts on biodiversity;
- developing sound tools for transition to sustainable food and agriculture; and
- environmental performance of large ruminant supply chains.

7.4 Concluding Remarks

Growing concern about the potential impacts of climate change has translated into concerted efforts to: (i) improve our understanding of how and why GHG emissions arise along food supply chains; and (ii) identify the best ways of reducing these emissions. There have a been a large number of papers and technical reports published on these topics over the last 10 years and (biophysical and economic) modelling has made a significant contribution to these. Life cycle analysis provides a more complete picture of the GHG emissions arising from food production, and can help to avoid the displacement of emissions. While the modelling of agricultural GHG emissions (whether on farm or along the supply chain) and mitigation measures remains complex, there are a growing number of initiatives seeking to fill data gaps and address methodological challenges.

References

Basset-Mens, C., van der Werf, H.M.G., Durand, P. and Leterme, P. (2004) Implications of uncertainty and variability in the life cycle assessment of pig farming systems. *Paper presented to the International Environmental Modelling and Software Society International Conference, iEMSs 2004, Complexity and Integrated Resources Management, Osnabruck, Germany*. Available at: http://iemss.logismi.co:8080/xmlui/handle/iemss/4203 (accessed 2 March 2016).

Beauchemin, K.A., Janzen, H., Little, S.M., McAllister, T.A. and McGinn, S.M. (2010) Life cycle assessment of greenhouse gas emissions from beef production in western Canada: a case study. *Agricultural Systems* 103, 371–379.

BSI (2011) *PAS 2050:2011: Specification for the Assessment of the Life Cycle Greenhouse Gas Emissions of Goods and Services*. BSI (British Standards Institution), London.

BSI (2012a) *PAS 2050-1:2012: Assessment of Life Cycle Greenhouse Gas Emissions from Horticultural Products – Supplementary Requirements for the Cradle to Gate Stages of GHG Assessments of Horticultural Products Undertaken in Accordance with PAS 2050*. BSI (British Standards Institution), London.

BSI (2012b) *PAS 2050-2:2012: Assessment of Life Cycle Greenhouse Gas Emissions – Supplementary Requirements for the Application of PAS 2050:2011 to Seafood and Other Aquatic Food Products*. BSI (British Standards Institution), London.

Casey, J.W. and Holden, N.M. (2005) Analysis of greenhouse gas emissions from the average Irish milk production system. *Agricultural Systems* 86, 97–114.

Casey, J.W. and Holden, N.M. (2006) Quantification of GHG emissions from suckler-beef production in Ireland. *Agricultural Systems* 90, 79–98.

Cederberg, C. and Flysjö, A. (2004) *Environmental Assessment of Future Pig Farming Systems – Quantifications of Three Scenarios from the FOOD 21 Synthesis Work*. SIK-rapport 723, SIK – The Swedish Institute for Food and Biotechnology [now the Food and Bioscience unit of the SP Technical Research Institute of Sweden], Borås, Sweden.

Cederberg, C., Sonesson, U., Henriksson, M., Sund, V. and Davis, J. (2009) *Greenhouse Gas Emissions from Swedish Production of Meat, Milk and Eggs from 1990 and 2005*. SIK-report 793, SIK – The Swedish Institute for Food and Biotechnology [now the Food and Bioscience unit of the SP Technical Research Institute of Sweden], Borås, Sweden.

Christie, K.M., Rawnsley, R.P. and Eckard, R.J. (2011) A whole farm systems analysis of greenhouse gas emissions of 60 Tasmanian dairy farms. *Animal Feed Science and Technology* 166/167, 653–662.

Crosson, P., Shalloo, L., O'Brien, D., Lanigan, G.J., Foley, P.A., Boland, T.M. and Kenny, D.A. (2011) A review of whole farm systems models of greenhouse gas emissions from beef and dairy cattle production systems. *Animal Feed Science and Technology* 166/167, 29–45.

Curran, M.A. (2006) *Life Cycle Assessment: Principles and Practice*. Document EPA/600/R-06/060, May 2006, National Risk Management Research Laboratory, Office of Research and Development, US

Environmental Protection Agency, Cincinnati, Ohio. Available at: http://nepis.epa.gov/Exe/ZyPDF.cgi/P1000L86.PDF?Dockey=P1000L86.PDF (accessed 2 March 2016).

Dalgaard, R.L. (2007) The environmental impact of pork production from a life cycle perspective. PhD thesis, University of Aarhus, Aarhus and Aalborg University, Aalborg, Denmark.

Dekker, S.E.M., de Boer, I.J.M. and Arnink, A.J.A. (2008) Environmental hotspot identification of organic egg production. In: Nemecek, T. and Gaillard, G. (eds) *Proceedings of the 6th International Conference on the Life Cycle Assessment in the Agri-food Sector – Towards a Sustainable Management of the Food Chain, November 12–14, 2008, Zurich, Switzerland*. Agroscope Reckenholz-Tänikon Research Station (ART), Zurich, Switzerland, pp. 381–389.

Edenhofer, O., Pichs-Madruga, R., Sokona, Y., Kadner, S., Minx, J.C. *et al.* (2014) Technical Summary. In: Edenhofer, O., Pichs-Madruga, R., Sokona, Y., Farahani, E., Kadner, S. *et al.* (eds) *Climate Change 2014: Mitigation of Climate Change. Contribution of Working Group III to the Fifth Assessment Report of the Intergovernmental Panel on Climate Change*. Cambridge University Press, Cambridge, UK and New York, pp. 33–107.

Eory, V., Topp, C.F.E., Moran, D. and Butler, A. (2014a) Assessing uncertainty in the cost-effectiveness of agricultural greenhouse gas mitigation. *Contributed paper prepared for presentation at the 88th Annual Conference of the Agricultural Economics Society, AgroParisTech, Paris, France 9–11 April 2014*. Available at: http://ageconsearch.umn.edu/bitstream/170784/2/Vera_Eory_Eoryetal_full.pdf (accessed 2 March 2016).

Eory, V., MacLeod, M., Shrestha, S. and Roberts, D. (2014b) Linking an economic and a life-cycle analysis biophysical model to support agricultural greenhouse gas mitigation policy. *German Journal of Agricultural Economics* 63, 133–142.

Eriksson, I.S., Elmquist, H., Stern, S. and Nybrant, T. (2005) Environmental systems analysis of pig production – the impact of feed choice. *The International Journal of Life Cycle Assessment* 10, 143–154.

European Commission Joint Research Centre (2015) Low Carbon Farming. Available at: http://ies-webarchive-ext.jrc.it/mars/mars/Projects/LC-Farming.html (accessed 3 March 2016).

FAO (2015a) GLEAM model in brief. Food and Agriculture Organization of the United Nations, Rome. Available at: http://www.fao.org/gleam/en/ (accessed 3 March 2016).

FAO (2015b) Global Livestock Environmental Assessment Model (GLEAM): Applications. Food and Agriculture Organization of the United Nations, Rome. Available at: http://www.fao.org/gleam/model-description/applications/en/ (accessed 3 March 2016).

FAO (2015c) Livestock Environmental Assessment and Performance (LEAP) Partnership: Resources. Food and Agriculture Organization of the United Nations, Rome. Available at: http://www.fao.org/partnerships/leap/resources/resources/en/ (accessed 3 March 2016).

Flysjö, A., Henriksson, M., Cederberg, C., Ledgard, S. and Englund, J.-E. (2011) The impact on various parameters on the carbon footprint of milk production in New Zealand and Sweden. *Agricultural Systems* 104, 459–469.

Foley, P.A., Crosson, P., Lovett, D.K., Boland, T.M., O'Mara, F.P. and Kenny, D.A. (2011) Whole-farm systems modelling of greenhouse gas emissions from pastoral suckler beef cow production systems. *Agriculture, Ecosystems and Environment* 142, 222–230.

Gerber, P., Steinfeld, H., Henderson, B., Opio, C., Mottet, A., Dijkman, J., Falcucci, A. and Tempio, G. (2013) *Tackling Climate Change through Livestock: A Global Assessment of Emissions and Mitigation Opportunities*. Food and Agriculture Organization of the United Nations, Rome.

Hagemann, M., Ndambi, A., Hemme, T. and Latacz-Lohmann, U. (2012) Contribution of milk production to global greenhouse gas emissions: an estimation based on typical farms. *Environmental Science and Pollution Research* 19, 390–402.

Halberg, N., Hermansen, J.E., Kristensen, I.S., Eriksen, J., Tvedegaard, N. and Petersen, B.M. (2010) Impact of organic pig production systems on CO_2 emission, C sequestration and nitrate pollution. *Agronomy for Sustainable Development* 30, 721–731.

Henriksson, M., Flysjö, A., Cederberg, C. and Swensson, C. (2011) Variation in carbon footprint of milk due to management differences between Swedish dairy farms. *Animal* 5, 1474–1484.

Hristov, A.N., Oh, J., Lee, C., Meinen, R., Montes, F. *et al.* (2013) *Mitigation of Greenhouse Gas Emissions in Livestock Production – A Review of Technical Options for non-co_2 Emissions*. FAO Animal Production and Health Paper No. 177, Food and Agriculture Organization of the United Nations, Rome.

IDF (2010) *A Common Carbon Footprint Approach for Dairy. The IDF Guide to a Standard Lifecycle Assessment Methodology for the Livestock Sector*. Bulletin of the International Dairy Federation 445/2010, IDF, Brussels.

IPCC (2006a) *2006 IPCC Guidelines for National Greenhouse Gas Inventories. Volume 1, General Guidance and Reporting*. Edited by Eggleston, H.S., Buendia, L., Miwa, K., Ngara, T. and Tanabe, K. Prepared by the National Greenhouse Gas Inventories Programme, Institute for Global Environmental Strategies (IGES), Hayama, Kanagawa, Japan. Available at: http://www.ipcc-nggip.iges.or.jp/public/2006gl/index.html (accessed 2 March 2016).

IPCC (2006b) *2006 IPCC Guidelines for National Greenhouse Gas Inventories, Volume 4, Agriculture, Forestry and Other Land Use*. Edited by Eggleston, H.S., Buendia, L., Miwa, K., Ngara, T. and Tanabe, K. Prepared by the National Greenhouse Gas Inventories Programme, Institute for Global Environmental Strategies (IGES), Hayama, Kanagawa, Japan. Available at: http://www.ipcc-nggip.iges.or.jp/public/2006gl/index.html (accessed 2 March 2016).

ISO (2013) *ISO/TS 14067:2013(en) Greenhouse Gases — Carbon Footprint of Products — Requirements and Guidelines for Quantification and Communication*. International Standards Organization, Geneva, Switzerland.

Jones, A.K., Jones, D.L. and Cross, P. (2014) The carbon footprint of UK sheep production: current knowledge and opportunities for reduction in temperate zones. *Journal of Agricultural Science* 152, 288–308.

Kool, A., Blonk, H., Ponsioen, T., Sukkel, W., Vermeer, H., de Vries, J. and Hoste, R. (2009) *Carbon Footprints van Conventioneel en Biologisch Varkensvlees. Uitgebreide Samenvatting. Analyse van Typische Productiesystemen in Nederland, Denemarken, Engeland en Duitsland.* [*Carbon Footprints of Conventional and Organic Pork. Detailed Summary. Assessment of Typical Production Systems in the Netherlands, Denmark, England and Germany*]. Blonk Milieu Advies, Gouda and Wageningen UR, Wageningen, The Netherlands. Available at: http://blonkconsultants.nl/upload/pdf/Carbon%20footprint%20pork%20samenvatting.pdf (accessed 2 March 2016).

Kristensen, T., Mogensen, L., Knudsen, M.T. and Hermansen, J.E. (2011) Effect of production system and farming strategy on greenhouse gas emissions from commercial dairy farms in a life cycle approach. *Livestock Science* 140, 136–148.

Lafontaine, M., Saad, R., Gilbert, D. and Margni, M. (2012) Environmental LCA of milk in Canada In: Corson, M.S. and van der Werf, H.M.G. (eds) *Proceedings of the, 8th International Conference on Life Cycle Assessment I the Agri-food Sector (LCA Food 2012), October 1–4, 2012, Saint-Malo, France*. INRA, Rennes, France, pp. 323–328. Available at: https://colloque4.inra.fr/var/lcafood2012/storage/fckeditor/file/Proceedings/Proceedings%20-%20LCA%20Food%202012%20-%20ISBN.pdf (accessed 2 March).

Ledgard, S.F., Lieffering, M., Coup, D. and O'Briens, B. (2011) Carbon footprinting of New Zealand lamb from the perspective of an exporting nation. *Animal Frontiers* 1, 40–45.

Leinonen, I., Williams, A.G., Wiseman, J., Guy, J. and Kyriazakis, I. (2012a) Predicting the environmental impacts of chicken systems in the United Kingdom through a life-cycle assessment: broiler production systems. *Poultry Science* 91, 8–25.

Leinonen, I., Williams, A.G., Wiseman, J., Guy, J. and Kyriazakis, I. (2012b) Predicting the environmental impacts of chicken systems in the United Kingdom through a life-cycle assessment: egg production systems. *Poultry Science* 91, 26–40.

Leip, A., Weiss, F., Wassenaar, T., Perez, I., Fellmann, T. *et al.* (2010) *Evaluation of the Livestock Sector's Contribution to the EU Greenhouse Gas Emissions (GGELS) – Final Report – Administrative Arrangements AGRI-2008-0245 and AGRI-2009-0290*. European Commission Joint Research Centre, Institute for Environment and Sustainability, Climate Change Unit (TP 290), Ispra, Italy. Available at: http://ec.europa.eu/agriculture/analysis/external/livestock-gas/full_text_en.pdf (accessed 2 March 2016).

Leip, A., Weiss, F., Lesschen, J.P. and Westhoek, H. (2014) The nitrogen footprint of food products in the European Union. *The Journal of Agricultural Science* 152, 20–33.

Lesschen, J.P., van den Berg, M., Westhoek, H.J., Witzke, H.-P. and Oenema, O. (2011) Greenhouse gas emission profiles of European livestock sectors. *Animal Feed Science and Technology* 166/167, 16–28.

MacLeod, M., Moran, D., Eory, V., Rees, R.M., Barnes, A. *et al.* (2010) Developing greenhouse gas marginal abatement costs curves for agricultural emissions from crops and soils in the UK. *Agricultural Systems* 103, 198–209.

MacLeod, M., Gerber, P., Mottet, A., Tempio, G., Falcucci, A. *et al.* (2013) *Greenhouse Gas Emissions from Pig and Chicken Supply Chains – A Global Life Cycle Assessment*. Food and Agriculture Organization of the United Nations, Rome.

MacLeod, M., Eory, V., Gruère, G. and Lankoski, J. (2015) *Cost-Effectiveness of Greenhouse Gas Mitigation Measures for Agriculture: A Literature Review*. Organisation for Economic Co-operation and Development,

Paris. Available at: http://www.oecd-ilibrary.org/agriculture-and-food/cost-effectiveness-of-greenhouse-gas-mitigation-measures-for-agriculture_5jrvvkq900vj-en (accessed 2 March 2016).
Mollenhorst, H., Berentsen, P.B.M. and De Boer, I.J.M. (2006) On-farm quantification of sustainability indicators: an application to egg production systems. *British Poultry Science* 47, 405–417.
Nielsen, N., Jørgensen, M. and Bahrndorff, S. (2011) *Greenhouse Gas Emission from the Danish Broiler Production Estimated via LCA Methodology*. AgroTech/Knowledge Centre for Agriculture, Aarhus, Denmark.
Ogino, A., Orito, H., Shimada, K. and Hirooka, H. (2007) Evaluating environmental impacts of the Japanese beef cow-calf system by the life cycle assessment method. *Animal Science Journal* 78, 424–432.
Opio, C., Gerber, P., Mottet, A., Falcucci, A., Tempio, G., MacLeod, M., Vellinga, T., Henderson, B. and Steinfeld, H. (2013) *Greenhouse Gas Emissions from Ruminant Supply Chains – A Global Life Cycle Assessment*. Food and Agriculture Organization of the United Nations, Rome.
Pelletier, N. (2008) Environmental performance in the US broiler poultry sector. *Agricultural Systems* 98, 67–73.
Pelletier, N., Lammers, P., Stender, D. and Pirog, R. (2010) Life cycle assessment of high- and low-profitability commodity and deep-bedded niche swine production systems in the Upper Midwestern United States. *Agricultural Systems* 103, 599–608.
Prudêncio da Silva, V., van der Werf, H.M., Soares, S.R. and Corson, M.S. (2014) Environmental impacts of French and Brazilian broiler chicken production scenarios: an LCA approach. *Journal of Environmental Management* 133, 222–231.
Ripoli-Bosch, R., de Boer, I.J.M., Bernues, A. and Vellinga, T. (2010) Greenhouse gas emissions throughout the life cycle of Spanish lamb-meat: a comparison of three production systems. In: Bernués, A., Boutonnet, J.P., Casasús, I., Chentouf, M., Gabiña, D., Joy, M., López-Francos, A., Morand-Fehr, P. and Pacheco, F. (eds) *Economic, Social and Environmental Sustainability in Sheep and Goat Production Systems, Proceedings of the 7th International Seminar of the Sub-Network on Production Systems of the FAO-CIHEAM Inter-Regional Cooperative Research and Development Network on Sheep and Goats, 2010/11/10-12, Zaragoza (Spain). Options Méditerranéennes: Série A. Séminaires Méditerranéens* No. 100, 125–130. Abstract available at: http://om.ciheam.org/article.php?IDPDF=801493 (accessed 2 March 2016).
Schils, R.L.M., Olesen, J.E., del Prado, A. and Soussana, J.F. (2007) A review of farm level modelling approaches for mitigating greenhouse gas emissions from ruminant livestock systems. *Livestock Science* 112, 240–251.
Smith, P., Bustamante, M., Ahammad, H., Clark, H., Dong, H. *et al.* (2014) Chapter 11: Agriculture, forestry and other land use (AFOLU). In: Edenhofer, O., Pichs-Madruga, R., Sokona, Y., Farahani, E., Kadner, S. *et al.* (eds) *Climate Change 2014: Mitigation of Climate Change. Contribution of Working Group III to the Fifth Assessment Report of the Intergovernmental Panel on Climate Change*. Cambridge University Press, Cambridge, UK and New York, pp. 811–922.
Thoma, G., Nutter, D., Ulrich, R., Maxwell, C., Frank, J. and East, C. (2011) *National Life Cycle Carbon Footprint Study for Production of US Swine. Report commissioned by the National Pork Board*. Applied Sustainability Center, University of Arkansas, Fayetteville, Arkansas. Available at: https://porkcdn.s3.amazonaws.com/sites/all/files/documents/NPB%20Scan%20Final%20-%20May%202011.pdf (accessed 2 March 2016).
Thoma, G., Popp, J., Nutter, D., Shonnard, D., Ulrich, R. *et al.* (2013) Greenhouse gas emissions from milk production and consumption in the United States: a cradle-to-grave life cycle assessment circa 2008. *International Dairy Journal* 31(Supplement 1), S3–S14.
Thomassen, M.A., Dalgaard, R., Heijungs, R. and de Boer, I. (2008) Attributional and consequential LCA of milk production. *International Journal of Life Cycle Assessment* 13, 339–349.
US EPA (2016) Climate Change Website: The Social Cost of Carbon. US Environmental Protection Agency, Washington, DC, Available at: https://www3.epa.gov/climatechange/EPAactivities/economics/scc.html (accessed 2 March 2016).
Velthof, G.L., Oudendag, D., Witzke, H.R., Asman, W.A.H., Klimont, Z. and Oenema, O. (2009) Integrated assessment of nitrogen losses from agriculture in EU-27 using MITERRA-EUROPE. *Journal of Environmental Quality* 38, 402–417.
Vergé, X.P.C., Dyer, J.A., Desjardins, R.L. and Worth, D. (2009a) Greenhouse gas emissions from the Canadian pork industry. *Livestock Science* 121, 92–101.
Vergé, X.P.C., Dyer, J.A., Desjardins, R. and Worth, D. (2009b) Long-term trends in greenhouse gas emissions from the Canadian poultry industry. *The Journal of Applied Poultry Research* 18, 210–222.
Webb, N., Broomfield, M., Brown, P., Buys, G., Cardenas, L. *et al.* (2014) *UK Greenhouse Gas Inventory, 1990 to 2012 Annual Report for Submission under the Framework Convention on Climate Change*. Ricardo-AEA Ltd., Didcot, UK.

Weiss, F. and Leip, A. (2012) Greenhouse gas emissions from the EU livestock sector: a life cycle assessment carried out with the CAPRI model. *Agriculture, Ecosystems and Environment* 149, 124–134.
Wiedemann, S.G. and McGahan, E.J. (2011) *Environmental Assessment of an Egg Production Supply Chain using Life Cycle Assessment. Final Project Report.* A report for the Australian Egg Corporation Limited. AECL Publication No. 1FS091A, North Sydney, New South Wales, Australia. Available at: https://www.aecl.org/assets/RD-files/Outputs-2/1FS091A-Final-Report.pdf (accessed 2 March 2016).
Wiedemann, S.G., McGahan, E.J., Grist, S. and Grant, T. (2010) *Environmental Assessment of Two Pork Supply Chains using Life Cycle Assessment.* Project Nos PRJ-003176 and PRJ-004519, RIRDC Publication No 09/176, Rural Industries Research and Development Corporation, Barton, Australian Capital Territory, Australia. Available at: https://rirdc.infoservices.com.au/downloads/09-176.pdf (accessed 2 March 2016).
Wiedemann, S.G., McGahan, E. and Poad, G. (2012) *Using Life Cycle Assessment to Quantify the Environmental Impact of Chicken Meat Production.* RIRDC Project Nos PRJ-004596 and PRJ-007445, RIRDC Publication No. 12/029, Rural Industries Research and Development Corporation, Barton, Australian Capital Territory, Australia. Available at: https://rirdc.infoservices.com.au/downloads/12-029 (accessed 2 March 2016).
Williams, A.G., Audsley, E. and Sandars, D.L. (2006) Determining the environmental burdens and resource use in the production of agricultural and horticultural commodities. Defra project report IS0205. Silsoe Research Institute, Silsoe, Cranfield University, Bedford and Department for Environment, Food and Rural Affairs, London. Available at: http://randd.defra.gov.uk/Document.aspx?Document=IS0205_3958_EXE.doc (accessed 2 March 2016).
World Resources Institute (2014) *Greenhouse Gas Protocol. GHG Protocol Agricultural Guidance: Interpreting the Corporate Accounting and Reporting Standard for the Agricultural Sector.* World Resources Institute, Washington, DC. Available at: http://www.ghgprotocol.org/files/ghgp/GHG%20Protocol%20Agricultural%20Guidance%20%28April%2026%29_0.pdf (accessed 2 March 2016).

8 Moving Beyond the Farm: Representing Farms in Regional Modelling

Jinxiu Ding,[1]* Bruce A. McCarl[2] and Weiwei Wang[3]
[1]*Xiamen University, Xiamen, Fujian, China;* [2]*Texas A&M University, College Station, Texas, USA;* [3]*University of Illinois Urbana-Champaign, USA*

8.1 Introduction

Farm-level modelling has been commonly used to study detailed questions about behaviour, policy effects and technology at the farm level. Such models are useful for understanding farm processes, detailed resource availability and practice sustainability (Bender and McCarl, 1992; van Calker *et al.*, 2007). However, there are cases where farm-level modelling can be misleading owing to case study specificity, locality of impact and a lack of coverage of interactions between alternative farms, resources and markets. Hence, it can be important to move beyond the farm itself and represent farms in regional modelling.

At the regional level, insights usually focus on welfare, markets, interactions and the environment. Typically, farm management decisions influence crop and livestock production, nutrient runoff and water quantity or quality. The consequences of these processes can affect markets in terms of prices and quantities. Regional modelling is useful for analysing the effects of policies and programmes on farm production systems, including marketed quantities, prices and trade, plus environmental loadings from agriculture. For example, policy analysis on the efficient allocation of water likely needs to be done at the regional level because of the heterogeneous access to water resources that farms have, and water trading between farms. Also, the analysis of policy and environmental influences on emission loadings to water, air and soil is not generally a single farm issue, especially for non-point pollution. Regional modelling does have some limitations though, i.e. typically, the models are more aggregate in character, and they lose detailed depictions of machinery, labour and resource constraints, as well as technology considerations that influence farm responses. There are also large issues in the selection of representative farms, market wedges between farm and regional prices, and endogeneity of prices and input markets.

This chapter discusses farm-based regional modelling, and covers the reasons for using this type of approach to represent farms in regional models. Additionally, two regional modelling case studies are provided.

8.2 Review of Regional Modelling

Econometric approaches estimate equations that give the behaviour of the system concerned based on statistical data regression over historical data.

*E-mail: jinxiuding@gmail.com

Adams *et al.* (1999) indicated that an approach based on historical data provides a greater description of the differences in actual crop yields based on observed phenomena, but that the phenomena to be represented must be within the range of historical fluctuations or at least reasonably inferred therefrom. Oczkowski and Bandara (2013) used econometric estimation to examine how agricultural land allocation was influenced by climate and economic factors. They estimated acreage response elasticities for rainfall, prices and land holdings.

Econometric estimation results developed from farm-level data can be used in a regional analysis to project farm behaviour. Mushtaq *et al.* (2014) tested the hypotheses of farm-level structural adjustments using econometric modelling and integrated these with a regional computable general equilibrium (CGE) model to investigate the impacts of water-use policy and climatic change on the rice industry. However, econometric modelling depends on large sets of data and observations of the full range of the phenomena to be studied, which are not easily obtained at the farm level. Usually, county-level data or state-level data will be used as a substitute, depending on the nature and scope of the research problem.

Modelling decision processes based on decision rules is an approach in which farmer behaviour is represented as the interplay between a disposition to act, an inventory of available resources and the external environment (Shucksmith, 1993). Bontkes and van Keulen (2003) described the decision making of farmers by decision rules and combined it with a regional model to simulate agricultural development in the south-west of Mali. They chose the cropping pattern based on hypothetical decision rules, involving food preferences, food requirements, expected yields, etc.

Mathematical programming deals with the analysis of optimization problems. It optimizes some measures of satisfaction by selecting values for a set of variables, given a set of goals and constraints (McCarl and Spreen, 2014). It has been commonly applied in farm planning and regional analysis. Such models can offer unique advantages over other methods. First, they can address the individually diverse but highly interlinked nature of the agricultural sector. Secondly, they can bring detailed micro-level databases to bear in the analysis of such policy issues as pricing, employment, investment decisions, comparative advantage and risk analysis (Hazell and Norton, 1986). In the following sections, we will review the methods of representing farms in regional modelling under mathematical programming.

8.2.1 Overview of aggregation issues under mathematical programming

Many regional mathematical programs have been formulated that contain a set of representative farms and not a full representation of each and every farm in the region. This raises the issue of aggregation. Generally, it is impossible to explicitly represent every farm in most regional models, so the number of representative farms to use, and the form and structure of each farm type, must be selected. Also, aggregation is involved in terms of representing the seasonality of operations and the complexity of the sets of production possibilities. There are three basic approaches used to address aggregation:

- the derivation of conditions for the elimination of aggregation bias, which involves the conditions under which firms may be aggregated into representative firms;
- the evaluation of strategies to reduce or minimize bias, which involves approaches such as farm type grouping (Frick and Andrews, 1965; Buckwell and Hazell, 1972) or exploration of the aggregation principles implicit in the Dantzig–Wolfe decomposition (McCarl, 1982; Önal and McCarl, 1989, 1991); and
- the derivation of bounds on maximum bias caused by aggregation, which deals with the aggregation of activities and/or constraints within a particular linear programming model (Kuhn and Cullen, 1975; Geoffrion, 1977).

Conditions for the elimination of aggregation bias were first derived by Day (1963). He developed conditions for zero aggregation error that required firms to be virtually identical. Others developed strategies to minimize bias by grouping related firms. For example, Frick and Andrews (1965) presented an approach based on common limiting resources, while Buckwell and Hazell

(1972) developed an approach using cluster analysis. Jayasuriya (2004) used representative farm models to denote the key characteristics of irrigation farming, with the data on the key features collected using a local consensus data approach (representing farm size, irrigation entitlements, crop areas, etc.). Louhichi *et al.* (2008) developed representative farms defined by farm size, intensity, land use and specialization.

The above aggregation procedures are largely bottom-up approaches based on the assumption that detailed farm level information is available. Without this resolution of farm-level information, a top-down approach has been introduced by McCarl (1982), who suggested a method to aggregate over regions in sector models by requiring the crop mix to be a convex combination of historical crop mixes. The main assumption here is that there is a duality between solving an aggregate (regional) model with the full detail of all the farm models and an aggregate, less detailed model, which is constrained to the convex combination of the chosen decisions by the farms. A detailed application of this method can be found in two case studies presented in this chapter.

8.2.2 Representing farms in farm-based regional modelling: conceptual issues

Regional models allow the representation of interactions among farms in an area in which agricultural entities compete for space and resources (Bontkes and van Keulen, 2003). Regional models are able to measure policy or programme effects on:

- farm production decisions across the region as they differ by type of farm, including crop mix, livestock mix, input use, on-farm consumption/off farm sales, farm income and factor returns;
- regional market activities and trade, such as regional food prices, production and processing, exports, imports, factor usage and factor prices;
- regional economic characteristics, including total farm income, welfare, government cost and balance of payments; and
- environmental attributes, including emissions to water, air and soil as well as water use, land use and vegetative cover.

A regional model is often a regionally representative multi-farm model that includes supply and demand sources from all farms in a producing region such as a county or a state. It is generally designed to capture the interactions between alternative farms, resources and markets. In such a model, some assumptions are called for, for example:

- there are markets for purchased inputs and products;
- input and output prices are the same between farms or different by a constant amount; and
- regionally specific factors, such as land, labour and water, are fixed but can be tradable among farms.

Generally, most regional model structures contain the following elements:

- A description of producers' economic behaviour, that is, the objective(s) that they pursue in choosing land use, irrigation water use and output composition. Often this is profit maximization (as in the Texas High Plains case study below), but it may involve the maximization of surpluses of regional consumers and producers (as in the Edwards case study).
- Production functions, or technology sets for the different producers as depicted for the representative farms in the region. These functions relate agricultural yields to inputs, and are generally differentiated by farming zone, commodity produced and whether the farms are irrigated or dry land (e.g. livestock vs annual crops vs perennial crops, irrigated vs non-irrigated, etc.).
- A definition of the resource endowments and access conditions held by each type of producer. These include land, irrigation water and farm machinery.
- A specification of the commodity market environment in which the producer operates. Here, whether the regional producers are price takers or price makers has to be specified according to the scale of the regional market. For example, modelling crop production in a small region with a small market share implies a price-taker assumption, but modelling, say, the US corn belt would require price-endogenous approaches

(McCarl and Spreen, 1980). Moreover, the import and export of commodities in this region may also be considered.

- A specification of the factor market environment in which the producer operates. Here, whether the regional producers have fixed or tradable factors, with models sometimes involving cases with high transaction cost-bearing factors (like a water market – McCarl *et al.*, 1999), with water return flows and diversions affecting downstream use (McCarl *et al.*, 2013), or with upward sloping factor prices as more, say, labour is purchased, groundwater is extracted or land is developed (see the second case study below).

8.3 The Core Model

The core model needs a specification of the policy environment of the region, for example, the rates of input and/or output subsidy, and whether the land or water use regulations would affect producers' decisions and final profits.

When representing farms in regional modelling, an individual farm can technically be represented by:

$$\begin{aligned} \text{Maximize} \quad & \mathbf{px} - \mathbf{cz} \\ \text{Subject to} \quad & \mathbf{x} - \mathbf{Ay} = 0 \\ & \mathbf{By} - \mathbf{z} = 0 \\ & \mathbf{Dy} \le \mathbf{e} \\ & \mathbf{x}, \mathbf{y}, \mathbf{z} \ge 0 \end{aligned}$$

where $\mathbf{p}$ is the vector of product prices, $\mathbf{c}$ is the vector of purchased input prices, $\mathbf{x}$ is the vector of quantities produced by commodity on the farm, $\mathbf{y}$ is the vector of production activity levels on the farm, $\mathbf{z}$ is the vector of input purchases by input type, $\mathbf{A}$ is the matrix of product yields per unit of production of $\mathbf{y}$, $\mathbf{B}$ is the matrix of input usages per unit of production of $\mathbf{y}$, $\mathbf{D}$ is the matrix of fixed input uses per unit of production of $\mathbf{y}$ and $\mathbf{e}$ is the vector of fixed input endowments on the farm. In this farm model, the objective function comprises profit maximization. The first inequality constraint reflects an upper limit on the availability of resources or fixed factor endowments.

Now, when we move to the regional model, we introduce multiple farms:

$$\begin{aligned} \text{Maximize} \quad & \mathbf{p}\sum_i \mathbf{x}_i - \mathbf{c}\sum_i \mathbf{z}_i \\ \text{Subject to} \quad & \mathbf{x}_i = \mathbf{A}_i\mathbf{y}_i \quad \text{for each farm } i \\ & \mathbf{B}_i\mathbf{y}_i = \mathbf{z}_i \quad \text{for each farm } i \\ & \mathbf{D}_i\mathbf{y}_i \le \mathbf{e}_i \quad \text{for each farm } i \\ & \mathbf{x}_i, \mathbf{y}_i, \mathbf{z}_i \ge 0 \quad \text{for each farm } i \end{aligned}$$

where $\mathbf{x}_i$, $\mathbf{y}_i$, $\mathbf{z}_i$ and $\mathbf{e}_i$ are the vectors across the farms of produced quantities, production activity levels, purchased input and fixed input endowments for each farm *i*; and $\mathbf{A}_i$, $\mathbf{B}_i$ and $\mathbf{D}_i$ are the matrices of farm-specific product yields, input usages and fixed input usage rates (per unit of production) for each farm *i*, respectively.

In practice, three types of knowledge at increasing levels of analytical complexity can be aggregated. These are:

- individual data, such as farm size or the area planted per crop;
- response functions, such as supply or demand equations; and
- systems models, such as programming approximations of farm types.

The aggregation of individual data depends on a consistent classification framework. Individual data can be aggregated as long as they refer to strictly the same quality of function. For example, irrigation water endowments for farmer A and irrigation water for farmer B when they withdraw from the same location. Requirements for the aggregation of response functions are rather stringent. In order to meet these requirements, response functions may be defined in simple linear terms of only price and income, omitting other variables that would not be significant in every individual equation (Theil, 1954). System model aggregation is even more difficult, and can only be done at a certain cost in terms of representativeness and detail.

8.3.1 Limitations and challenges of the regional model

There are several challenges and limitations involved in building a regional model, and these are briefly described below.

Loss of farm-level accuracy

When modelling at the farm level, there are often relatively detailed depictions available of the constraints and technology considerations that influence farm crop choice. Modellers can contact the subject farmers and obtain data on price expectations, yields, resource endowments, input applications, working rates, etc. However, moving up to the regional and possibly national level generally means loss of this intimacy. There are generally not the resources or time to obtain as much farm-specific information or to do as good a job of depicting the full production possibilities set and the associated constraints. Generally, a more aggregated approach is taken that uses statistical or expert opinion data and a much less detailed production representation.

Aggregation of farms

When focusing on a large region instead of a farm, the time available and the data that would be required do not permit attempts to depict individual production systems for all of the farms. Nor, typically, if it was done once, would it be affordable to maintain such a data set over time. Thus, modellers usually choose to depict farm production by subregion, such as a county with different farm types therein (see the case studies in Section 8.5). This situation raises the classic concerns of aggregation and bias reduction; namely, how can a representation of production choices in the region be developed that gives an accurate portrayal of how the region might respond to changes in the item to be studied? That representation should be designed so that it exhibits minimal differences between what would happen if there was a full regional representation and the more compact region representation that is chosen for use. Such a choice is obviously hampered by the fact that we never know the full characteristics of the region.

Endogeneity of prices in product and input markets

When dealing with models of large regions one often has to wrestle with the issue of endogenous prices. For example, in a small region, while changes in the amount of agricultural commodities produced are typically insufficient to change the nationally determined agricultural commodity prices, the alterations in agricultural water demand and productivity that they involve might affect local water and land rental prices, because they might affect local markets for hay and pasture grazing. Decisions also alter water withdrawals; they may influence aquifer water stocks and river flows, which would influence the scarcity of water for industrial, municipal and environmental interests, as well as affecting water market prices. Regional modelling studies may find it desirable to deal with endogenous prices for land, labour and water, for instance, and possibly for some commodities.

Calibration of the model

In a regional setting, producers' budgets give data on the production activities that were carried out in face of potentially higher or lower national prices. Data explaining the wedge between local prices and national or regional prices are generally not available in a systematic fashion. Because of this, regional modellers may need to deal with disparate prices and model calibration to best simulate the real market in the way of representing regionally varying prices. Two approaches have been developed. The first was adopted by Fajardo *et al.* (1981), who calibrated a model of Nicaraguan agriculture by imposing the production economic optimality criterion of marginal revenue = marginal cost for all activities and adding a calculated value to cost to make this equation balance.

The second approach is commonly known as positive quadratic programming (PQP) or positive mathematical programming (PMP), as explained in Howitt (1995). In PMP, given the prevailing national or regional average prices, the model is solved with additional constraints that hold the solution to observed acreage levels. The resultant shadow prices of the added constraints indicate that the model solution would not replicate the observed acreage if they were deleted. In turn, PQP/PMP users divide the shadow prices by the acreage level and add the result in a quadratic fashion to the production costs, which causes the model to calibrate at that point. This procedure, however, alters the marginal cost when the solution varies from the shadow price allocation made, adding to the cost when going above it and taking away from the cost when going below it.

Generally, both of these procedures compensate for missing costs and omitted constraints as a result of factors such as: (i) unknown transactions costs encountered to move goods through the marketing channel which are not reported in the crop budgets and may differ by farming location; (ii) differences in the quality and timing of the commodities sold; and (iii) omitted seasonality in resource constraints (as explored in Baker and McCarl, 1982).

Other limitations

Regional models can yield results on the effects of changes in parameters on market and resource competition, although they can be weak in capturing farm-level details of production decision making and detailed resource availability; for example, Bender and McCarl (1992) found that omitting detailed resource timing tends to bias results. Furthermore, regional models can fail to provide generalizable results if decision making is at the national or global level and includes multiple heterogeneous regions. Generally, the selection of model scale and analysis should be determined by the goals of the study and the scope of the issue concerned.

8.4 Review of Regional Modelling Applications

Regional modelling has been used to analyse environmental change, resource allocation management, sustainability and policy evaluation. Some issues cannot be effectively examined at the farm level because of regional resource competition/usage interrelationships and unequal access to resources across a region, e.g. water management in the Edwards aquifer region of Texas (McCarl *et al.*, 1999), or unequal depths of aquifer water supplies (Wang *et al.*, 2011). Also, regions commonly contain different types of farms that need to be represented with intermediate commodities and localized demands that are important, e.g. cow-calf, cattle finishing and livestock feed producing units. Finally, changes in commodity and resource prices can also make a regional analysis give different results from a farm analysis. Regional modelling has been recently applied to a wide variety of issues, including climate effects, water resources, environmental loadings from agriculture and policy analysis, as we will discuss below.

8.4.1 Climate effects

The effects of climate variability and change have primarily been examined in a regional or larger context, as reasonable adaptation to climate change often entails shifts in regional production and resources (Adams *et al.*, 1990; Giupponi *et al.*, 2006) and climate effects are not regionally uniform (Bates *et al.*, 2008). The components involved in climate change assessment include:

- climate scenarios, e.g. emission control scenarios from general circulation models (GCMs);
- simulations of the effects of climate changes on crops, e.g. regional crop yields, regional irrigation, crop water use;
- hydrological simulations, e.g. effects on irrigation water supply;
- effects on livestock growth, fecundity, death loss, grazing land requirements;
- yield simulation for range and hay production, e.g. alterations in hay and pasture feed supply and effects on animal unit monthly grazing supply;
- non-agricultural resource competition, e.g. changes in municipal and industrial demands, and in environmental flow requirements; and
- economics, e.g. shifts in commodity demands, market conditions, production costs, pest control costs and trade with other regions.

Chen *et al.* (2001b) applied a regional hydro-economic model (EDSIM, see Section 8.5.1) to investigate the effect of climate change on the water-dependent economy in the Texas Edwards Aquifer region. Regional climate change projections were drawn from Intergovernmental Panel on Climate Change (IPCC) 2001 scenarios from the Canadian Climate Centre (CCC) model and the Hadley Climate Center (HAD) model. Also at a regional level, Chen *et al.* (2001a) used an empirical model of the US agricultural sector (ASM) to investigate the economic impacts from potential frequency and strength shift in El Niño Southern Oscillation (ENSO). Soboll *et al.* (2011) emphasized the importance of integrated regional modelling and scenario design when examining the effects of global climate change on future water demand. They showed that integrated modelling and scenario development provided a good way to discuss complex human–environment-interactions with stakeholders. McCarl *et al.* (2013) examined the regional effects of climate change in Egypt, modelling crop and livestock implications as well as effects on regional water supply, return flow and municipal use.

8.4.2 Water resources

Water demand has been increasing. Agriculture accounts for more than two thirds of global water use, with regional shares as large as 90% (Morrison *et al.*, 2009). Water resources are generally available in a heterogeneous fashion to farms in a region due to depth to water table, proximity to rivers, up and down stream locations, etc. Also, in some cases, water can be traded into markets. From this perspective, the analysis of efficient water allocation may need to be addressed at a regional scale.

There have been a number of studies on water allocation and efficiency studies at the regional level. The earlier studies are differentiated in their coverage of water users and water types. For water management, some studies have integrated a hydrological component into an economic model to investigate: (i) the economic effect of improvement in water allocation (Dillon, 1991; Ward and Lynch, 1997; Rosegrant *et al.*, 2000); (ii) minimum in-stream requirements in support of the environment (Gillig *et al.*, 2001; Cai, 2009); (iii) use for hydroelectric purposes (McCarl and Parandvash, 1988); and (iv) regional interdependency of surface water flows, return flows, municipal uses and environmental flows (Booker and Young, 1994; McCarl *et al.*, 2013). Some studies have focused on surface water, some on groundwater and others on the conjunctive use of surface water and groundwater. Surface water is mainly discussed in water allocation efficiency, e.g. the application of inter-basin water transfers in Texas (Cai, 2009), and the performance of intrastate and interstate market institutions for water resource in the Colorado River (Booker and Young, 1994). Groundwater studies have focused on recharge, discharge, pumping, aquifer elevation and spring flow (Gisser and Sanchez, 1980; Dillon, 1991; Keplinger *et al.*, 1998; Almas, 2008). Conjunctive use of groundwater and surface water was initiated by Burt (1964) and further investigated in several studies (Gillig *et al.*, 2001; Pulido-Velazquez *et al.*, 2004; Cai, 2009). Daneshmand *et al.* (2014) applied an integrated hydrological, socio-economic and environmental approach to access conjunctive water use during drought in the Zayandehrood water basin in Iran. They found that conjunctive use would preserve water supply reduction under 10% of irrigation demand during a drought.

8.4.3 Environmental loadings from agriculture

As the population grows, the demand for food increases, which stimulates the development of agricultural production and also increases the environmental loadings from agriculture, for example:

- input usage, e.g. increased nitrogen fertilizer usage;
- emission loadings to water, air and soil, e.g. pesticide use;
- water use, e.g. groundwater depletion; and
- land use, e.g. deforestation.

An integrated economic–hydrological water model was applied by Rosegrant *et al.* (2000) to account for the interactions between water, food production, economic welfare and the environmental consequences in the Maipo River basin in Chile. Pattanayak *et al.* (2005) developed estimates of national and regional water quality co-effects from greenhouse gas (GHG) emission offset strategies in US agriculture. They linked a national agricultural sector model to a national water quality model to simulate the effect of GHG mitigation strategies. They found that GHG price incentives could stimulate conversion from agricultural land to forested land, with co-effects of water quality increase.

8.4.4 Policy analysis

A proposed policy can have a regional focus or have differential effects across a region (Bontkes and van Keulen, 2003). Judez and de Miguel (2001) analysed the effects of the European Union (EU) agricultural policy 'Agenda 2000' in Spain. Hamilton *et al.* (1985) investigated the implications of aggregate response assumptions on the economic effects of air quality change on agriculture production in the US Corn Belt. Das *et al.* (2010) estimated the effects of water conservation policies in the form of an extraction tax and a quota, and found that neither policy significantly inhibited groundwater use in the Texas Aquifer region.

8.5 Application

Here we cover two regional water-based studies in some detail. Both studies build up from farm considerations but involve different market and other assumptions.

8.5.1 Edwards Aquifer management

The Edwards Aquifer (EA) is the major water source for more than 2 million people in south-central Texas around San Antonio and provides much of the base flow to the Guadalupe River. The EA water supply depends on precipitation-based recharge, which is highly influenced by climate variability and adversely affected by drought. Several studies have already been conducted in the EA and a model named EDSIM (the Edwards Aquifer Simulation Model) has been developed (Dillon, 1991; McCarl *et al.*, 1993; Keplinger, 1996; Williams, 1996; McCarl *et al.*, 1999). EDSIM is a regional, farm-based economic and hydrological simulation model that simulates water and land allocation within a perfectly competitive economy under uncertainty in recharge. The model has been used to study dry-year irrigation suspension (Keplinger *et al.*, 1998), climate change effects (Chen *et al.*, 2001b), regional water planning (Gillig *et al.*, 2001), ENSO effects (Chen *et al.*, 2001b, 2005), elevation-dependent management (Chen *et al.*, 2006) and, most recently, decadal climate variability (Ding, 2014).

Herein, we describe the most recent version of the EDSIM model and its use in a study on the value of decadal climate variability (DCV) information. Specifically, we study three DCV phenomena, the Pacific Decadal Oscillation (PDO), the Tropical Atlantic Gradient (TAG), and the West Pacific Warm Pool (WPWP). As each DCV phenomenon has a positive phase and a negative phase, we have eight DCV phase combinations that are ordered as PDO, TAG and WPWP with a positive sign for a positive phase and a negative sign for a negative phase.

EDSIM model structure

The most recent version of EDSIM simulates agricultural, municipal and industrial water use, plus irrigated versus dry land cropping, livestock herd size, grassland use, land conversion, pumping cost and spring flow with recharge varying by DCV phase combinations. It optimizes consumers' and producers' surpluses, simulating the economic allocation of water subject to pumping limits. Equation (1) shows the objective function with uncertain recharge as influenced by DCV information (see equation 1 at bottom of the page):

A transition probability for movements between DCV phases is represented by prob_DCV_d, which is calculated based on DCV (d), and $\text{prob}_{r/d}$ is the probability of a recharge state r given a DCV phase combination. The function k denotes the cost of developing new irrigated land (IRRLAND) in a county (p) and lift zone (z). The function h is the net benefit from agricultural production (CROPPROD), livestock production (LIVEPROD) and water use in the agricultural (AGWATER), municipal (MUN) and industrial (IND) sectors for livestock (l) or crop (c) with irrigation strategy (s) in month (m). GRASSUSE is the acreage of grassland used by livestock. The h function has the integral under municipal and industrial (M&I) demand functions plus farm profits being a producers' and consumers' surplus-maximizing

$$\text{Max}: \sum_d \text{prob\_DCV}_d \times \left\{ \begin{array}{l} k(\text{IRRLAND}_{pzd}) + \sum_{r/d} \text{prob}_{r/d} \times h(\text{CROPPROD}_{pzrcsd}, \text{AGWATER}_{pzrmd}, \\ \text{LIVEPROD}_{pzrld}, \text{GRASSUSE}_{pzrd}, \text{MUN}_{prmd}, \text{IND}_{prmd}) \end{array} \right\} \quad (1)$$

approach (as discussed in McCarl and Spreen, 1980). This involves the assumption that the farms are price takers but that the water price involves the M&I demand.

Constraints on land conversion permit land movement between irrigated cropping, dry land cropping and grassland, with the grassland movements being longer term and the dry land movements being shorter term. Irrigated land use (IRRLAND) cannot exceed the initial irrigated land available less the irrigated land converted to dry land and grassland. Likewise, grassland use (GRASSUSE) is limited to the available initial grassland plus land converted to grassland from irrigated land.

DCV information is entered in three ways in the EDSIM model. First, we include eight combinations of DCV phases and the corresponding historical DCV probabilities are applied (for discussions on DCV phenomena see Mehta *et al.*, 2012; Ding, 2014; Fernandez, 2013). Secondly, econometrically estimated DCV impacts on crop yields were added into the model. Thirdly, the model is formulated with and without allowing crop and livestock mixes to adapt to DCV phase combination information.

Due to time and farm-level data limitation, in the EDSIM model we depict farm production by county and pumping lift zone within county, which raises the problem of aggregation. We use three representative farms per county – an irrigated one, a dry land one and a livestock grazing one. We allow the transfer of land between these and model interrelationships at the aquifer level across the irrigated entities. We also model M&I water use.

To alleviate aggregation issues, we use a top-down representation approach, requiring the representative farm crop mix to be a convex combination of historical crop/livestock mixes. We separately set up the crop mix constraints for the irrigated and dry land farms. The crop mix constraint is defined in Equation (2). Cropland use (CROPROD) is a convex combination of historical crop mixes (cropmixdata) for crops (c) and mix possibilities (x) in county (p). Livestock production is also constrained to be a convex combination of historical livestock mixes.

For the DCV study, we allow crop/livestock mixes to vary across stochastic conditions depending on knowledge of DCV phase and phase strength information. Following Fernandez (2013), we have three cases to discuss here (see equation 2 at bottom of the page):

- the base scenario case, in which the crop and livestock mix is selected without DCV information;
- the transition probability case, when the DCV information to set up the crop and livestock mix is known; and
- the perfect information case, when both the DCV phase combination and phase strength information are known in advance.

where d, r and z are defined as for Equation (1).

Other constraints are defined in the same way as stated in McCarl *et al.* (1999) except for adding the DCV phases. For instance, pumping cost is a linear function of aquifer lift, and both ending water level and spring flow level are a function of recharge level, initial water level and total water use (AGWATER + MUN + IND), respectively.

Simulation results

In this section, we mainly report results relative to the economic benefits gained with disseminated knowledge about DCV phases and their effects, much like the value of El Niño information studies as reported in Adams *et al.* (1999) or Chen *et al.* (2005). We also report on regional crop mix adaptation and inter-sectoral water allocation.

$$\sum_s \mathrm{CROPPROD}_{pzrcsd} = \begin{cases} \sum_x \mathrm{cropmixdata}_{pcx}\mathrm{CROPMIX}_{px} \text{ without DCV information, for all } p,z,r,c,d \\ \sum_x \mathrm{cropmixdata}_{pcx}\mathrm{CROPMIX}_{pxd} \text{ with DCV information, for all } p,z,r,c,d \\ \sum_x \mathrm{cropmixdata}_{pcx}\mathrm{CROPMIX}_{pxrd} \text{ with DCV and phase strength information, for all } p,z,r,c,d \end{cases} \quad (2)$$

Table 8.1. Comparison of economic benefits for alternative forecasting cases (unit: US$106) under different combinations of variation of the Pacific Decadal Oscillation (PDO), Tropical Atlantic Gradient (TAG) and West Pacific Warm Pool (WPWP).

	PDO– TAG– WPWP–	PDO– TAG+ WPWP–	PDO– TAG– WPWP+	PDO+ TAG+ WPWP–	PDO– TAG+ WPWP+	PDO+ TAG+ WPWP+	PDO+ TAG– WPWP–	PDO+ TAG– WPWP+	Mean
Agriculture									
Base scenario	207.6	213.9	202.8	203.5	206.6	211.7	200.1	197.3	205.4
Transition probability	0.73	0.82	0.30	1.13	0.36	1.36	0.02	0.78	0.68
Perfect information	42.17	44.05	43.55	40.08	45.07	42.94	37.28	32.91	41.72
Livestock									
Base scenario	105.8	111.9	113.4	101.4	104.8	107.0	121.3	97.3	106.6
Transition probability	0.56	0.12	0.32	–0.47	0.45	0.12	0.25	0.20	0.21
Perfect information	–1.80	–1.80	–1.59	–2.57	–2.11	–1.98	–1.67	–0.92	–1.91
Municipal and Industrial (M&I)									
Base scenario	768.9	763.6	765.1	748.1	739.8	753.6	761.9	716.1	750.7
Transition probability	0.11	0.16	0.03	0.11	0.13	0.01	0.04	0.38	0.12
Perfect information	0.21	0.37	0.31	0.57	0.47	0.26	0.24	1.12	0.43
Totals									
Base scenario	1082.2	1089.4	1081	1052.9	1051.2	1072.3	1083.3	1010.7	1062.7
Transition probability	1.40	1.09	0.65	0.77	0.94	1.48	0.31	1.36	1.01
Perfect information	40.58	42.62	42.27	38.09	43.43	41.22	35.85	33.11	40.25

We consider two cases, the value of perfect information and the value of transition probabilities to next year's DCV state given information on the current year's state. We find that total benefits increase under perfect information of the DCV phase when compared with both the base scenario without DCV information and also relative to the transition probability information (see Table 8.1). The potential welfare gains from an adaptation in the crop and livestock mix with the knowledge of DCV and phase strength information vary from US$33.11 million to US$43.43 million a year, depending on the initial phase of the DCV combination. The average economic value of the perfect DCV forecast is US$40.25 million a year in the EA region. The value given transition probability information is much smaller, US$1.01 million a year, as the uncertainty in effects is substantial.

The results of crop mix adaptation are shown in Tables 8.2 and Table 8.3. Table 8.2 reports the changes in crop mix under transition probability relative to the base scenario with a pumping limit. The production of winter wheat, sorghum hay and watermelon display a negative acreage shift for most of the DCV scenarios. Under perfect information, acreage shifts in

Table 8.2. Crop mix adaptation under transition probability relative to base scenario with a 400,000 acre ft pumping limit, showing the % change under different combinations of variation in the Pacific Decadal Oscillation (PDO), Tropical Atlantic Gradient (TAG) and West Pacific Warm Pool (WPWP).

	PDO− TAG− WPWP−	PDO− TAG+ WPWP−	PDO− TAG− WPWP+	PDO+ TAG+ WPWP−	PDO− TAG+ WPWP+	PDO+ TAG+ WPWP+	PDO+ TAG− WPWP−	PDO+ TAG− WPWP+
Cotton	−16.73	−13.99	63.30	396.37	281.32	156.78	−21.26	−0.26
Hay	−30.19	−38.30	−10.30	−33.13	−23.34	−10.20	2.34	−4.98
Maize	18.60	29.37	−0.32	−4.81	−2.48	3.20	4.68	−6.74
Oats	17.33	26.20	−16.08	4.27	−27.66	−25.33	−6.30	−11.56
Onion	−7.51	1.07	−8.23	4.80	−14.05	−6.80	−7.61	−3.89
Peanuts	−24.68	−30.45	0.81	−0.79	5.78	9.95	10.77	14.36
Sorghum	−21.44	−12.55	−0.94	61.67	1.98	−7.01	−10.93	−15.41
Sorghum hay	−42.58	−45.77	18.57	35.40	42.31	66.67	−41.20	52.74
Watermelon	−14.43	−25.85	−2.73	−26.22	−11.21	−9.13	2.98	−9.24
Winter wheat	−39.75	−40.93	−19.41	−20.23	−37.08	−31.62	5.35	−13.55

Table 8.3. Crop mix adaptation under perfect information relative to base scenario with a 400,000 acre ft pumping limit, showing the % change under different combinations of variation in the Pacific Decadal Oscillation (PDO), Tropical Atlantic Gradient (TAG) and West Pacific Warm Pool (WPWP).

	PDO− TAG− WPWP−	PDO− TAG+ WPWP−	PDO− TAG− WPWP+	PDO+ TAG+ WPWP−	PDO− TAG+ WPWP+	PDO+ TAG+ WPWP+	PDO+ TAG− WPWP−	PDO+ TAG− WPWP+
Cabbage	−1.57	−1.57	−3.10	−1.78	−0.47	−1.05	−4.67	−3.10
Carrot	0.04	0.04	0.08	0.05	0.01	0.03	0.13	0.08
Cotton	46.90	60.42	246.40	858.50	909.06	383.77	66.82	60.84
Hay	−36.53	−37.67	−5.39	−22.71	−21.36	−2.97	23.38	26.68
Lettuce	29.41	29.41	60.00	34.12	9.41	20.00	89.41	60.00
Maize	61.10	55.30	30.07	18.60	20.79	25.33	45.48	13.30
Oats	20.36	6.62	−29.98	5.90	−21.30	−26.19	−16.79	−18.25
Onion	17.71	14.31	1.61	26.86	14.95	7.33	1.95	−5.84
Peanuts	−14.32	−15.89	34.01	9.32	21.15	24.54	52.69	24.33
Sorghum	−17.42	−26.95	−10.21	66.91	31.25	−12.43	−20.45	−13.25
Sorghum hay	−20.73	−33.87	47.86	22.36	43.59	78.91	−15.02	4.11
Watermelon	−23.67	−34.02	−0.78	−16.63	−14.06	−10.44	17.36	−3.47
Winter wheat	−48.63	−44.52	−29.42	−17.71	−35.39	−27.77	−7.73	5.16

maize, cotton, carrot and lettuce are all positive for all DCV phase combinations (see Table 8.3).

Although they are not shown here, we also have results showing that livestock mix, and land use in irrigated land, dry land and grassland are sensitive. We can also see some results on inter-sectoral water allocation from Figs 8.1 and 8.2. With perfect information on the DCV phase, the amount of agricultural water use is greater than the values under both the cases of the base scenario and transition probability, no matter whether the pumping limit is constrained or not (see Fig. 8.1). However, under perfect information, M&I water use decreases relative to the case of transition probability (see Fig. 8.2), implying that more water would flow from the M&I sector to the agricultural sector when the information on DCV and phase strength is announced.

Main findings

The analysis shows that information on DCV phases has substantial economic value and that the crop and livestock mix adjusts with DCV information. It also shows that a regional model can be used to examine inter-sectoral water allocation.

8.5.2 Economic and groundwater use implications of climate change in the Ogallala Aquifer region

Sustainable water use is a major concern in the Texas Ogallala Aquifer region. In the last 50 years, the aquifer has exhibited depletion, with its water stock decreasing by approximately 33%. Climate change is also a potential factor that could exacerbate the situation. Regional projections of a warmer and regionally drier future (Bates *et al.*, 2008; Seager *et al.*, 2013) suggest increases in water demand which, in turn, could hasten depletion. Agricultural producers can adapt to the depletion and climate change influences by adjustments in production, including deficit

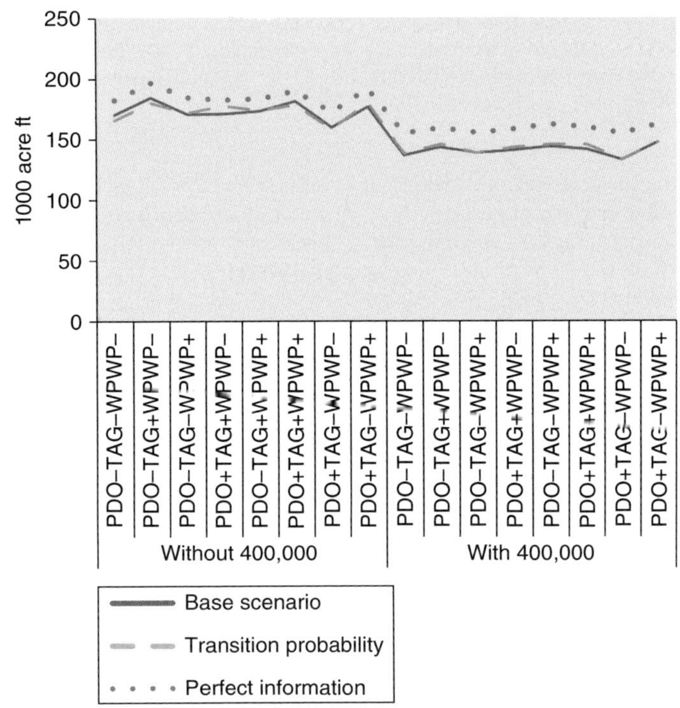

Fig. 8.1. Agricultural water use under alternative forecasting cases (unit: 103 acre feet): different combinations of variation of the Pacific Decadal Oscillation (PDO), Tropical Atlantic Gradient (TAG) and West Pacific Warm Pool (WPWP). With or without 400,000 indicates with or without a 400,000 acre ft pumping limit.

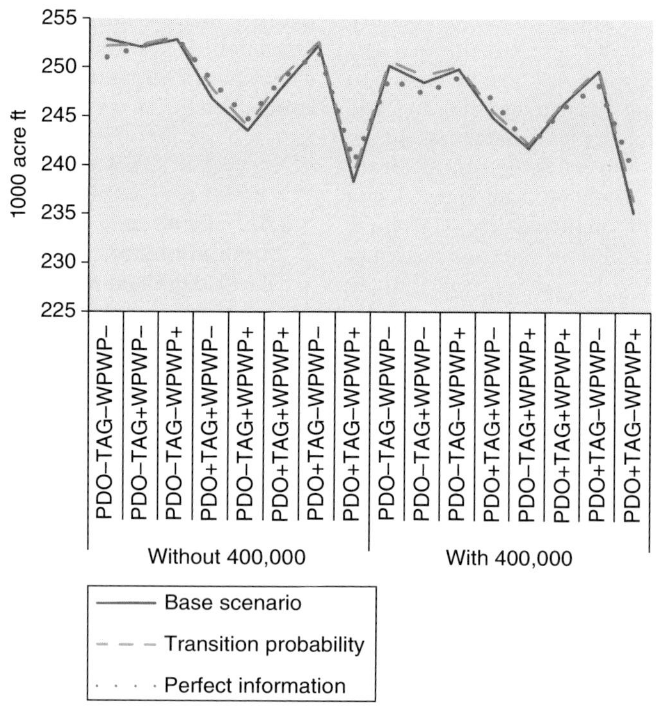

Fig. 8.2. Municipal and industrial water use under alternative forecasting cases (unit: 103 acre feet): different combinations of variation of the Pacific Decadal Oscillation (PDO), Tropical Atlantic Gradient (TAG) and West Pacific Warm Pool (WPWP). With and without 400,000 indicates with or without a 400,000 acre ft pumping limit.

irrigation, changing the crop mix or moving out of irrigated cropping to dry land or pasture.

Three counties in this region – Dallam, Hartley and Sherman – are the focus of our study, as they have been identified as critical groundwater depletion areas (Marek *et al.*, 2009). The questions we study are how climate change will affect regional economics and groundwater use, and what strategies can be used to adapt to this change.

Structure of the model

To examine future land and water use as affected by depletion and climate change, we construct a dynamic agricultural/hydrological model. The model assumes that farmers maximize the net present value of income by choosing land use (between irrigated and dry land cropping), water application rates, the conversion of land to grazing livestock, and crop mix, while facing constraints on land, water levels and production inputs. The hydrological component determines water stock, depth to water and pumping cost over time, based on the initial stock and cumulative withdrawals. Water level endowments and changes are heterogeneous within and across counties so that there are representative irrigated farms by county with differing initial water and water depletion characteristics. The model also has a representative dry land farm for each initial pumping lift and a livestock grazing operation.

The basic model extends previous farm-level studies (Feng, 1992; Terrell, 1998; Johnson, 2003; Das, 2004) to account for the following features that are important in analysing the region:

- The irrigated farms within each county are grouped into five representative farms which have different characteristics in terms of depth to water table and saturated thickness (distance from the surface of the water in the aquifer to the Triassic clay bottom).

- Dry land cropping farms and a grazing farm are included for each county and initial depth to water case.
- There is endogenous land conversion between each of the irrigated cropping farms and the dry land cropping and livestock grazing farms.
- There are constraints on crop mix reflecting rotations that are modelled separately for dry land and irrigated crops to solve the aggregation problem, as discussed above.
- Climate change scenarios alter yields and irrigation requirements, reflecting a generally hotter and drier climate in conformity with climate change projections.
- The irrigation strategies for each representative farm allow the water applied per acre to be an endogenous variable, including a water-use response function allowing deficit irrigation.
- There is an explicitly dynamic 50 year representation of annual cropping, water use, pumping cost and aquifer status under each farm in terms of depth to water and remaining saturated thickness.

Thus, the model represents five representative farms with heterogeneous pumping conditions within each county. The county lands are divided based on a 2008 saturated thickness contour map developed by the Center for Geospatial Technology, Texas Tech University. Cropping conditions (yields, non-pumping costs, etc.) are assumed to be homogenous in a county owing to a lack of sub-county information. Algebraically, the objective function and net returns are defined as:

$$\max \text{NPV} = \sum_{t=1}^{T} (1+r)^{-(t-1)} \text{NR}_t \quad (3)$$

$$\text{NR}_t = \sum_z \sum_i \sum_l \left[P_{it} Y_{zilt} \text{ACR}_{zilt} - C_{zilt} \right] + \sum_z \text{NB}_t (\text{RANGELAND}_{zt}) \quad (4)$$

where NPV is the net present value of net returns to land use, r is the discount rate, and NR_t is the net revenue in period t. The subscript z represents pumping zones, i represents the crops grown and l is the land type (irrigated land or dry land). P_{it} is the exogenous price of crop i at time t; Y_{zilt} is the yield of crop i under irrigation strategy l in zone z in time period t; ACR_{zilt} is the acreage of crop i under irrigation type l in zone z at time t; and C_{zilt} is the cost of production of ACR_{zilt}. That cost includes operating cost, harvest cost, equipment cost, and the energy and labour cost of pumping, as further defined below. NB_t is the net benefit of livestock grazing on rangeland use in zone z at time t, RANGELAND_{zt} is the amount of grazing land in zone z at time t. T is the time horizon considered (40 years).

Irrigated crop yields are a function of water use, allowing the possibility of deficit irrigation (applying an amount of water below that which would maximize physical yields). Conceptually this function is:

$$Y_{zilt} = f(\text{WUS}_{zit}), \quad l = \text{irrigated land} \quad (5)$$

where WUS_{zit} is the water use per acre of crop i in zone z at time t. The crop production function $f(\cdot)$ describes the crop yield response to water applied using the functions estimated by Guerrero *et al.* (2011) for the major regional crops of maize, cotton, wheat and sorghum. For dry land crops, 5 year (2005–2010) average county level yields are used in all zones.

Cattle grazing is an alternative land use. The maximum amount of cattle that can graze in each zone is restricted by the equation:

$$\text{QCL}_{zt} \leq \text{RANGELAND}_{zt}/\text{gr} \quad (6)$$

where the variable RANGELAND_{zt} is the amount of land for grazing, and the parameter 'gr' denotes the amount of grazing land required per unit of cattle. QCL_{zt} is the number of cattle that are produced in zone z at time t, which is an endogenous variable, as is the amount of rangeland available.

To prevent extreme specification in land use caused by our aggregation of farms, the acreage of dry land and irrigated crops is modelled separately and required to be a convex combination (weighted average) of historical dry land and irrigated crop mix following McCarl (1982), as shown in Equations (7) and (8).

$$\text{ACR}_{zilt} = \sum_j W_{zljt} \times \text{hiscrop}_{zilj} \quad (7)$$

$$\sum_i \sum_l \text{ACR}_{zilt} \leq \text{CROPLAND}_{zt} \quad (8)$$

where subscript j stands for the observed crop mixes drawn from historical year 1990–2010;

hiscrop_{zilj} is the ratio of historically observed acreage patterns of crop *i* in the *j*th mix for land type *l* (dry land or irrigated) in zone *z* to the total acreage under all crops for each land type *l* in zone *z*. W_{zljt} represents the weight assigned to each historical crop mix, which varies with land type (*l*), zone (*z*) and time (*t*); it is defined as a variable to be endogenously determined by the model. Notice that here the historical crop mix (hiscrop_{zilj}) is represented as the proportion of land each crop takes rather than the real acreage; therefore we drop the traditional convexity equation which requires the summation of all the weights (W_{zljt}) to be less than or equal to 1. Instead, we take constraint shown in Equation (7), which makes these mixes a convex combination of cropland available. CROPLAND_{zt} is the total available cropland in zone *z* at time *t*, which is an endogenous variable subject to total land balance as discussed in Equation (10) below. Equation (8) restricts cropland acreage in each zone by irrigation type to no more than land allocated to that irrigation type.

Notice that the use of this approach may not fully reflect future production possibilities under unprecedented changes in conditions such as climate change and biofuel feedstock production (Chen and Önal, 2012). Therefore, we follow the methodology taken in Beach and McCarl (2010) to accommodate such effects and phase out the effects of crop mixes: (i) the historical crop mix approach is fully applied until 2030; and (ii) after year 2030, a 5% deviation is allowed from the observed historical mixes and this is increased at 5% annually until the year 2050, gradually phasing out the mix. Equation (7′) replaces Equation (7) and reflects such adjustments:

$$\text{ACR}_{zilt} = (1-\theta)\times\left(\sum_j W_{zljt}\times \text{hiscrop}_{zilj}\right) + \theta\times \text{ANYMIX}_{zilt} \qquad (7')$$

where

$$\theta = \begin{cases} 0, & t \le 2030 \\ \dfrac{t-2030}{2050-2030}, & 2030 < t \le 2050 \\ 1, & t > 2050 \end{cases}$$

and

$$\text{ANYMIX}_{zilt} \ge 0 \text{ for all } z, i, l, t \qquad (9)$$

The cropland acreage ACR_{zilt} is the weighted combination of two parts: the first a 'fully historical mix (hiscrop_{zilt}) and the second 'any mix' (ANYMIX_{zilt}). The crop mix then phases out over time *t*. In the first 20 years (2010–2030), θ equals 0 and the mixes hold. After year 2050, θ is equal to 1 and the crop mixes are inactivated. Between 2030 and 2050, the percentage of the land obeying the mix diminishes. Equation (9) imposes non-negativity on ANYMIX_{zilt}.

We have a total land constraint that restricts total land use for the sum of cropping plus pasture:

$$\text{RANGELAND}_{zt} + \text{CROPLAND}_{zt} \le \text{availland}_z \qquad (10)$$

where the total land allocation to range and crops in zone *z* at time *t* cannot exceed total available land (availland_z). That benefit of livestock production is expressed in terms of the rental rate for rangeland. Land conversion is allowed from cropland to rangeland at a one-time cost of establishing grass at about US$60/acre (Sullivan and Rinehart, 2010).

In the hydrological component, irrigation cost and aquifer depletion potential are determined by the annual characteristics of: (i) pumping lift; (ii) saturated thickness; (iii) specific yield (the volume of water that can be extracted from a unit volume of saturated ground); and (iv) cumulative water withdrawals now and in the past (Hornbeck and Keskin, 2011). As water use increases, levels decline and pumping costs increase. The hydrological component reflects these conditions through five equations following Das *et al.* (2010):

$$\text{ST}_{zt} = \text{ST}_{zt-1} - (\text{TW}_{zt} - \text{rech}_z)/(\text{overall}_z \times s_z) \qquad (11)$$

$$\text{LIFT}_{zt} = \text{LIFT}_{zt-1} + (\text{TW}_{zt} - \text{rech}_z)/(\text{overall}_z \times s_z) \qquad (12)$$

$$\sum_i \text{WUS}_{zit} \le 0.37 \times \left(\frac{\text{wy}_z}{\text{aw}_z}\right) \times \left(\frac{\text{ST}_{zt}}{\text{ist}_z}\right)^2 \qquad (13)$$

$$\text{TW}_{zt} = \sum_i \text{ACR}_{zilt} \times \text{WUS}_{zit}, \quad \text{where } l = \text{irrigated land} \qquad (14)$$

$$\text{TW}_{zt} \le \text{ST}_{zt} \times \text{overall}_z \times s_z \qquad (15)$$

Equation (11) computes the change in saturated thickness, where: ST_{zt} is the saturated thickness at zone z in time t, TW_{zt} is the total water pumping at zone z in time t; $rech_z$ is the annual recharge rate in zone z; $overall_z$ is the surface area of the aquifer under zone z; and s_z is specific yield (%). The annual recharge rate is set as 0 because the southern portion of the Ogallala Aquifer has no significant recharge. Equation (12) relates lift to water pumping, where $LIFT_{zt}$ is the pumping lift at zone z in time t with the data for $t = 0$ reflecting the initial lift.

Equation (13) limits the pumped water use per acre (WUS_{zit}) to an amount that diminishes at the rate of the squared ratio of the current saturated thickness and initial saturated thickness following Hughes and Wyatt (1969), where: ist_z is the initial saturated thickness at zone z; wy_z is the average initial well yield (in gal/min) for zone z; aw_z is the average number of wells per irrigated acre in zone z; and 0.37 (acre feet/gal/min) is a factor based on the assumption of 2000 pumping hours in a growing season. The data show that well yield and well numbers are changing over time, as are saturated thickness and climate. Equation (14) equates total water use to acreage times per acre use and Equation (15) restricts water use to be less than or equal to the volume of groundwater available in time t, which equals the saturated thickness times the aquifer area times the specific yield.

The next two equations compute the pumping costs following Almas *et al.* (2006) and Das *et al.* (2010):

$$PC_{zit} = WUS_{zit} \times ef \times (LIFT_{zt} + 2.31 \times psi) \times enp/pe \quad (16)$$

$$TPC_{zit} = FC_z + PC_{zit} + LMR_{zt} + LC_{zit} \quad (17)$$

Equation (16) expresses the energy cost of pumping (PC_{zit}) per acre for crop i in zone z given that a water use of WUS_{zit} is the representative energy use factor (ef) for electricity; psi represents the irrigation system operating pressure (pounds/in^2 pumping head); enp is the energy price; pe is the pumping engine efficiency; and the factor 2.31 (ft) is the height of a column of water that will exert a pressure of 1 pound/in^2. Equation (17) computes the total pumping cost (TPC_{zit}) per acre for crop i at zone z in time t, where: FC_z is the annualized fixed costs for the irrigation system; LMR_{zit} is the lubrication, maintenance and repair costs; and LC_{zit} is the labour cost per acre for irrigation.

Collectively, this model can adapt to the evolving climate and aquifer on a farm-by-farm basis in each year by reducing water use (shifting crops or employing deficit irrigation), changing the crop mix (planting fewer water-using crops), moving to dry-land farming or moving land into grass.

Results

Here we focus on the results for the climate change effects on agricultural production and water use. We use two climate change scenarios:

- MRI-CGCM (developed by Meteorological Research Institute, Japan) which projects the smallest increase in mean temperature and largest increase in mean precipitation; and
- GFDL-CM2.1 (developed by the Geophysical Fluid Dynamics Laboratory, USA) which depicts the largest increase in mean temperature and largest decrease in mean precipitation.

Crop yields and water use are altered to account for the climate change effects based on crop simulations from Beach *et al.* (2009).The results for acreages of selected major crops relative to the baseline in selected zones of Dallam County are presented in Figs 8.3 and 8.4. In Zone 2 (Fig. 8.3), which has lower water availability, the MRI-CGCM scenario tends to initially increase higher value crops such as irrigated maize and wheat acreage, with a later turn towards dry land wheat and sorghum and a small shift to rangeland. The results under GFDL2.1 scenario show decreased maize acreage over the entire period. This is mainly because reduced per acre maize returns arise as the lower water productivity (the ratio of crop yields to unit of water input) makes farmers less likely to extend heavier water use for maize relative to the 'no climate change' case. The advantage of dry land wheat shows up particularly in the long run as farmers adjust their production to less water-intensive crops. In the water-rich Zone 4 of Dallam County, there is less land use change relative to the water poorer zone. Enough access to water allows

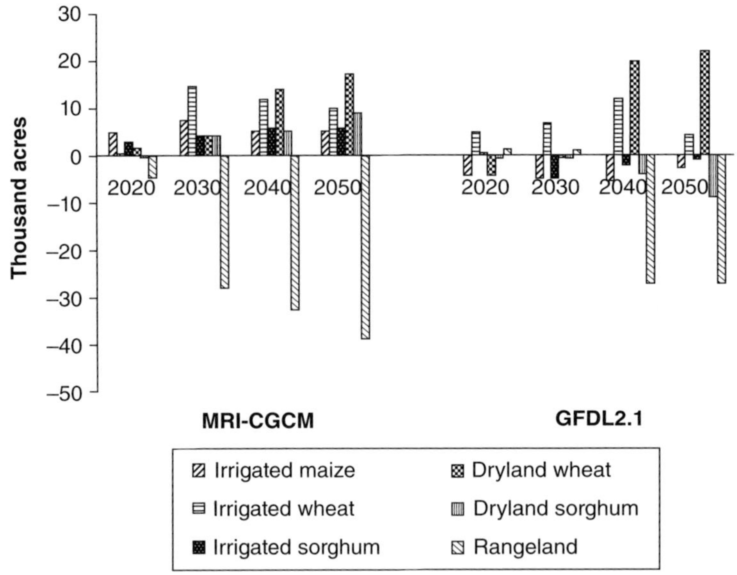

Fig. 8.3. Modelled crop mix changes under climate change in selected years relative to 2010 for Zone 2, a relatively water-poor area in Dallam County. Texas, USA. MRI-CGCM (from the Meteorological Research Institute, Japan) projects the smallest increase in mean temperature and largest increase in mean precipitation; GFDL 2.1 (from the US Geophysical Fluid Dynamics Laboratory) depicts the largest increase in mean temperature and largest decrease in mean precipitation.

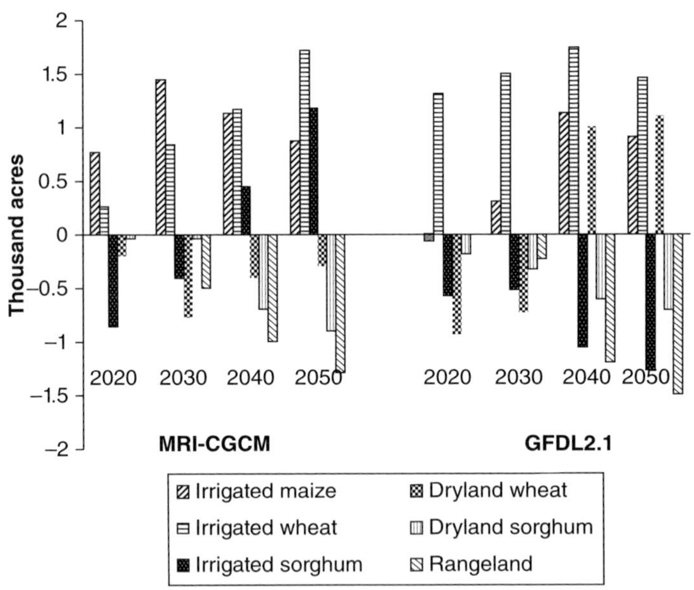

Fig. 8.4. Modelled crop mix changes under climate change in selected years relative to 2010 for Zone 4, a relatively water-rich area in Dallam County, Texas, USA. MRI-CGCM (from the Meteorological Research Institute, Japan) projects the smallest increase in mean temperature and largest increase in mean precipitation; GFDL 2.1 (from the US Geophysical Fluid Dynamics Laboratory) depicts the largest increase in mean temperature and largest decrease in mean precipitation.

farmers to adapt to climate change by adjusting with increased irrigation at the intensive margin instead of shifting land to more drought-resistant relative low-value crops. This shows the importance of using multiple representative farms to depict heterogeneous responses to phenomena such as climate change.

As shown in Table 8.4, climate change increases regional income under MRI-CGCM, while runs under GFDL2.1 lead to reduced income. This is mainly due to the projected higher (lower) yields for most major crops under MRI-CGCM (GFDL2.1). Table 8.5 presents per-acre net present value (NPV) of income in Dallam County. There we find that MRI-CGCM increases per-acre NPV in all five zones but that GFDL2.1 decreases NPV.

Table 8.4. Percentage change in net present value of agricultural income in three counties in Texas, USA under two climate change scenarios (MRI-CGCM and GFDL2.1; see text) relative to the baseline case.

County	Baseline (US$ million)	MRI-CGCM (%)	GFDL2.1 (%)
Dallam	2.25	19.12	−7.50
Hartley	2.61	22.93	−4.98
Sherman	2.09	13.80	−9.57

Table 8.5. Percentage change in per-acre net present value by zone Dallam County, Texas, USA under two climate change scenarios (MRI-CGCM and GFDL2.1; see text) relative to the baseline case,.

Zone	Baseline (US$/acre)	MRI-CGCM (%)	GFDL2.1 (%)
1	1664	5.13	−12.90
2	1870	13.39	−9.25
3	2441	21.52	−9.00
4	2989	20.95	−4.41
5	3669	24.91	−3.25

Implications

This case study constructs and uses a regional model to estimate climate change impacts on agricultural production in the Texas Ogallala Aquifer region. It depicts a spatially explicit representative farm-level forward-looking dynamic farm management showing a water use, crop mix, land conversion and livestock herd solution, with multiple representative farms depicting hydrological heterogeneity. This enhances the credibility and nature of regional results on land/water allocation and predictions of the effects on farmers' income. By applying this model, we find that the impacts of climate change vary with water endowments and GCM projections, and that some combinations have a positive effect on farm income due to case-specific adaptations such as changes in crop mix and deficit irrigation.

8.6 Summary and Conclusions

This chapter describes a multi-farm regional modelling approach and its characteristics, and discusses why one might want to use such an approach. Summarizing regional-level modelling is useful for investigating broader issues about the regional implications for new technology or agricultural policy when there are market and resource interactions between the farms. Aggregation procedures for forming the representative farms are illustrated. Two case studies in the Edward and Ogallala Aquifer regions of Texas are included to illustrate applications of such regional models. In these two areas, water issues and resource substitution possibilities create the need to examine interactions among farms and therefore use a regional model.

References

Adams, R.M., Rosenzweig, C., Peart, R.M., Ritchie, J.T., McCarl, B.A. *et al.* (1990) Global climate change and US agriculture. *Nature* 345, 219–224.

Adams, R.M., Chen, C., McCarl, B.A. and Weiher, R.F. (1999) The economic consequences of ENSO events for agriculture. *Climate Research* 13, 165–172.

Almas, L.K. (2008) Economic optimization models to project income and hydrological changes: a case of groundwater management in Texas. In: Bayer, A. (ed.) *Proceedings International Conference on Policy Modelling – ECOMOD 2–4 July 2008 Berlin*. EcoMod Press [no further details available].

Almas, L.K., Colette, W.A. and Park, S.C. (2006) Economic optimization of groundwater resources in the Texas panhandle. *Paper presented at the 2006 Annual Meeting, Southern Agricultural Economics Association, Orlando, Florida, February 5–8, 2006.* Available at: http://ageconsearch.umn.edu/handle/35321 (accessed 4 March 2016).

Baker, T.G. and McCarl, B.A. (1982) Representing farm resource availability over time in linear programs: a case study. *North Central Journal of Agricultural Economics* 4, 59–68.

Bates, B.C., Kundzewicz, Z.W., Wu, S. and Palutikof, J.P. (2008) *Climate Change and Water*. Technical Paper IV, Intergovernmental Panel on Climate Change, IPCC Secretariat, Geneva, Switzerland.

Beach, R.H. and McCarl, B.A. (2010) *U.S. Agricultural and Forestry Impacts of the Energy Independence and Security Act: FASOM Results and Model Description*. Final report prepared for the US Environmental Agency, Washington, DC.

Beach, R.H., Zhen, C., Thomson, A., Rejesus, R.M., Sinha, P., Lentz, A.W., Vedenov, D.V. and McCarl, B.A. (2009) *Climate Change Impacts on Crop Insurance*. Report prepared for the USDA Risk Management Agency, Kansas City, Kansas.

Bender, D.A. and McCarl, B.A. (1992) Energy crop selection using linear programming. In: Peart, R.M. and Brook, R.C. (eds) *Analysis of Agricultural Energy Systems*. Elsevier, Amsterdam, pp. 57–82.

Bontkes, T.S. and van Keulen, H. (2003) Modelling the dynamics of agricultural development at farm and regional level. *Agricultural Systems* 76, 379–396.

Booker, J.F. and Young, R.A. (1994) Modeling intrastate and interstate markets for Colorado River water resources. *Journal of Environmental Economics and Management* 26, 66–87.

Buckwell, A.E. and Hazell, P.B.R. (1972) Implications of aggregation bias for the construction of static and dynamic linear programming supply models. *Journal of Agricultural Economics* 23, 119–134.

Burt, O.R. (1964) Optimal resource use over time with an application to ground water. *Management Science* 11, 80–93.

Cai, Y. (2009) Water scarcity, climate change, and water quality: three economic essays. PhD dissertation, Texas A&M University, College Station, Texas.

Chen, X. and Önal, H. (2012) Modeling agricultural supply response using mathematical programming and crop mixes. *American Journal of Agricultural Economics* 94, 674–686.

Chen, C., McCarl, B.A. and Adams, R.M. (2001a) Economic implications of potential ENSO frequency and strength shifts. *Climatic Change* 49, 147–159.

Chen, C., Gillig, D. and McCarl, B.A. (2001b) Effects of climatic change on a water dependent regional economy: a study of the Texas Edwards Aquifer. *Climatic Change* 49, 397–409.

Chen, C., Gillig, D., McCarl, B.A. and Williams, R. (2005) ENSO impacts on regional water management: case study of the Edwards Aquifer (Texas, USA). *Climate Research* 28, 175–181.

Chen, C., McCarl, B.A. and Williams, R.L. (2006) Elevation dependent management of the Edwards Aquifer: linked mathematical and dynamic programming approach. *Journal of Water Resources Planning and Management* 132, 330–340.

Daneshmand, F., Karimi, A., Nikoo, M.R., Bazargan-Lari, M.R. and Adamowski, J. (2014) Mitigating socio-economic-environmental impacts during drought periods by optimizing the conjunctive management of water resources. *Water Resources Management* 28, 1517–1529.

Das, B. (2004) Towards a comprehensive regional water policy model for the Texas high plains. *Selected paper presented at the Annual Meetings of the Southern Agricultural Economics Association, Tulsa, Oklahoma, February 14–18, 2004*. Available at: http://ageconsearch.umn.edu/bitstream/34638/1/sp04da08.pdf (accessed 4 March 2016).

Das, B.R., Willis, D.B. and Johnson, J. (2010) Effectiveness of two water conservation policies: an integrated modeling approach. *Journal of Agricultural and Applied Economics* 42, 695–710.

Day, R.H. (1963) On aggregating linear programming models of production. *Journal of Farm Economics* 45, 797–813.

Dillon, C.R. (1991) An economic analysis of Edwards Aquifer water management. PhD dissertation, Texas A&M University, College Station, Texas.

Ding, J. (2014) Three essays on climate variability, water and agricultural production. PhD dissertation, Texas A&M University, College Station, Texas.

Fajardo, D., McCarl, B.A. and Thompson, R.L. (1981) A multicommodity analysis of trade policy effects: the case of Nicaraguan agriculture. *American Journal of Agricultural Economics* 63, 23–31.

Feng, Y. (1992) Optimal intertemporal allocation of ground water for irrigation in the Texas high plains. PhD dissertation, Texas Tech University, Lubbock, Texas.

Fernandez, M. (2013) Decadal climate variability: economic implications in agriculture and water in the Missouri river basin. PhD dissertation, Texas A&M University, College Station, Texas.

Frick, G.E. and Andrews, R.A. (1965) Aggregation bias and four methods of summing farm supply functions. *Journal of Farm Economics* 47, 696–700.

Geoffrion, A.M. (1977) A priori error bounds for procurement commodity aggregation in logistics planning models. *Naval Research Logistics Quarterly* 24, 201–212.

Gillig, D., McCarl, B.A. and Boadu, F. (2001) An economic, hydrologic, and environmental assessment of water management alternative plans for the south central Texas region. *Journal of Agricultural and Applied Economics* 33, 59–78.

Gisser, M. and Sanchez, D.A. (1980) Competition versus optimal control in groundwater pumping. *Water Resources Research* 16, 638–642.

Giupponi, C., Jakeman, A.J., Karssenberg, D. and Hare, M.P. (2006) *Sustainable Management of Water Resources: An Integrated Approach*. Edward Elgar Publishing, Cheltenham, UK.

Guerrero, B., Amosson, S., Johnson, J. and Weinheimer, J. (2011) Economic evaluation of water conservation strategies for the North Plains groundwater conservation district. *Selected Paper presented at the Universities Council on Water Resources Annual Conference, Boulder, Colorado, July 11–14*. [No further details available.]

Hamilton, S.A., McCarl, B.A. and Adams, R.M. (1985) The effect of aggregate response assumptions on environmental impact analyses. *American Journal of Agricultural Economics* 67, 407–413.

Hazell, P.B.R. and Norton, R.D. (1986) *Mathematical Programming For Economic Analysis in Agriculture*. Macmillan, New York.

Hornbeck, R. and Keskin, P. (2011) *Farming the Ogallala Aquifer: Short-run and Long-run Impacts of Groundwater Access*. Working Paper, Department of Economics, Harvard University, Cambridge, Massachusetts.

Howitt, R.E. (1995) Positive mathematical programming. *American Journal of Agricultural Economics* 77, 329–342.

Hughes, W.S. and Wyatt, L.H. (1969) *Projected Economic Life of Water Resources, Subdivision Number 1, High Plains Underground Water Reservoir*. Texas Agricultural Experiment Station, College Station, Texas.

Jayasuriya, R.T. (2004) Modelling the regional and farm-level economic impacts of environmental flows for regulated rivers in NSW, Australia. *Agricultural Water Management* 66, 77–91.

Johnson, J.W. (2003) Regional policy alternatives in response to depletion of the Ogallala Aquifer. PhD dissertation, Texas Tech University, Lubbock. Texas.

Judez, L. and de Miguel, J.M. (2001) Modelling crop regional production using positive mathematical programming. *Mathematical and Computer Modelling* 35, 77–86.

Keplinger, K.O. (1996) An investigation of dry year options for the Edwards Aquifer. PhD dissertation, Texas A&M University, College Station, Texas.

Keplinger, K.O., McCarl, B.A., Chowdhury, M.E. and Lacewell, R.D. (1998) Economic and hydrologic implications of suspending irrigation in dry years. *Journal of Agricultural and Resource Economics* 23, 191–205.

Kuhn, H.W. and Cullen, D. (1975) Aggregation in network models for transportation planning. Final Report Mathematic Inc., Princeton, New Jersey.

Louhichi, K., Belhouchette, H., Wery, J., Therond, O. and Flichman, G. (2008) Impact assessment of the 2003 CAP reform and the Nitrate Directive on the arable farming system in the Midi-Pyrénées region: Bio-economic modelling at field, farm and regional levels. *Paper prepared for the 109th EAAE Seminar the CAP after the Fischler Reform: National Implementations, Impact Assessment and the Agenda for Future Reforms. Viterbo, Italy, November 20–21, 2008*. Available at: http://ageconsearch.umn.edu/bitstream/44826/2/3.3.3_Louhichi.pdf (accessed 4 March 2016).

Marek, T., Amosson, S., Bretz, F., Guerrero, B. and Kotara, R. (2009) *Panhandle Regional Water Plan, Task 2 Report: Agricultural Water Demand Projections*. A Water Planning Report, Texas AgriLife Experiment Station, Lubbock Texas.

McCarl, B.A. (1982) Cropping activities in agricultural sector models: a methodological proposal. *American Journal of Agricultural Economics* 64, 768–772.

McCarl, B.A. and Parandvash, G.H. (1988) Irrigation development versus hydroelectric generation: can interruptible irrigation play a role? *Western Journal of Agricultural Economics* 13, 267–276.

McCarl, B.A. and Spreen, T.H. (1980) Price endogenous mathematical programming as a tool for sector analysis. *American Journal of Agricultural Economics* 62, 87–102.
McCarl, B.A. and Spreen, T.H. (2014) Applied Mathematical Programming using Algebraic Systems. Available at: http://agecon2.tamu.edu/people/faculty/mccarl-bruce/books.htm (accessed 4 March 2016).
McCarl, B.A., Jordan, W.R., Williams, R.L., Jones, L.L. and Dillon, C.R. (1993) *Economic and Hydrologic Implications of Proposed Edwards Aquifer Management Plans*. Technical Report No. 158, Texas Water Resources Institute, Texas A&M University, College Station, Texas. Available at: http://agecon2.tamu.edu/people/faculty/mccarl-bruce/papers/496.pdf (accessed 30 June 2014).
McCarl, B.A., Dillon, C.R., Keplinger, K.O. and Williams, R.L. (1999) Limiting pumping from the Edwards Aquifer: an economic investigation of proposals, water markets, and spring flow guarantees. *Water Resources Research* 35, 1257–1268.
McCarl, B.A., Musumba, M., Smith, J.B., Kirshen, P., Jones, R., *et al.* (2013) Implications of climate change for Egypt's agricultural sector. *Mitigation and Adaptation Strategies for Global Change* 20, 1097–1109.
Mehta, V.M., Rosenberg, N. and Mendoza, K. (2012) Simulated impacts of three decadal climate variability phenomena on dryland corn and wheat yields in the Missouri River basin. *Agricultural and Forest Meteorology* 152, 109–124.
Morrison, J., Morikawa, M., Murphy, M. and Schulte, P. (2009) *Water Scarcity and Climate Change: Growing Risk for Businesses and Investors*. A Ceres Report commissioned from the Pacific Institute. Ceres, Boston, Massachusetts and Pacific Institute, Oakland, California. Available at: http://www.ceres.org/resources/reports/water-scarcity-climate-change-risks-for-investors-2009 (accessed 4 March 2016).
Mushtaq, S., Cockfield, G., White, N. and Jakeman, G. (2014) Modelling interactions between farm-level structural adjustment and a regional economy: a case of the Australian rice industry. *Agricultural Systems* 123, 34–42.
Oczkowski, E. and Bandara, Y. (2013) Modelling agricultural land use allocation in regional Australia. *Australian Journal of Agricultural and Resource Economics* 57, 422–440.
Önal, H. and McCarl, B.A. (1989) Aggregation of heterogeneous firms in mathematical programming models. *European Review of Agricultural Economics* 16, 499–513.
Önal, H. and McCarl, B.A. (1991) Exact aggregation in mathematical programming sector models. *Canadian Journal of Agricultural Economics* 39, 319–334.
Pattanayak, S.K., McCarl, B.A., Sommer, A.J., Murray, B.C., Bondelid, T., Gillig, D. and DeAngelo, B. (2005) Water quality co-effects of greenhouse gas mitigation in US agriculture. *Climatic Change* 71, 341–372.
Pulido-Velazquez, M., Jenkins, M.W. and Lund, J.R. (2004) Economic values for conjunctive use and water banking in Southern California. *Water Resources Research* 40(3): W03401.
Rosegrant, M.W., Ringler, C., McKinney, D.C., Cai, X., Keller, A. and Donoso, G. (2000) Integrated economic–hydrologic water modeling at the Basin Scale: the Maipo river basin. *Agricultural Economics* 24, 33–46.
Seager, R., Ting, M., Li, C., Nail, N., Cook, B., Nakamura, J. and Liu, H. (2013) Projections of declining surface-water availability for the southwestern United States. *Nature Climate Change* 3, 482–486.
Shucksmith, M. (1993) Farm household behaviour and the transition to post-productivism. *Journal of Agricultural Economics* 44, 466–478.
Soboll, A., Elbers, M., Barthel, R., Schmude, J., Ernst, A. and Ziller, R. (2011) Integrated regional modelling and scenario development to evaluate future water demand under global change conditions. *Mitigation and Adaptation Strategies for Global Change* 16, 477–498.
Sullivan, P. and Rinehart, L. (2010) Converting cropland to perennial grassland. Report Prepared for the National Sustainable Agriculture Information Service, National Center for Appropriate Technology (NCAT), Butte, Montana. Available at: https://attra.ncat.org/attra-pub/download.php?id=108 (accessed 4 March 2016).
Terrell, B. (1998) Economic impacts of the depletion of the Ogallala Aquifer: an application to the Texas High Plains. MS thesis, Texas Tech University, Lubbock, Texas.
Theil, H. (1954) *Linear Aggregation of Economic Relations*. North-Holland, Amsterdam.
van Calker, K.J., Berentsen, P.B.M., de Boer, I.J.M., Giesen, G.W.J. and Huirne, R.B.M. (2007) Modelling worker physical health and societal sustainability at farm level: an application to conventional and organic dairy farming. *Agricultural Systems* 94, 205–219.

Wang, W., Park, S.C., McCarl, B.A. and Amosson, B. (2011) Economic and groundwater use implications of climate change and bioenergy feedstock production in the Ogallala Aquifer region. *Selected paper prepared for presentation at the Agricultural and Applied Economics Association's 2011 AAEA and NAREA Joint Annual Meeting, Pittsburgh. Pennsylvania, July 24–26, 2011.* Available at: http://ageconsearch.umn.edu/bitstream/103642/1/AAEA-OAP.pdf (accessed 4 March 2016).

Ward, F.A. and Lynch, T.P. (1997) Is dominant use management compatible with basin-wide economic efficiency? *Water Resources Research* 33, 1165–1170.

Williams, R.L. (1996) *Drought management and the Edwards Aquifer: an economic inquiry.* PhD dissertation, Texas A&M University, College Station, Texas.

9 Farm-level Microsimulation Models

Cathal O'Donoghue*
Rural Economy and Development Programme (REDP), Teagasc, Athenry, Ireland

9.1 Introduction

In many parts of the world, agriculture is one of the sectors that is most influenced by public policy. Given the importance of food production in the production of goods that are essential for survival, and the characteristics of food production as a biology-based sector that is prone to risk and volatility, faces pressures in meeting the food requirements of the growing world population (particularly in the face of climate and environmental constraints) and has impacts upon the wider environment (as one of the most significant land uses), it is unsurprising that a modelling field has developed to look at these issues at farm level. In this chapter, we will focus specifically on the microsimulation modelling literature applied at the farm level.

Microsimulation modelling utilizes micro-units of analysis such as individuals, households, firms and farms for *ex ante* analysis based on data from surveys or administrative data sets. The simulation-based methodology utilizes computer programmes to simulate the effects of changes in public policy, and economic or social changes, on the micropopulation of interest.

The methodology allows the design of public policy to be evaluated and improved on a computer before rolling out often costly programmes to the general population. To some extent, the methodology can be regarded as a computer-based laboratory for running policy experiments. The aim, thus, is to help to improve the effectiveness and efficiency of public programmes. The field spans both users in government – who employ the method in the design and evaluation of public policy – and mainstream academic researchers.

As a field, microsimulation modelling has its roots in a proposal by Guy Orcutt (Orcutt, 1957; Orcutt *et al.*, 1961). Although this field has existed since the late 1950s, progress was relatively limited until the 1990s due to challenges in computing power and data requirements, except for pockets of development in the UK, Sweden and the USA. However, the advent of the personal computer in the 1980s and the availability of micro-data has allowed development on the field to take off, and it has since grown very rapidly, with the development and use of models in many countries. Whether formally called microsimulation modelling or not, micro-based *ex ante* simulation-based analysis is now used extensively around the world for policy analysis and design.

Much of the microsimulation literature focuses on tax and social policy (O'Donoghue, 2014), though there has been a growing application in the area of farm-level microsimulation (Richardson *et al.*, 2014). As farm-level simulation modelling is also technically farm-

*E-mail: Cathal.ODonoghue@teagasc.ie

level microsimulation modelling, much of the literature covered in this book could be referred to as microsimulation modelling. Ciaian *et al.* (2013) and Klein and Narayanan (2008) have provided a recent review of the farm-level simulation modelling literature. These farm-level models utilize a variety of methodologies and include models such as FARMIS (Osterburg *et al.*, 2001), FSSIM (Louhichi *et al.*, 2010), the FAPRI-Ireland Farm Level Model (Hennessy and Rehman, 2008) and the FLIPSIM model developed by Richardson and Nixon (1986).

Farm-level simulation modelling has historically developed as a parallel field to microsimulation modelling in that relatively few farm-level papers appear in microsimulation conferences or journals or vice versa. Nevertheless, the objectives are fundamentally similar: micro-level simulation of policy and economic change. A differentiating feature of microsimulation modelling relative to other micro-level simulation modelling approaches, such as agent-based modelling (Niazi and Hussain, 2011), which simulates the actions and interactions of autonomous agents, social simulation modelling (Troitzsch *et al.*, 1996) or farm-level simulation modelling (Ciaian *et al.*, 2013), has been a focus on the static incidence of policy. Although the methodology often incorporates partial behavioural response, typically on the basis of econometric estimation (O'Donoghue, 2014), the behavioural analysis is typically not of the same breadth as in these other modelling areas. Typically, in modelling the trade-off between the precision of policy simulation and behavioural complexity, the former is prioritized. Where significant behaviour is endogenous, as in the case of dynamic microsimulation models, the behavioural equations are often of a reduced form rather than structural (Li and O'Donoghue, 2013).

While farm-based micro-level simulation modelling differs from microsimulation-based modelling for other units of analysis – such as the household or the firm – in that incomes partially derive from biological processes, farms do also have specific business structures. So, in concept, and in terms of profit, output and costs, farm-based models are not too different from firm-level models (Richardson *et al.*, 2014). However, the policy environment is typically quite different.

As mainstream farm-level simulation modelling has been described in better detail elsewhere in the volume, the focus of this chapter will be on models with a farm-level unit of analysis from within the microsimulation literature, largely (though not uniquely) focusing on static incidence rather than behavioural responses to policy change that affect agriculture. We will focus on the dimensions of microsimulation that have been applied at the farm level and can be found in the *Handbook of Microsimulation Modelling* (O'Donoghue, 2014). These include: hypothetical farm models; static incidence models; behavioural response models; dynamic inter-temporal models; macroeconomic impact models; spatial impact models; and environmental impact models. Even though the objectives and types of these modelling approaches are similar, there has been relatively little mutual learning between them. This chapter aims to communicate some of the areas in common and highlight differences from those in other fields.

9.1.1 Modelling complexity

As a modelling framework, microsimulation modelling is a mechanism of abstracting from reality to help us to understand complexity better. Figure 9.1 outlines potential sources of complexity in a static, single time period, microsimulation model.

In the context of policy design and evaluation, complexity can take the form of:

- the structure of the population;
- the behavioural response to the policy; and
- the structure of the policy.

These levels of complexity themselves interact with each other, resulting in a degree of complexity that is difficult to disentangle without recourse to a model.

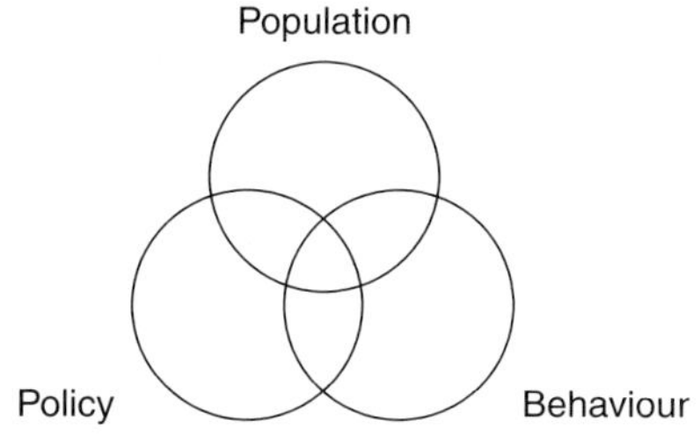

Fig. 9.1. Sources of complexity in policy design and evaluation

Consider first of all the dimension of population complexity. The first dimension of complexity of a population is whether an analysis takes place on a population with limited or extensive heterogeneity. Many analyses focus merely on the impact of policy on typical families, abstracting almost entirely from population complexity, such as the IFCN (International Farm Comparison Network)[1] hypothetical farm models based upon typical farms (Hemme *et al.*, 2000). None the less, data-based microsimulation models do contain a degree of heterogeneity at the farm level (e.g. Hynes *et al.*, 2009a). Other models focusing on agriculture contain an individual or household unit of analysis (Ferreira *et al.*, 2013).

The second dimension of complexity is policy complexity. This relates to the range of different policy or socio-economic impacts and the degree of policy complexity involved; many microsimulation models try to replicate the fine detail of legislation in modelling policy to the different types of policy modelled (Shrestha *et al.*, 2007). Different geopolitical contexts may influence the nature of the policy complexity. For example, many OECD (Organisation for Economic Co-operation and Development) countries have very significant agricultural and environmental policies targeted at the farm level (Kruseman *et al.*, 2008a,b), and these may be different to those simulated in a developing country, where market and macroeconomic impacts may be more important (Boccanfuso and Savard, 2007).

The third dimension of complexity is behaviour. Many policies are explicitly aimed at influencing behaviour, as in the case of attempts to improve work incentives or in environmental policy. Models that abstract from behavioural response are known as static microsimulation models.

For models that incorporate either inter-temporal dimensions or spatial dimensions, the level of complexity is increased still further by the addition of time complexity (inter-temporal models) or place complexity (spatial models) (Fig. 9.2). Policies that depend upon long-term economic changes utilize inter-temporal or dynamic models. Land use and spatially targeted policy or spatially targeted socio-economic effects require spatial models.

9.2 Applications of Farm-level Microsimulation Modelling

In this section, we provide an overview of a selection of farm-level microsimulation models that have been reported in the literature. We consider the types of models that have been developed to understand the farm-level implications for the five dimensions of complexity discussed above: population, policy, behaviour, time and place (space/environment). In order to bridge the gap between the microsimulation field and the farm-level simulation field, we will follow the thematic structure of the areas covered in the *Handbook of Microsimulation Modelling* (O'Donoghue, 2014) and describe relevant analyses that have been undertaken at the farm level.

9.2.1 Hypothetical analyses

Abstracting from population complexity in order to enhance understanding, a variety of models have been developed internationally that have simulated the effects of biological, market and

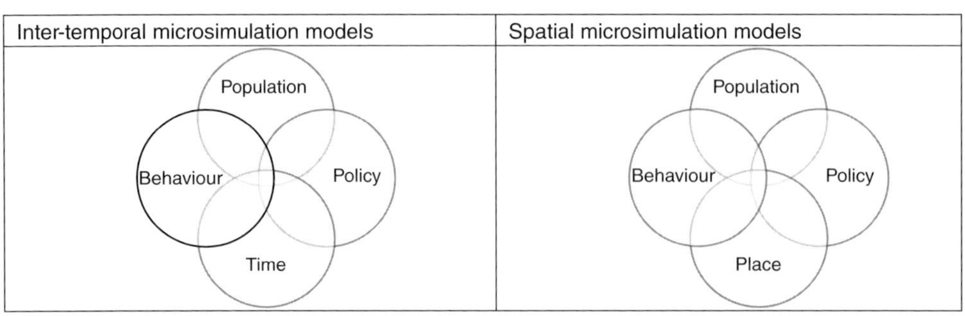

Fig. 9.2. Enhanced complexity in inter-temporal and spatial microsimulation models.

policy changes at farm level. In addition to deliberately abstracting from complexity, these models are also used where microdata does not exist (Zander *et al.*, 2007).

Representative or hypothetical farms have been used for policy analysis since the 1960s – see, for example, Plaxico and Tweeten (1963) and Sharples (1969). At a biological level, bioeconomic simulation models of farm systems have been developed to assess the economic impact of alternative management practices and technologies at the farm level (Shalloo *et al.*, 2004; Crosson *et al.*, 2006).

As a means of understanding the impact of different systems, and combined with differences in local market prices for inputs, outputs and other production factors, the IFCN (Hemme *et al.*, 2000) has been able to simulate farm-level profits to compare farm systems across the world at a synthetic farm level (Thorne and Fingleton, 2006; Prochorowicz and Rusielik, 2007). Such hypothetical farm analyses have also been used for policy analysis purposes (Doucha and Vaněk, 2006; McCormack *et al.*, 2014).

9.2.2 Static modelling

As the requirements of agricultural policy have become more complicated, in terms of agricultural and environmental objectives and the constraints imposed by, for instance, WTO (World Trade Organization) considerations or public finance issues, there has been an increased need for more farm-level simulation modelling. Static models focus on the distributional incidence of policy reforms. As static models, they simulate the 'day after effect' before behavioural change occurs.

While much of the related literature focuses on *ex post* static redistributional analysis of agricultural policy reform (e.g. Keeney, 2000; Severini and Tantari, 2013), the microsimulation literature focuses on *ex ante* analysis. An illustration is given by Donnellan *et al.* (2013a), who simulated the impact of a range of alternative reforms to the Common Agricultural Policy (CAP) on a nationally representative sample of farms from the Teagasc National Farm Survey – the Irish variant of the European Union (EU) Farm Accountancy Data Network (FADN).

Another dimension of interest to analysts is the distributional incidence on farms of the impact of market changes. Even though many organizations undertake market outlooks for agricultural commodities prices and farm incomes (e.g. OECD-FAO, 2013), it is relatively rare that the distributional impact of these changes is analysed. The Teagasc *Outlook on Economic Prospects for Agriculture*, which is undertaken annually in Ireland, is an exception in that it provides data on the impact of market changes for farms in different terciles by commodity sector (Donnellan *et al.*, 2013b).

9.2.3 Behavioural modelling

The objectives of many policies and their implications for market price change involve behavioural response. While most non-farm microsimulation models focus on population representativeness and policy accuracy rather than on the inclusion of behavioural response, behavioural responses that are often included in microsimulation models relate to consumer demand, policy participation or take-up and labour supply (O'Donoghue, 2014). Agent-based models tend to have a greater focus on behaviour, but with less detail in relation to population or policy heterogeneity.

For farm simulation models, much of the literature described in this book and in Ciaian *et al.*, (2013) focuses on farm-level production, consumption and investment decisions. The inclusion of demand systems that incorporate price responses for farm-level inputs such as feed and fertilizer is more commonplace in these farm-level models (Burrell, 1989; Peeters and Surry, 1997). Rather than reproducing these accounts here, we focus on methodology that draws specifically upon the methods used in the wider microsimulation literature.

As in the case of many other households reliant on income from entrepreneurial activity, farm households often rely on income from other sources. Other labour income is important, but though there is a large econometric literature in relation to off-farm employment (Huffman and Lange, 1989), the microsimulation literature is relatively sparse. Callan and Van Soest (1993) simulated the female off-farm labour supply elasticity relative to wage rates in Ireland using a structural labour supply model. Van Leeuwen and Dekkers (2013) used a spatial microsimulation methodology to understand the determinants of off-farm labour supply in the Netherlands.

Behavioural models in the microsimulation literature often utilize conditional logit or random utility maximizing models to estimate and simulate the behavioural impact of policy reform (Aaberge and Colombino, 2014). While microdata incorporates the observed characteristics associated with the actual choices made, this literature utilizes a microsimulation model to simulate counterfactual characteristics of choices not made. Recent examples of the application of these methods in the farm-level literature include the participation of farmers in agri-environmental schemes (Murphy *et al.*, 2014) and forestry planting decisions (Ryan *et al.*, 2015). Murphy *et al.* (2014) simulated the counterfactual output, cost, labour and subsidy impact of participating (or not) in the Rural Environmental Protection Scheme (a voluntary agri-environmental scheme that operated in Ireland from 1994 to 2009), and used the panel data available to estimate a utility function for scheme participation. Ryan *et al.* (2015) utilized a similar method, but over a longer term perspective, to compare the net present value of market and subsidy income flows as well as the impacts on leisure and wealth of alternative land uses in relation to agriculture and forestry.

9.2.4 Dynamic modelling

Dynamic microsimulation models are those that incorporate the inter-temporal dimensions of markets and policy. In other areas of microsimulation, these models are often used for the distributional analysis of pensions and education financing and *ex ante* analyses of retirement decision making (Li and O'Donoghue, 2013). Many of the models within the farm-level simulation modelling literature incorporate inter-temporal dimensions, particularly for investment decisions (Ciaian *et al.*, 2013).

Agriculture and Agri-Food Canada (AAFC) developed a farm-level dynamic microsimulation model, the Canadian Agriculture Dynamic Micro-Simulation Model (CADMS) (Galbraith *et al.*, 2011; Chan and Guerra, 2012). This model utilizes farm-level data and is used to estimate the current and future behaviour of farm businesses. It simulates the responses of farm-level income, wealth and financial indicators to scenarios involving proposed policy and market changes, the consequential structure and competitiveness of agricultural sub-sectors, and the adoption of innovations such as new crop varieties. Unlike other dynamic microsimulation models, it uses artificial neural networks rather than econometric methods for estimation and simulation.

9.2.5 Impact of macroeconomic change

Agricultural products are some of the largest globally traded commodities influenced by global markets, global weather and climate patterns, differential global demand trade agreements, tariffs and other trade constraints, so macroeconomic phenomena and policy can be influential in agricultural markets and have impacts at the micro level. As a result, a relatively large subcomponent of the macro to micro literature described here relates to agriculture and food. The methodologies applied include both CGE (computable general equilibrium) microsimulation (Boccanfuso and Savard, 2007) and partial equilibrium microsimulation (Hennessy and Thorne, 2006).

Some papers have focused on the top-down impacts of the liberalization of agricultural trade regimes, such as the impact of general trade liberalization (Chemingui and Thabet, 2009), potential WTO agreements (Breen and Hennessy, 2003; Morley and Piñeiro, 2004; Hennessy and Thorne, 2006), specific sectoral liberalization (Boccanfuso and Savard, 2008) and the distributional impact of agricultural price protection (Mabugu and Chitiga, 2009).

Given the importance of agricultural policy, a number of analyses have focused on the micro impact of these policies. Arndt *et al.* (2010) looked at the impact of biofuels policy on growth and poverty. Boccanfuso and Savard (2007) studied the impacts of cotton subsidies in Mali. Chitiga and Mabugu (2008) considered the impact of land redistribution policy in Zimbabwe. Cororaton and Corong (2006) analysed the impact of agricultural sector policies and poverty in the Philippines.

Tax policy has also been examined. For example, de Souza Ferreira Filho *et al.* (2010) modelled the effects of changes in taxes on food and agricultural inputs in Brazil, and Ahmed *et al.* (2007) modelled the effects of changes in agricultural income taxes in Pakistan.

As the impact of market price volatility, and particularly of agricultural commodity price volatility, becomes more important, there is increasing interest in understanding the micro-level impact of this volatility (Warr and Yusuf, 2009). This impact can be at farm, firm and household levels. Dartanto (2011) modelled the impact of soybean price on poverty. Ferreira *et al.* (2013) considered the impact of rising food prices on both farms and households in Brazil. Using a CGE model, Diao *et al.* (2009) modelled the economy-wide and household-level impacts of (highly pathogenic) avian flu in Nigeria, with the aim of designing a pro-poor risk-reduction strategy.

More generally, a number of studies have focused on the general distributional implications of growth in the agricultural sector. Pauw and Thurlow (2011) considered the situation in Tanzania, and Thurlow (2008) the situation in Mozambique.

9.2.6 Spatial models

As a sector that largely depends on the local environment in terms of soils and weather, there is significant spatial heterogeneity in agriculture. It is, therefore, important to understand this spatial heterogeneity so as to be able to better target policy interventions. In particular, the spatial distribution of agricultural income and the consequential impact of policy reform – such as CAP reform – are important in targeting (for example) specific localized interventions or in developing localized rural development interventions.

The challenge in understanding the spatial distribution of farm incomes and policy reform is one of data. It may be possible to simulate the spatial pattern of farm direct payments using administrative data, as in the case of Bergmann *et al.* (2011), who looked at the distributional equality of CAP payments at a spatial scale in Scotland, and Donnellan *et al.* (2013a), who undertook a distributional analysis of farm payments at an aspatial scale in Ireland, but these data sets often lack contextual information, thus limiting the depth of analysis possible. Typically, data from censuses of agriculture and administrative data provide spatial information on the structure of agriculture, but they provide no income or farm structure data. In contrast, farm survey data contain excellent farm income and structural data, but have weak spatial dimensions. In order to combine the strengths of both types of data, data imputation/enhancement methods known as spatial microsimulation (Hermes and Poulsen, 2012; O'Donoghue *et al.*, 2013) have been developed. Small-area statistical analysis can be used for this purpose (see Ghosh and Rao, 1994). However, here we are interested not only in inter-spatial variation in incomes but also intra-spatial area variation of incomes. So we require a method that maintains both spatial variability and micro-level variability.

Spatial microsimulation (Clarke, 1996) is a potential methodology for achieving both of these dimensions within its data enhancement process. There is an extensive literature on this that is summarized in O'Donoghue *et al.* (2014a) and covers many different policy areas, utilizing various methodologies that were reviewed earlier by Hermes and Poulsen (2012). The methodology has been applied in a number of instances within agriculture and rural development. Ballas *et al.* (2006) utilized iterative proportional fitting to examine CAP reform as part of the Luxembourg agreement. Hynes *et al.* (2009b) developed a model of spatial farm incomes utilizing simulated annealing, which has been used to examine the impact of changes in the CAP (Shrestha *et al.*, 2007). This forms part of the Simulation Model of the Irish Local Economy (SMILE) (O'Donoghue *et al.*, 2013). O'Donoghue (2013) extended farm-focused models to include wider household income sources so as to be able to assess the wider economic sustainability of farm households. Clancy *et al.* (2013) utilized the SMILE model in Ireland to assess the optimal spatial location for the growth of willow and miscanthus for biomass production. van Leeuwen (2010) applied a spatial microsimulation method to look at farm and non-farm households in a rural setting in the Netherlands.

O'Donoghue *et al.* (2014a) and Hermes and Poulsen (2012) described a number of potential methodologies to develop spatial microsimulation models. Potential options include:

- iterative proportional fitting;
- deterministic reweighting;

- combinatorial optimization; and
- quota sampling.

The deterministic approach to reweighting national sample survey data is an attempt to fit small area statistics tables or benchmarks for each small area without the use of random sampling procedures (Ballas *et al.*, 2005). Iterative proportional fitting (IPF) may be used to generate cross-tabulated control totals at the small area level. These are compared with similar cross-tabulated totals from the survey data to produce weights.

An alternative mechanism for generating weights for spatial microdata is to use a regression-based reweighting method. An example is GREGWT, which is a generalized regression weighting (reweighting) algorithm written by the Australian Bureau of Statistics (ABS). GREGWT was developed to reweight the ABS survey data to constraints from other Australian data sources (see Tanton *et al.*, 2011).

Another alternative approach to generating spatially disaggregated microdata is the use of combinatorial optimization-based simulated annealing methods, which can be used to reweight an existing microdata sample to fit small-area population statistics. For instance, aspatial microdata sets can be reweighted to estimate the micropopulation at a local spatial scale (Williamson *et al.*, 1998). This method differs from IPF primarily in that it reweights or samples from a micro data set until a new micro data set is generated that reflects the characteristics of the small area. In an agricultural context, this method has been applied to the spatial modelling of agri-environmental policy (Hynes *et al.*, 2009b).

There are a number of challenges faced in developing farm-level spatial microsimulation models using the methods described above. These are avoiding the income smoothing concern of the weighting methodology, having improved computational efficiency and adjusting to improve the spatial heterogeneity of stocking rates. O'Donoghue *et al.* (2014b) have employed a method known as quota sampling to deal with these issues.

Quota sampling (QS) is a probabilistic reweighting methodology that operates in a similar fashion to simulated annealing (SA), whereby survey data are reweighted according to key constraining totals for each small area, with amendments made in the sampling procedure in order to improve computational efficiency. In a manner similar to SA, QS selects observations at random and considers whether they are suitable for selection for a given small area based on conformance with aggregate totals for each small area characteristic. Unlike SA, QS only assigns units that conform to aggregate constraint totals and, once a unit is deemed selected, it is not replaced; this is the main computational improvement.

9.2.7 Environmental analysis

In response to the demand from policy makers for information on the impact of farming and farm policies on the environment, many models have been expanded and further enhanced to focus on the impact of policy scenarios on environmental as well as economic indicators. See Thomas (2013) for a review of farm-level environmental impact models. The wider farm simulation modelling literature contains examples of farm-level models developed for this purpose, such as the AROPAj model (De Cara and Jayet, 2011). Here we focus purely on those from within the microsimulation literature.

There have been a number of explicitly environment-focused microsimulation models that have simulated the distributional incidence of environmental policy and issues at farm level. Berntsen *et al.* (2003) modelled the incidence of potential environmental taxes on farm-level nitrogen emissions, Hynes *et al.* (2009b, 2013b) simulated the effect of farm-level taxes on methane emissions from cattle and Doole *et al.* (2013) examined the distributional impact of a farm-level CAP and trade instruments. Kruseman *et al.* (2008a,b) modelled the impact of tightening environmental policy on ammonia, nitrate and phosphate emissions, and Villot (1998) modelled a tax related to sulfur emissions.

Given the potential impact of agricultural practices on the environment, a number of microsimulation models have simulated the pollution impacts of agriculture. Potter *et al.* (2009) modelled carbon sequestration from cropland in the USA, Dijk *et al.* (1996) modelled nutrient flow in Dutch agriculture and Ramilan *et al.* (2007, 2011) modelled the environmental and economic efficiency of dairy farms. Farm-based microsimulation models have also simulated greenhouse gas emissions (Hynes *et al.*, 2009a, 2013b), while Potter *et al.* (2009) modelled carbon sequestration in the soil on cropland and Kimura *et al.* (2012) applied a microsimulation model to simulate crop risk management choices under climate change.

From a biological perspective, there have been studies that model the impact at the hypothetical farm systems level of changes to management practices and technology. Lovett *et al.* (2006), O'Brien *et al.* (2010 and Crosson *et al.* (2011) all looked at the effect of farm and management practices on greenhouse gas emissions from dairy or beef production systems. Lindgren and Elmquist (2005) evaluated the economic and environmental impact of alternative farm management practices on an arable farm.

Microsimulation models have also been used to model biodiversity-related issues, including wildlife–recreation interactions (Bennett *et al.*, 2009) and wild bird conservation (Hynes *et al.*, 2013a), and participation in Agri-Environmental Protection Schemes (Hynes *et al.*, 2008, 2013a; Hynes and Garvey, 2009; Kelley *et al.*, 2013).

A small literature has developed that describes the use of farm-level simulation models to model marginal abatement cost (MAC) curves for agriculture. Chyzheuskaya *et al.* (2014) developed a dairy farm farm-level MAC for Ireland, and Doole (2012) modelled a farm-level nitrogen MAC curve.

9.3 Conclusions and Future Directions

Microsimulation models have been developed since the late 1950s to simulate the micro-level impact of policy and economic change on an *ex ante* basis. Over the past two decades, a distinct subfield has developed in the use of these models at the farm level. This field has developed in parallel to and in partnership with the farm-level simulation field and the agent-based modelling field, which often have similar objectives but sometimes use different methodologies. In this short paper, we have reviewed the literature on farm-level microsimulation models to provide a link to wider farm simulation modellers who may be interested in utilizing and adapting some of these methodologies within their own disciplines.

The chapter has considered the application of the different microsimulation modelling methodologies – hypothetical, static, behavioural, dynamic, macro to micro, spatial and environmental – to farm-level and agricultural sector issues.

Future research directions for the field include deepening our understanding of the interaction between the environment and agricultural productivity as more georeferenced data becomes available and, vice versa, enhancing our knowledge of the impact of agriculture on the environment.

Most bioeconomic systems models simulate farm-level changes using single farms. However, given the heterogeneity of responses on different farms, and different systems and efficiency, the development of bioeconomic systems models using heterogeneous data as in the case of Janssen *et al.* (2010) can help us to understand the potential impact across agricultural sectors of market, practice and technological changes.

Future developments could help us to understand better the impact of extension or policy interventions from an *ex ante* perspective, as well as the intersection between farms and the wider rural economy – another area that is also underdeveloped.

Note

[1] The IFCN is a worldwide association of agricultural scientists, advisors and farmers, and an international network that applies internationally harmonized methods of data collection, analysis and modelling in order to provide policy advice. Within this framework, agricultural farms and production systems are defined that are typical for their region, their economic situations are analysed and the farms are projected into the future.

References

Aaberge, R. and Colombino, U. (2014) Labour supply models. In: O'Donoghue, C. (ed.) *Handbook of Microsimulation Modelling.* Contributions to Economic Analysis, Volume 293, Emerald Group Publishing, Bingley, UK, pp. 167–221.

Ahmed, V., Abbas, A. and Ahmed, S. (2007) Taxation reforms: a CGE-microsimulation analysis for Pakistan. In: *6th PEP Research Network General Meeting, Sheraton Lima Hotel, Paseo de la Republica 170, Lima, Peru, June 14–16, 2007*. Partnership for Economic Policy, Global Office, Nairobi, Kenya. Available at: http://www.pep-net.org/sites/pep-net.org/files/typo3doc/pdf/files_events/ahmed-pa.pdf (accessed 7 March 2016).

Arndt, C., Benfica, R., Tarp, F., Thurlow, J. and Uaiene, R. (2010) Biofuels, poverty, and growth: a computable general equilibrium analysis of Mozambique. *Environment and Development Economics* 15, 81–105.

Ballas, D., Clarke, G.P. and Wiemers, E. (2005) Building a dynamic spatial microsimulation model for Ireland. *Population, Space and Place* 11, 157–172.

Ballas, D., Clarke, G.P. and Wiemers, E. (2006) Spatial microsimulation for rural policy analysis in Ireland: the implications of CAP reforms for the national spatial strategy. *Journal of Rural Studies* 22, 367–378.

Bennett, V.J., Beard, M., Zollner, P.A., Fernández-Juricic, E., Westphal, L. and LeBlanc, C.L. (2009) Understanding wildlife responses to human disturbance through simulation modelling: a management tool. *Ecological Complexity* 6, 113–134.

Bergmann, H., Noack, E.M. and Thomson, K.J. (2011) The distribution of CAP payments – redistributional injustice or spatially adapted policy? *Paper presented to 85th Annual Conference of the Agricultural Economics Society, April 18–20, 2011, Warwick University, Coventry, UK*. Available at: http://ageconsearch.umn.edu/handle/108958 (accessed 7 March 2016).

Berntsen, J., Petersen, B., Jacobsen, B., Olesen, J. and Hutchings, N. (2003) Evaluating nitrogen taxation scenarios using the dynamic whole farm simulation model FASSET. *Agricultural Systems* 76, 817–839.

Boccanfuso, D. and Savard, L. (2007) Impacts analysis of cotton subsidies on poverty: a CGE macro-accounting approach. *Journal of African Economies* 16, 629–659.

Boccanfuso, D. and Savard, L. (2008) Groundnut sector liberalization in Senegal: a multi-household CGE analysis. *Oxford Development Studies* 36, 159–186.

Breen, J. and Hennessy, T. (2003) *The Impact of the MTR and WTO reform on Irish Farms*. The Irish Agriculture and Food Development Authority, Teagasc, Athenry, Ireland.

Burrell, A. (1989) The demand for fertiliser in the United Kingdom. *Journal of Agricultural Economics* 40, 1–20.

Callan, T. and Van Soest, A. (1993) Female labour supply in farm households: farm and off-farm participation. *Economic and Social Review* 24, 313–334.

Chan, B. and Guerra, P. (2012) Modeling farm inventory behaviour using artificial neural network in CADMS. *Paper presented to the European Meeting of the International Microsimulation Association, Dublin, May 17–19, 2012*.

Chemingui, M.A. and Thabet, C. (2009) Agricultural Trade Liberalisation and Poverty in Tunisia: Microsimulation in a General Equilibrium Framework. Aussenwirtschaft: *Schweizerische Zeitschrift für Internationale Wirtschaftsbeziehungen/The Swiss Review of International Economic Relations*, 64, 71.

Chitiga, M. and Mabugu, R. (2008) Evaluating the Impact of Land Redistribution: A CGE Microsimulation Application to Zimbabwe. *Journal of African Economies* 17(4), 527–549.

Ciaian, P., Espinosa, M., Gomez y Paloma, S., Heckelei, T., Langrell, S. *et al.* (2013) *Farm Level Modelling of CAP: A Methodological Overview.* Institute for Prospective Technological Studies, European Commission Joint Research Centre. Publications Office of the European Union, Luxembourg.

Clancy, D., Breen, J., Morrissey, K., O'Donoghue, C. and Thorne, F. (2013) The location economics of biomass production for electricity generation. In: O'Donoghue, C., Hynes, S., Morrissey, K., Ballas, D. and Clarke, G. (eds) *Spatial Microsimulation for Rural Policy Analysis*. Springer, Berlin/Heidelberg, Germany, pp. 159–175.

Clarke, G.P. (ed.)(1996) *Microsimulation for Urban and Regional Policy Analysis*. Pion, London.

Corong, E.L. and Cororaton, C.B. (2006) *Philippine Tariff Reduction Program: Poverty Effect Insights from A CGE Analysis*. ResearchGate.net. Available at: https://www.researchgate.net/publication/237433106_Philippine_Tariff_Reduction_Program_Poverty_Effect_Insights_from_A_CGE_Analysis_i (accessed 14 July 2016).

Crosson, P., O'Kiely, P., O'Mara, F.P. and Wallace, M. (2006) The development of a mathematical model to investigate Irish beef production systems. *Agricultural Systems* 89, 349–370.

Crosson, P., Shalloo, L., O'Brien, D., Lanigan, G.J., Foley, P.A., Boland, T.M. and Kenny, D.A. (2011) A review of whole farm systems models of greenhouse gas emissions from beef and dairy cattle production systems. *Animal Feed Science and Technology* 166, 29–45.

Chyzheuskaya, A., O'Donoghue, C. and O'Neill, S. (2014) Using farm micro-simulation model to evaluate the impact of the nitrogen reduction mitigation measures on the farm income in Ireland. *International Journal of Agricultural Management* 3, 232–242.

Dartanto, T. (2011) Volatility of world soybean prices, import tariffs and poverty in Indonesia A CGE-microsimulation analysis. *Margin: The Journal of Applied Economic Research* 5, 139–181.

De Cara, S. and Jayet, P.A. (2011) Marginal abatement costs of greenhouse gas emissions from European agriculture, cost effectiveness, and the EU non-ETS burden sharing agreement. *Ecological Economics* 70, 1680–1690.

De Souza Ferreira Filho, J.B., dos Santos, C.V. and do Prado Lima, S.M. (2010) Case study: tax reform, income distribution and poverty in Brazil: an applied general equilibrium analysis. *International Journal of Microsimulation* 3(1), 114–117.

Diao, X., Alpuerto, V. and Nwafor, M. (2009) *Economywide Impact of Avian Flu in Nigeria – A Dynamic CGE Model Analysis*. Controlling Avian Flu and Protecting People's Livelihoods in Africa and Indonesia – HPAI [highly pathogenic avian influenza] Research Brief No. 15. Available at: http://r4d.dfid.gov.uk/PDF/Outputs/HPAI/IFPRI_ILRI_rbr15.pdf (accessed 7 March 2016).

Dijk, J., Leneman, H. and van der Veen, M. (1996) The nutrient flow model for Dutch agriculture: a tool for environmental policy evaluation. *Journal of Environmental Management* 46, 43–55.

Donnellan, T., Finn, J., Hanrahan, K., Hennessy, T., Keena, C. *et al.* (2013a) *Teagasc Submission Made in Response to the Department of Agriculture, Food and the Marine CAP Public Consultation Process*. Teagasc Working Group on CAP Reform, Teagasc, Athenry, Ireland. Available at: http://www.teagasc.ie/publications/2013/2985/DAFM_Submission_20131029.pdf (accessed 8 March 2016).

Donnellan, T., Hanrahan, K., Hennessy, T., Kinsella, A., McKeon, M., Moran, B. and Thorne, F. (2013b) *Outlook 2014: Economic Prospects for Agriculture* [also titled *Annual Review and Outlook 2014*]. Agricultural Economics and Farm Surveys Department, Irish Agriculture and Development Authority, Teagasc, Athenry, Ireland. Available at: http://www.teagasc.ie/publications/2013/3043/ReviewAndOutlook2014.pdf (accessed 8 March 2016).

Doole, G.J. (2012) Cost-effective policies for improving water quality by reducing nitrate emissions from diverse dairy farms: an abatement-cost perspective. *Agricultural Water Management* 104, 10–20.

Doole, G.J., Marsh, D. and Ramilan, T. (2013) Evaluation of agri-environmental policies for reducing nitrate pollution from New Zealand dairy farms accounting for firm heterogeneity. *Land Use Policy* 30, 57–66.

Doucha, T. and Vaněk, D. (2006) Interactions between agricultural policy and multifunctionality in Czech agriculture. In: Diakosavvas, D. (ed.) *The Development Dimension: Coherence of Agricultural and Rural Development Policies*. OECD Publishing, Paris, pp. 239–247.

Ferreira, F.H., Fruttero, A., Leite, P.G. and Lucchetti, L.R. (2013) Rising food prices and household welfare: evidence from Brazil in 2008. *Journal of Agricultural Economics* 64, 151–176.

Galbraith, C., Bakhshi, S., Kung, W. and Kjaer, P. (2011) Incorporating a farm-level balance sheet forecast into the Canadian agricultural dynamic microsimulation model. *Paper presented to the 3rd General Conference of the International Microsimulation Association. Microsimulation and Policy Design, Stockholm, Sweden, June 8th to 10th, 2011.* Available at: http://www.scb.se/Grupp/Produkter_Tjanster/Kurser/_Dokument/IMA/Incorporating%20a%20Farm-Level%20Balanc.pdf (accessed 7 March 2016).

Ghosh, M. and Rao, J.N.K. (1994) Small area estimation: an appraisal. *Statistical Science* 9, 55–76.

Hemme, T., Deblitz, C., Isermeyer, F., Knutson, R. and Anderson, D. (2000) The International Farm Comparison Network (IFCN) – objectives, organisation and first results on international competitiveness of dairy production. *Züchtungskunde* 72, 428–439.

Hennessy, T.D. and Rehman, T. (2008) Assessing the impact of the 'decoupling' reform of the Common Agricultural Policy on Irish farmers' off-farm labour market participation decisions. *Journal of Agricultural Economics* 59, 41–56.

Hennessy, T. and Thorne, F. (2006) *The Impact of the WTO Doha Development Round on Farming in Ireland*. Rural Economy Research Centre, Teagasc, Athenry, Ireland.

Hermes, K. and Poulsen, M. (2012) A review of current methods to generate synthetic spatial microdata using reweighting and future directions. *Computers, Environment and Urban Systems* 36, 281–290.

Huffman, W.E. and Lange, M.D. (1989) Off-farm work decisions of husbands and wives: joint decision making. *The Review of Economics and Statistics* 71, 471–480.

Hynes, S. and Garvey, E. (2009) Modelling farmers' participation in agri-environmental schemes using panel data: an application to the rural environmental protection scheme in Ireland. *Journal of Agricultural Economics* 60, 546–562.

Hynes, S., Farrelly, N., Murphy, E. and O'Donoghue, C. (2008) Modelling habitat conservation and participation in agri-environmental schemes: a spatial microsimulation approach. *Ecological Economics* 66, 258–269.

Hynes, S., Morrissey, K., O'Donoghue, C. and Clarke, G. (2009a) A spatial microsimulation analysis of methane emissions from Irish agriculture. *Journal of Ecological Complexity* 6, 135–146.

Hynes, S., Morrissey, K., O'Donoghue, C. and Clarke, G. (2009b) Building a static farm level spatial microsimulation model for rural development and agricultural policy analysis in Ireland. *International Journal of Agricultural Resources, Governance and Ecology* 8, 282–299.

Hynes, S., Farrelly, N., Murphy, E. and O'Donoghue, C. (2013a) Conservation and rural environmental protection schemes. In: O'Donoghue, C., Hynes, S., Morrissey, K., Ballas, D. and Clarke, G. (eds) *Spatial Microsimulation for Rural Policy Analysis*. Springer, Berlin, Heidelberg, Germany, pp. 123–141.

Hynes, S., Morrissey, K. and O'Donoghue, C. (2013b) Modelling greenhouse gas emissions from agriculture. In: O'Donoghue, C., Hynes, S., Morrissey, K., Ballas, D. and Clarke, G. (eds) *Spatial Microsimulation for Rural Policy Analysis*. Springer, Berlin, Heidelberg, Germany, pp. 143–157.

Janssen, S., Louhichi, K., Kanellopoulos, A., Zander, P., Flichman, G., Hengsdijk, H., Meuter, E., Andersen, E., Belhouchette, H., Blanco, M. *et al.* (2010) A generic bio-economic farm model for environmental and economic assessment of agricultural systems. *Environmental Management* 46, 862–877.

Keeney, M. (2000) The distributional impact of direct payments on Irish farm incomes. *Journal of Agricultural Economics* 51, 252–265.

Kelley, H., Rensburg, T.M.V. and Yadav, L. (2013) A micro-simulation evaluation of the effectiveness of an Irish grass roots agri-environmental scheme. *Land Use Policy* 31, 182–195.

Kimura, S., Antón, J. and Cattaneo, A. (2012) Effective risk management policy choices under climate change: an application to Saskatchewan crop sector. Selected poster prepared for presentation at the *International Association of Agricultural Economists (IAAE) Triennial Conference*, Foz do Iguaçu, Brazil, 18–24 August, 2012. Available at: http://ageconsearch.umn.edu/bitstream/126736/2/Kimura.pdf (accessed 7 March 2016).

Klein, K. and Narayanan, S. (2008) Farm level models: a review of developments, concepts and applications in Canada. *Canadian Journal of Agricultural Economics* 40, 351–368.

Kruseman, G., Blokland, P.W., Bouma, F., Luesink, H., Mokveld, L. and Vrolijk, H. (2008a) *Micro-simulation as a Tool to Assess Policy Concerning Non-point Source Pollution: The Case of Ammonia in Dutch Agriculture*. LEI Wageningen UR, The Hague. Also available as: Contributed paper, European Association of Agricultural Economists, 107th Seminar, January 30–February 1, 2008, Sevilla, Spain; at: http://ageconsearch.umn.edu/handle/6661 (accessed 7 March 2016).

Kruseman, G., Blokland, P.W., Luesink, H. and Vrolijk, H. (2008b) Ex-ante evaluation of tightening environmental policy: the case of mineral use in Dutch agriculture. In: *European Association of Agricultural Economists, 2008 International Congress, August 26–29, 2008, Ghent, Belgium*. Available at: http://ageconsearch.umn.edu/handle/43644 (accessed 7 March 2016).

Li, J. and O'Donoghue, C. (2013) A survey of dynamic microsimulation models: uses, model structure and methodology. *International Journal of Microsimulation* 6, 3–55.

Lindgren, U. and Elmquist, H. (2005) Environmental and economic impacts of decision-making at an arable farm: an integrative modelling approach. *Ambio* 34, 393–401.

Louhichi, K., Kanellopoulos, A., Janssen, S., Flichman, G., Blanco, M., Hengsdijk, H., Heckelei, T., Berentsen, P., Lansink, A.O. and van Ittersum, M. (2010) FSSIM, a bio-economic farm model for simulating the response of EU farming systems to agricultural and environmental policies. *Agricultural Systems* 103, 585–597.

Lovett, D.K., Shalloo, L., Dillon, P. and O'Mara, F.P. (2006) A systems approach to quantify greenhouse gas fluxes from pastoral dairy production as affected by management regime. *Agricultural Systems* 88, 156–179.

Mabugu, R. and Chitiga, M. (2009) Is increased agricultural protection beneficial for South Africa? *Economic Modelling* 26, 256–265.

McCormack, M., O'Donoghue, C. and Hynes, S. (2014) Trends in CAP over time: a hypothetical farm analysis. *Paper presented to the Agricultural Economics Society of Ireland, Dublin March 14th, 2014*. Available at: http://www.aesi.ie/aesi2014/mmccormack.pdf (accessed 8 June 2014).

Morley, S. and Piñeiro, V. (2004) *The Effect of WTO and FTAA on Agriculture and the Rural Sector in Latin America*. DSGD Discussion Paper No. 3, Development Strategy and Governance Division, International Food Policy Research Institute (IFPRI), Washington, DC.

Murphy, G., O'Donoghue, C., Hynes, S. and Murphy, E. (2014) Modelling the participation decision in agri-environmental schemes. *Paper presented to the European Association of Agricultural Economists Conference, Ljubljana, August 26–29, 2014*. Available at: http://www.bioecon-network.org/pages/17th_2015/Murphy.pdf (accessed 7 March 2016).

Niazi, M. and Hussain, A. (2011) Agent-based computing from multi-agent systems to agent-based models: a visual survey. *Scientometrics* 89: 479.

O'Brien, D., Shalloo, L., Grainger, C., Buckley, F., Horan, B. and Wallace, M. (2010) The influence of strain of Holstein–Friesian cow and feeding system on greenhouse gas emissions from pastoral dairy farms. *Journal of Dairy Science* 93, 3390–3402.

O'Donoghue, C. (2013) Modelling farm viability. In: O'Donoghue, C., Hynes, S., Morrissey, K., Ballas, D. and Clarke, G. (eds) *Spatial Microsimulation for Rural Policy Analysis*. Springer, Berlin, Heidelberg, Germany, pp. 177–191.

O'Donoghue, C. (ed.) (2014) *Handbook of Microsimulation Modelling.* Contributions to Economic Analysis, Volume 293, Emerald Group Publishing, Bingley, UK.

O'Donoghue, C., Hynes, S., Morrissey, K., Ballas, D. and Clarke, G. (eds) (2013) *Spatial Microsimulation for Rural Policy Analysis*. Springer, Berlin, Heidelberg, Germany.

O'Donoghue, C., Morrissey, K. and Lennon, J. (2014a) Spatial microsimulation modelling: a review of applications and methodological choices. *International Journal of Microsimulation* 7, 26–75.

O'Donoghue, C., Grealis, E., Farrell, N., Loughrey, J., Hanrahan, K., Hennessy, T. and Meredith, D. (2014b) Modelling the spatial distributional effect of Common Agricultural Policy reform. *Presentation to the Agricultural Economics Society of Ireland, Dublin, March* 14, 2014. Available at: http://www.aesi.ie/aesi2014/codonoghue.pdf (accessed 8 June 2014).

OECD-FAO (2013) *OECD–FAO Agricultural Outlook 2013–2022. OECD Publishing*, Paris/Food and Agriculture Organization of the United Nations, Rome.

Orcutt, G.H. (1957) A new type of socio-economic system. *Review of Economics and Statistics* 58, 773–797 (reprinted in *International Journal of Microsimulation* 1, 3–9).

Orcutt, G., Greenberg, M., Korbel, J. and Rivlin, A. (1961) *Microanalysis of Socioeconomic Systems: A Simulation Study*. Harper and Row, New York.

Osterburg, B., Offermann, F. and Kleinhanss, W. (2001) A sector consistent farm group model for German agriculture. In: Heckelei, T., Witzke, H.P. and Henrichsmeyer, W. (eds) *Agricultural Sector Modelling and Policy Information Systems: Proceedings of the 65th European Seminar of the European Association of Agricultural Economists (EAAA), March 29–31, 2000, Bonn, Germany*. Verlag Vauk, Kiel, Germany, pp. 152–160.

Pauw, K. and Thurlow, J. (2011) Agricultural growth, poverty, and nutrition in Tanzania. *Food Policy* 36, 795–804.

Peeters, L. and Surry, Y. (1997) A review of the arts of estimating price-responsiveness of feed demand in the European Union. *Journal of Agricultural Economics* 48, 379–392.

Plaxico, J.S. and Tweeten, L.G. (1963) Representative farms for policy and projection research. *Journal of Farm Economics* 45, 1458–1465.

Potter, S.R., Atwood, J.D., Lemunyon, J. and Kellogg, R.L. (2009) A national assessment of soil carbon sequestration on cropland: a microsimulation modeling approach. In: Lal, R. and Follett, R.F. (eds) *Soil Carbon Sequestration and the Greenhouse Effect*, 2nd edn. SSSA Special Publication 57, Soil Science Society of America, Madison, Wisconsin, pp. 1–11.

Prochorowicz, J. and Rusielik, R. (2007) Relative efficiency of oilseed crops production in the selected farms in Europe and the world in 2005. *Acta Scientiarum Polonorum Oeconomia* 6, 57–62.

Ramilan, T., Scrimgeour, F.G., Levy, G. and Romera, A.J. (2007) Modelling economic impact of agri-environmental policy on dairy farms–a catchment perspective. In: Oxley, L. and Kulasiri, D. (eds) *MODSIM 2007 International Congress on Modelling and Simulation.* Modelling and Simulation Society of Australia and New Zealand, *December* 2007, pp. 1089–1095. Available at: http://www.mssanz.org.au.previewdns.com/MODSIM07/papers/17_s18/ModellingEconomic_s18_Ramilan_.pdf (accessed 7 March 2016).

Ramilan, T., Scrimgeour, F. and Marsh, D. (2011) Analysis of environmental and economic efficiency using a farm population micro-simulation model. *Mathematics and Computers in Simulation* 81, 1344–1352.

Richardson, J.W. and Nixon, C.J. (1986) *Description of FLIPSIM V: A General Firm Level Policy Simulation Model.* Bulletin B No. 1528, Texas Agricultural Experiment Station, Texas A&M University, College Station, Texas.

Richardson, J.W., Hennessy, T. and O'Donoghue, C. (2014) Farm level models. In: O'Donoghue, C. (ed.) *Handbook of Microsimulation Modelling.* Contributions to Economic Analysis Volume 293, Emerald Group Publishing, Bingley, UK, pp. 503–534.

Ryan, M., O'Donoghue, C., Hynes, S. and Upton, V. (2015) Land use change from agriculture to forestry: a structural model of the income leisure choices of farmers. In: *International Conference of Agricultural Economists: ICAE 29th, Milan, Italy 2015, Università degli Studi di Milano August 8–14. Agriculture in an Interconnected World*. Available at: http://ageconsearch.umn.edu/bitstream/212285/2/Ryan-Land%20Use%20Change%20from%20Agriculture%20to%20Forestry-820.pdf (accessed 7 March 2016).

Severini, S. and Tantari, A. (2013) The effect of the EU farm payments policy and its recent reform on farm income inequality. *Journal of Policy Modeling* 35, 212–227.

Shalloo, L., Dillon, P., Rath, M. and Wallace, M. (2004) Description and validation of the Moorepark dairy system model. *Journal of Dairy Science* 87, 1945–1959.

Sharples, J.A. (1969) The representative farm approach to estimation of supply response. *American Journal of Agricultural Economics* 51, 353–361.

Shrestha, S., Hennessy, T. and Hynes, S. (2007) The effect of decoupling on farming in Ireland: a regional analysis. *Irish Journal of Agricultural and Food Research* 46, 1–13.

Tanton, R., Vidyattama, Y., Nepal, B. and McNamara, J. (2011) Small area estimation using a reweighting algorithm. *Journal of the Royal Statistical Society: Series A (Statistics in Society)* 174, 931–951.

Thomas, A. (2013) Linking farm level models with environmental impact models. In: Ciaian, P., Espinosa, M., Gomez y Paloma, S., Heckelei, T., Langrell, S. *et al.* (eds) *Farm Level Modelling of CAP: A Methodological Overview*. Institute for Prospective Technological Studies, European Commission Joint Research Centre. Publications Office of the European Union, Luxembourg, pp. 53–60.

Thorne, F.S. and Fingleton, W. (2006) Examining the relative competitiveness of milk production: an Irish case study (1996–2004). *Journal of International Farm Management* 3(4), 49–61.

Thurlow, J. (2008) *Agricultural Growth Options for Poverty Reduction in Mozambique. Preliminary Report Prepared for Mozambique's Ministry of the Agriculture and Strategic Analysis and Knowledge Support System (SAKSS).* ReSAKSS Working Paper No. 20, Regional Strategic Analysis and Knowledge Support System, International Food Policy Research Institute, Washington, DC.

Troitzsch, K.G., Mueller, U., Gilbert, N. and Doran, J.E. (eds) (1996) *Social Science Microsimulation*. Springer, Berlin.

van Leeuwen, E.S. (2010) Microsimulation of rural households. In: *Urban-Rural Interactions: Towns as Focus Points in Rural Development*. Physica-Verlag (Springer), Heidelberg, Germany, pp. 115–135.

van Leeuwen, E. and Dekkers, J. (2013) Determinants of off-farm income and its local patterns: a spatial microsimulation of Dutch farmers. *Journal of Rural Studies* 31, 55–66.

Villot, X.L. (1998) *The Effects of a Sulphur Tax Levied on the Spanish Electricity Industry*. Documentos de Trabajo, Facultad de Ciencias Económicas y Empresariales, Universidad Nacional Educación a Distancia, Madrid.

Warr, P. and Yusuf, A.A. (2009) International food prices and poverty in Indonesia. *Paper presented to 54th National Conference of the Australian Agricultural and Resource Economics Society, Adelaide, February 2010*. Available at: http://ageconsearch.umn.edu/bitstream/59259/2/Warr,%20Peter.pdf (accessed 7 March 2016).

Williamson, P., Birkin, M. and Rees, P.H. (1998) The estimation of population microdata by using data from small area statistics and samples of anonymised records. *Environment and Planning A* 30, 785–816.

Zander, K., Thobe, P. and Nieberg, H. (2007) Economic impacts of the adoption of the Common Agricultural Policy on typical organic farms in selected new Member States. *Jahrbuch der Österreichischen Gesellschaft für Agrarökonomie* 16, 85–96.

10 Scaling Up and Out: Agent-based Modelling to Include Farmer Regimes

Andrew P. Barnes,[1]* Eleonore Guillem[2] and Dave Murray-Rust[3]
[1]Land Economy, Environment and Society (LEES) Research Group, Scotland's Rural College (SRUC), Edinburgh, UK; [2]Université Aix-Marseille, Marseille, France; [3]University of Edinburgh, Edinburgh, UK

10.1 Introduction

The needs of policy makers have become more complex as policy itself embraces increasingly intractable problems related to maintaining or enhancing production and the natural environment. The extension of policies to accommodate environmental and social goals, alongside ensuring economic development within farming communities, has led the agricultural economic modeller to seek ways to accommodate these effects within a robust modelling framework. Augmenting economic processes with those of other complex ecological and social systems is clearly a non-trivial task; nevertheless, some understanding of the consequences of the full primary and secondary effects of reform benefits the policy maker and informs future land-use decision making.

The idea of scaling has been mostly discussed in the realm of technology adoption and development, but scaling up and out are useful concepts for the farm-oriented discussion here. Cook and Fujisaka (2004) point out that the process of 'scaling out' is extending scope by repeating a process at one scale to other individuals at a similar scale, for example a farm-level model parameterized for a range of farmers (Shrestha *et al.*, 2014). Conversely, 'scaling up' occurs when the dimension of a process increases to accommodate agents, for example agents other than farmers, including non-human agents.

The inclusion of these other processes requires the modeller to apply techniques that adequately represent a spatial scale beyond the economy of a single farm. However, aggregating these farm-level models themselves is fraught with bias and can be a contentious task. Specifically, the accuracy of the 'scaling' task is determined by a series of decisions the farm modeller may take, such as the use of a representative (or typical) farming type, or other factors which are ostensibly out of the control of the modeller, such as the precision and resolution of any regional data that may be included. In addition, 'scaling out' is compromised by the heterogeneity of the farms within the study landscape, the differences between the perceptions and objectives of the farmers in their decision making, and differences in land profiles. These issues implicitly require the integration of disciplines, models and data within the modelling process.

Moreover, much like the process of farm-level model building, some conceptual framework is needed to define how the results of the scaling process are to be understood and, more critically, the error rate at which results are estimated. Menter *et al.* (2004) pointed out that scale is

*E-mail: andrew.barnes@sruc.ac.uk

generally understood in terms of a hierarchy, yet different disciplines have different criteria for defining and measuring scale. In what follows, we refer to aggregating the results of farm-level models to a regional or country level as 'scaling up', and to extending farm-level models themselves to include wider ecological and social processes as 'scaling out'. This is an important distinction as it dictates the modelling approach used, the level of assumptions required within which to solve a model and the final acceptability of the results.

The scale of the farm is relatively distinct and reasonably well defined in government data sets, in contrast to the ecological, behavioural and social landscape in which the farm and farmer exist. Paucity of data and brevity in the modelling process tends to assign these within-farm non-production effects and post-farm gate effects as exogenous to the model itself. None the less, their influence, through determining scarcity of input and dictating marketing costs, is critical to providing a robust optimal solution.

Accordingly, this chapter focuses on the concept of scaling out and provides examples of mechanisms for refining and meeting the needs of policy makers in the multiple dimensional landscape of economic, environmental and sustainable development needs.

10.2 Review of Approaches to Scaling

A range of modelling approaches is available for the farm-level modeller to allow scaling outwards or upwards. Initially, and perhaps most parsimoniously, a first-stage decision is to determine the spatial scale of the decision-making unit and whether it could adequately obviate scaling issues – that is, to treat a region as a single-farm decision-making unit, or a set of homogenous units within a landscape or region. This is an attractive and popular approach, as it captures all of the resources and activity constraints available for the site of interest, and aggregate data at a regional level are available from government publications. Furthermore, this approach implicitly accounts for trade of inputs and outputs between actors within the overall resource, and for the material budget for this higher spatial unit, thus allowing simplicity in the characterization of the economic process underlying the farming economy of the region. Clearly, this approach imposes homogeneity within decision making and biophysical resource allocations in that it reduces the optimization problem to a single decision maker, and any disparity between the response of the farmer, the location and the quality of land is averaged out or, in some cases, discounted.

Other techniques have provided a more nuanced approach to scaling (O'Donoghue *et al.*, 2014), but this also has issues, as it aims to transfer the optimized resource allocation of a typical farm to the scaled distribution, and this may also be biased as it is dependent on how a farm is characterized by the farm-level modeller. Underlying the resource allocation problem is the complex decision maker. A wide range of literature has identified the range of motivating factors in determining a shift in decision making (Gasson, 1973). For example, one compelling argument is the role of intergenerational transfer and the identification of a successor in determining change within a farming environment (Barnes *et al.*, 2016).

Farmers, by applying specific farm strategies, are at the centre of the processes underlying land use and cover change (LUCC), and its subsequent effect on the provision and quality of ecosystem services. Accounting for non-rational behaviour is not uncommon in agricultural modelling, for instance Rehman *et al.* (2007) adjusted constraints to match environmental decisions to account for risk aversion. However, with increasing understanding of the ecological and social dimensions involved, an approach to modelling complex decisions is to disaggregate the decision process into discrete areas in which there is a benefit perceived by the farmer. The standard pillars of sustainability – economic viability, social benefit and environmental improvement – are common aspects for the decision maker (Faucheux and Froger, 1995) and, essentially, each decision maker assigns different levels of importance to these principles.

In order to adequately reflect the response to policy drivers, there are various requirements when empirically modelling farmer decision making: a strong conceptual framework, a variety of data and the characterization of a high number of individuals within the farming landscape.

Here, agent-based modelling (ABM) is a useful approach. ABM is a technique that has developed from computer modelling to incorporate other disciplines and complexity over different scales and organizational levels, and to reflect individual choices in higher level behavioural patterns (e.g. Brown, 2006). Only a few studies have used ABM incorporated into this aspect of LUCC models (e.g. Fernandez *et al.*, 2005; Brown and Robinson, 2006; Valbuena *et al.*, 2010; Chen *et al.*, 2011).

10.3 Core Model

In order to build an agent-based model, the core model must present the process of analysis in terms of the drivers and the different layers of the model. These are shown in Fig. 10.1.

A farm strategy is the result of a decision-making search and the weighted utility function. These lead to a set of regimes applied at the parcel level within a farm. A regime is a multi-year plan composed of a suite of management approaches applied to one field (e.g. crop rotation). The regimes can be intensively or extensively produced, or they can be associated with agri-environmental measures (e.g. grass margins). The decision-making process can be unpacked in terms of an initial search and an evaluation of a set of possible regimes in terms of their sustainable attributes (i.e. economic, social and environmental). The farmers apply the regime with the highest utility to a parcel when the previous regime has ended. The agent/farmer can plan for the future as well as make choices in the present (i.e. sacrifice today in order to obtain a better outcome in the future, or maximize the present). These choices lead to effects on production, through a module representing the biophysical attributes of landscape effects, relating to the mixture of regimes within each parcel at the farm level. This, in turn, is coupled to the relationship between landscape effects and ecosystem services, e.g. the provision of sites for nesting birds, and locking in carbon. For brevity, we focus on the farmer decision-making component of the model. The details of the agent-based model are discussed in more detail in Murray-Rust *et al.* (2011) and Guillem *et al.* (2015).

The farmer may choose a certain regime to meet profit-maximizing, societal and/or environmental goals. To accommodate heterogeneity within the farmer agents the following multi-attribute utility function is applied:

$$U_j = U_{E_j} + U_{S_j} + U_{N_j} \quad (1)$$

where U_j is the overall utility of a regime *j*, and U_{E_j}, U_{S_j} and U_{N_j} are, respectively, the economic, social and environmental partial utilities. These can be classified as:

- The economic score, *E*: this is the expected economic contribution of a given regime on a per hectare basis. The contribution is expressed as the difference existing between the farmer's average gross margin over the past 5 years and the expected gross margin estimated for a new regime *j*.

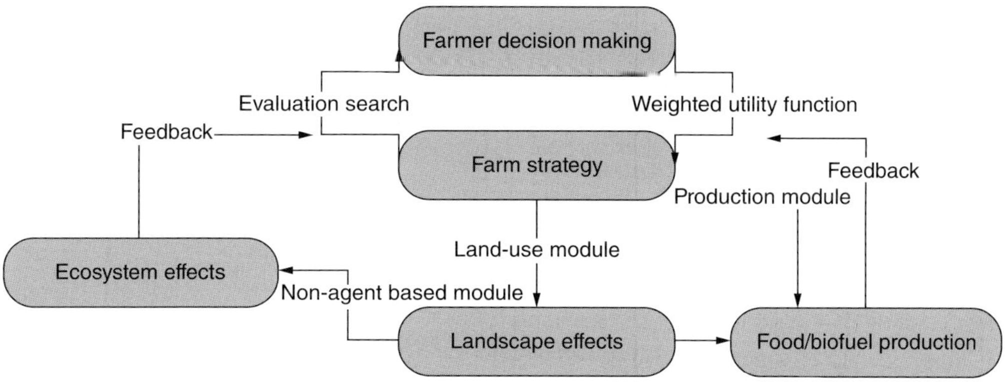

Fig. 10.1. Modelling architecture of an agent-based model, showing the different modules and information flow between the policy drivers.

- The environmental score, N: this score is characterized by three indicators: crop cover, nitrogen need and diversity. The first two indicators are applied at the management level. The third indicator, diversity, is applied at the regime level and is computed using a simple set of rules. Two aspects of rotation diversity are assessed: temporal diversity, which is the length of a crop rotation (in years), and biodiversity, which is the composition of the rotation in terms of plant groups. Longer rotations typically increase the diversity in insect species (Clergue *et al.*, 2005) and reduce the risk of crop failure due to single pest outbreak (Altieri, 1999).
- The social score, S: The social score is calculated from the mean of scores for access to green space (G) for the tradition (T) of a management i. Thus, the social score of a regime j over the time period t is calculated by Equation (2):

$$S_j = \frac{\sum(G_i + T_i)}{t_j} \tag{2}$$

10.4 Application

10.4.1 The Lunan catchment, east Scotland

The agent-based model described above is applied to the Lunan catchment in Scotland. Geographical information system (GIS) data were obtained for the catchment using IACS[1] (data provided by the Scottish government). These data give information at the parcel level on farmer ID, geographic coordinates and a timeline of land uses from 2001 to 2008. The model is calibrated against these historical data, which also serve to constrain the farmers in their choice of regime by representing the limitations in skills and machinery for specific crops.

A basic vegetation model is applied to all managements. This is configured using growth curves that represent the average harvestable biomass of a crop throughout a given year based on historical data. A technological coefficient is applied to the 2002 average yield of a given management for each subsequent year (Abildtrup *et al.*, 2006). These coefficients are scenario specific. Farm management actions are limited to fixed sowing and harvesting. The dates of actions are taken from the literature available for various years from SAC (the Scottish Agricultural Colleges, now SRUC, Scotland's Rural College). We assume that farmers with grassland over 5 years old and rough grazing are producing beef. In the lowlands, most beef farmers rear cattle from birth to 'finish', i.e. at 3 years old, so the time between start and harvest is set to 3 years.

To represent heterogeneity in farmers' decisions, and create a population of farmers within the catchment model, a typology of farmers was developed, based on the objectives and attitudes of Lunan farmers to make land-management decisions. This was achieved through a two-phase survey of these farmers. First, a postal survey of 123 farmers in the catchment was undertaken to determine and understand their attitudes towards farming and the environment. Using both factor and cluster analysis, a typology was created of these farmers to represent distinct groups of decision makers within the Lunan catchment (see Table 10.1).

A choice-based conjoint (CBC) approach was implemented (Adamowicz *et al.*, 1997) to evaluate quantitatively the non-use or passive-use aspects of a farmer's decision (e.g. social feedback, impact on the environment) as well as the economic aspects of decision making. CBC analysis is a technique used to derive the partial utilities of the different levels of one attribute of interest relative to others that are shown by a survey respondent. The responses were analysed using a hierarchical Bayes choice-based conjoint (HCBC) model to capture the preferences of individual respondents as well as groups of individuals, i.e. segment-level analysis (Orme and Howell, 2009).

In the conjoint experiment, respondents were assigned an attitudinal type from the first-stage postal survey and then presented with several choice tasks whereby they were asked to choose one land use option in each task. Each option is described by a combination of levels of attributes (see Table 10.2).

10.4.2 Evaluation of regimes

A farm strategy is the aggregation of regimes applied on all the parcels within a farm. We assume that farmers evaluate a set of possible regimes in

Table 10.1. Description of the behavioural types of farmers and their frequency of occurrence in the Lunan catchment in Scotland, based on the results of a postal survey. From Guillem *et al.*, 2012.

Type of farmer (% frequency)	Description of farmer type
Profit-oriented (38%)	These farmers maximize profit and do not hold strong positive attitudes towards the environment. Their intentions are to use the most profitable activities but without innovating. They seek information in an individualistic manner (from advisors, the media, etc.).
Multifunctionalist (25%)	These farmers are oriented equally towards profit maximization and environmental and social objectives. They have strong environmental values and participate in agri-environment schemes (AESs). Their activity is chosen for its profitability, but shows respect for the environment by reductions in chemical input. They intend to innovate if this fits with market and policy changes. They inform collectively (via open days, other farmers, etc.).
Traditionalist (36%)	The main orientation of these farmers is quality of lifestyle (both environmental and social). They are reluctant to innovate and participate in AESs, but wish to reduce their impact on the environment by adapting their management. They have local attachments and exchange information mainly with neighbouring farmers.

Table 10.2. Attributes and their levels used for constructing choice tasks by survey respondents.

	Levels of attribute involved (land use option)
Farm activity	Crop, livestock, non-food, manage environment
Social feedback	Positive, none, negative
Environment impact	Degrade, maintain, enhance
Change in income	Low (−10%), unchanged, high (+10%)

terms of their sustainable attributes (i.e. economic, social and environmental sustainability) and apply the regime with the highest utility to a parcel when the previous regime has ended. The performance of the regime across economic, environmental and social factors is assessed through a scoring system that corresponds with the attribute levels used in the CBC survey.

10.4.3 Model verification

To assess the plausibility of the performance of the model, a number of verification experiments were performed. At each step of model development, a series of 'unit tests' were applied to search for errors in the code (Beck, 2002). These unit tests are independent pieces of code that relate to a single equation in the software. Hence, it is possible to compare outputs from the model with expected results.

10.4.4 Results

Three of the ALARM-based socio-economic scenarios (Assessing LArge-scale Risks for biodiversity with tested Methods) (Bohunovsky *et al.*, 2011; Settele *et al.*, 2012; Spangenberg *et al.*, 2012) were adapted to the case study to analyse the effects of exogenous factors on the endogenous decision-making process. These were:

- BAMBU (Business-As-Might-Be-Usual), which represents approximately the current regional trends and can, therefore, be considered as a baseline for comparison with the other scenarios;
- GRAS (GRowth Applied Strategy), which is characterized by economic liberalism, free trade and international competitiveness; and
- SEDG (Sustainable European Development Goal) policies, which are oriented towards multifunctionality and local needs.

These scenarios provide the basis for running the modelling over the period 2025 to 2050. The results are discussed and presented in terms of the three farmer types.

The management styles differ between farmer types and across socio-economic scenarios. Figure 10.2, shows that most of the land is managed intensively, and the multifunctionalists apply agri-environmental measures and bioenergy on a larger area than do the two other farmer types. In the BAMBU (baseline) scenario

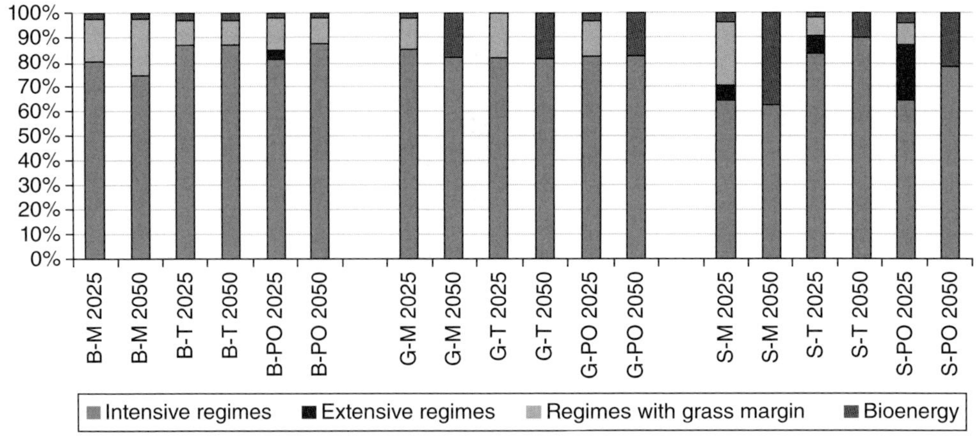

Fig. 10.2. Proportion of management styles in 2025 and 2050 for each of three socio-economic scenarios (B, BAMBU, Business-As-Might-Be-Usual; G, GRAS, Growth Applied Strategy; and S, SEDG, Sustainable European Development Goal) and farmer type (M, multifunctionalist; T, traditionalist; and PO, profit-oriented).

(the first set of bar graphs), while both profit-oriented and traditionalist farmers maintain a small number of regimes with grass margins across time, the multifunctionalists increase these. In addition, the profit-oriented farmers are the only type to apply a small area of extensive regime. The proportion of management styles is very similar between farmer types in the GRAS scenario (the second set of bar graphs): a small area of regimes with grass margins is replaced by bioenergy by 2050. The biggest difference between farmer types is found in the SEDG outputs (the third set of bar graphs). In 2025, the multifunctionalists apply more than 25% of their land to regimes including grass margins and about 10% to extensive regimes, the rest being intensive; in 2050, however, this is replaced by bioenergy crops. Only 18% of the land is managed extensively by the traditionalists or associated with agri-environmental measures. The traditionalists increase intensive management further by 2050, with a small area of land being allocated to bioenergy (about 10%). The profit-oriented farmers reduce regimes with intensive management by more than 35% in 2025, as do the multifunctionalists, but they prefer extensive management rather than agri-environmental measures. In spite of this, by 2050, this management becomes more intensive and 20% of the land is allocated to bioenergy crops.

The overall regime utility is largely dominated by the economic partial utility, in particular in the GRAS scenario (the second set of bar graphs) (Fig. 10.3). The environmental and social partial utilities are minor and similar in all scenarios because the scoring system was bounded. This reflects the economic adaptation of the different farmer types at the same time as maintaining the maximized environmental and social utilities permitted by the potential regime alternatives that were input to the model. The social utilities are negative in all cases, but the environmental utilities are positive or equal to zero. Profit-oriented farmers have higher environmental utilities than do the multifunctionalists and traditionalists, showing the adoption of farm strategies with neutral environmental scores that satisfy the profit-oriented farmers. This is in opposition to the multifunctionalists and traditionalists, for whom environmental utility increases as the scores attributed to regimes rise. The profit-oriented farmers show an adaptation to farm strategies, which increase profit in BAMBU and GRAS. In contrast, the multifunctionalists choose strategies that break even in BAMBU or lead to profit loss in GRAS, while not gaining environmental or social personal benefit. The traditionalists experience the smallest overall utility in both scenarios. In SEDG, the pattern is reversed between profit-oriented farmers who lose profit and traditionalists

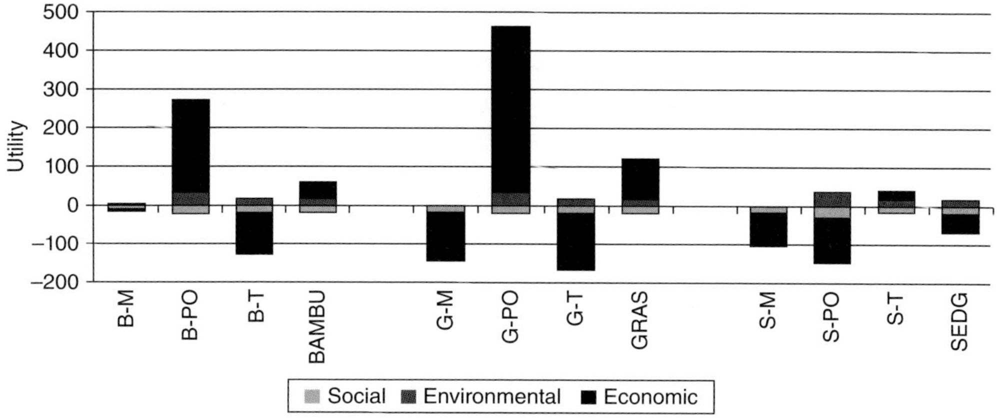

Fig. 10.3. Average partial utilities for economic, social and environmental attributes of farmer types (M, multifunctionalist; T, traditionalist; and PO, profit-oriented) for each of three socio-economic scenarios (B, BAMBU, Business-As-Might-Be-Usual; G, GRAS, Growth Applied Strategy; and S, SEDG, Sustainable European Development Goal) from 2008 to 2050. Each set of bars shows the overall partial utilities per scenario as well as the utilities broken down by farmer type.

who have managed to break even. The traditionalists maintain a positive utility, which emphasizes their characteristic of being satisfied without profit maximization.

10.4.5 Limitations

The main criticism of ABM, especially when applied to the social sciences, is a lack of empirical validation. Validation can be defined as 'the process by which a computer model is an accurate representation of the real world from the perspective of the model's intended applications' (Ormerod and Rosewell, 2009). The complexity of the processes being modelled and, conversely, the requirement to simplify the modelling of these processes, inherently give unobservable accuracy rates. This is compounded by the observable data, which may only offer some discrete representation of the process being modelled. This has implications for how outputs are reported in terms of a single result, and for the outcome of a number of runs made to assess sensitivities of different parameters to change. A range of approaches are offered to systematically validate ABM. For example, Bommell *et al.* (2014) used an iterative approach whereby inputs were calibrated from the observable data and then a simple model of livestock management was presented to livestock keepers facing drought conditions in Uruguay. Through an accessible front-end editor, the livestock keepers could then manipulate the actions of the agents within the model. Including participatory methodologies should capture some of the qualitative decision-making processes In addition, Gonzalès *et al.* (2013) argued for the use of online gaming platforms or the development of such game-playing models to reveal the rich data around farmer decision making to inform the rules of agent-based models.

10.5 Summary and Conclusions

Assuming homogeneity in decision making is a pragmatic solution to the problem of scaling out responses to reflect a population of interest. While nuances in farm structure and commodity focus can be captured by using national data sets, the underlying drivers for decision making can only be observed indirectly or require more detailed surveying of the farming population. Nevertheless, the granularity that this approach offers in terms of scaling out results allows a deeper understanding of policy response to future policy interventions and changes in market pressures.

Note

[1] The IACS is the Integrated Administration and Control System used by the European Union to manage and control payments made to farmers under the Common Agricultural Policy. As part of this system, farmers have the obligation to fill in a form in order to obtain their direct payments and other subsidies.

References

Abildtrup, J., Audsley, E., Fekete-Farkas, M., Giupponi, C., Gylling, M., Rosato, P. and Rounsevell, M. (2006) Socio-economic scenario development for the assessment of climate change impacts on agricultural land use: a pairwise comparison approach. *Environmental Science and Policy* 9, 101–115.

Adamowicz, W., Swait, J., Boxall, P., Louviere, J. and Williams, M. (1997) Perceptions versus objective measures of environmental quality in combined revealed and stated preference models of environmental valuation. *Journal of Environmental Economics and Management* 32, 65–84.

Altieri, M.A. (1999) The ecological role of biodiversity in agroecosystems. *Agriculture, Ecosystems and Environment* 74, 19–31.

Barnes, A.P., Sutherland, L.-A., Toma, L., Matthews, K. and Thomson, S. (2016) The effect of the Common Agricultural Policy reforms on intentions towards food production: evidence from livestock farmers. *Land Use Policy* 50, 548–558.

Beck, K. (2002) *Test Driven Development: By Example*. Addison-Wesley (Pearson Education), Boston, Massachussetts.

Bohunovsky, L., Jager, J. and Omann, I. (2011) Participatory scenario development for integrated sustainability assessment. *Regional Environmental Change* 11, 271–284.

Bommel, P., Dieguez, F., Bartaburu, D., Duarte, E., Montes, E. *et al.* (2014) A further step towards participatory modelling. Fostering stakeholder involvement in designing models by using executable UML. *Journal of Artificial Societies and Social Simulation* 17(1): 6.

Brown, D.G. (2006) Agent-based models. In: Geist, H. (ed.) *The Earth's Changing Land: An Encyclopedia of Land-Use and Land-Cover Change*. Greenwood Publishing Group, Westport, Connecticut, pp. 7–13.

Brown, D.G. and Robinson, D.T. (2006) Effects of heterogeneity in residential preferences on an agent-based model of urban sprawl. *Ecology and Society* 11(1): 46.

Chen, X., Lupi, F., An, L., Sheely, R., Vina, A. and Liu, J. (2011) Agent-based modelling of the effects of social norms on enrolment in payments for ecosystem services. *Ecological Modelling* 229, 16–24.

Clergue, B., Amiaud, B., Pervanchon, F., Lasserre-Joulin, F. and Plantureux, S. (2005) Biodiversity: function and assessment in agricultural areas: a review. *Agronomy for Sustainable Development* 25, 1–15.

Cook, S.E. and Fujisaka, S. (2004) Spatial dimension of scaling up and scaling out. In: Pachico, D. and Fujisaka, S. (eds) *Scaling Up and Scaling Out: Achieving Widespread Impact through Agricultural Research*. Economics and Impact Series 3, CIAT (Centro Internacional de Agricultura Tropical), Cali, Colombia, pp. 53–65.

Faucheux, S. and Froger, G. (1995) Decision-making under environmental uncertainty. *Ecological Economics* 15, 29–42.

Fernandez, L.E., Brown, D.G., Marans, R.W. and Nassauer, J.I. (2005) Characterizing location preferences in an exurban population: implications for agent-based modeling. *Environment and Planning B: Planning and Design* 32, 799–820.

Gasson, R. (1973) Goals and values of farmers. *Journal of Agricultural Economics* 24, 521–542.

Gonzalès, R., Cardille, J.A. and Parrott, L. (2013) Agent-based land-use models and farming games on the social web – fertile ground for a collaborative future? *Ecological Informatics* 15, 14–21.

Guillem, E.E., Barnes, A.P., Rounsevell, M. and Renwick, A. (2012) Refining perception-based farmer typologies with the analysis of past land use change. *Journal of Environmental Management* 110, 226–235.

Guillem, E.E., Murray-Rust, D., Robinson, D.T., Barnes, A. and Rounsevell, M.D.A. (2015) Modelling farmer decision-making to anticipate tradeoffs between provisioning ecosystem services and biodiversity. *Agricultural Systems* 137, 12–23.

Menter, H., Kaaria, S., Johnson, N. and Ashby, J. (2004) Scaling up. In: Pachico, D. and Fujisaka, S. (eds) *Scaling Up and Out: Achieving Widespread Impact through Agricultural Research.* Economics and Impact Series 3, CIAT, Cali, Colombia.

Murray-Rust, D., Dendoncker, N., Dawson, T., Acosta-Michlik, L., Karali, E., Guillem, E. and Rounsevell, M. (2011) Conceptualising the analysis of socio-ecological systems through ecosystem services and agent based modelling. *Journal of Land Use Science* 6, 83–99.

O'Donoghue, C., Morrissey, K. and Lennon, J. (2014) Spatial microsimulation modelling: a review of applications and methodological choices. *International Journal of Microsimulation* 7(1), 26–75.

Orme, B. and Howell, J. (2009) *Application of Covariates Within Sawtooth Software's CBC/HB Program: Theory and Practical Example.* Research Paper Series, Sawtooth Software, Sequim, Washington, DC.

Ormerod, P. and Rosewell, B.A. (2009) Validation and verification of agent-based models in the social sciences. In: Squazzoni, F. (ed.) *Epistemological Aspects of Computer Simulation in the Social Sciences*. Springer, Berlin, Heidelberg, Germany, pp. 130–140.

Rehman, T., McKemey, K., Yates, C.M., Cooke, R.J., Garforth, C.J., Tranter, R.B., Park, J.R. and Dorward, P.T. (2007) Identifying and understanding factors influencing the uptake of new technologies on dairy farms in SW England using the theory of reasoned action. *Agricultural Systems* 94, 281–293.

Settele, J., Carter, T.R., Kuhn, I., Spangenberg, J.H. and Sykes, M.T. (2012) Scenarios as a tool for large-scale ecological research: experiences and legacy of the ALARM project. *Global Ecology and Biogeography* 21, 1–4.

Shrestha, S., Hennessy, T., Abdalla, M., Forristal, D. and Jones, M. (2014) Determining responses of Irish dairy farms under climate change. *German Journal of Agricultural Economics* 63, 143–155.

Spangenberg, J.H., Bondeau, A., Carter, T.R., Fronzek, S., Jaeger, J. *et al.* (2012) Scenarios for investigating risks to biodiversity. *Global Ecology and Biogeography* 21, 5–18.

Valbuena, D., Verburg, P.H., Bregt, A.K. and Ligtenberg, A. (2010) An agent-based approach to model land-use change at a regional scale. *Landscape Ecology* 25, 185–199.

11 Catchment-level Modelling

Joana Guimarães Ferreira,[1]* Patrick Abbot[2] and Andrew P. Barnes[1]
[1]Land Economy, Environment and Society (LEES) Research Group, Scotland's Rural College (SRUC), Edinburgh, UK; [2]LTS International Ltd, Edinburgh, UK

11.1 Introduction

Agricultural policies are transitioning from programmes focused on subsidizing the commercial production of agricultural commodities into programmes that also seek to address the environmental impacts of that production and encourage the adoption of sustainable land management (SLM) practices (Antle *et al.*, 2010). This change recognizes that farms have important roles in providing ecosystem services and preserving natural values. With the rise in popularity of the ecosystem services approach in both research and policy, there is a need to model farms at scales that capture ecological processes beyond the traditional farm-level and administrative boundaries. Such processes include water quality (non-point and point-source pollution and its diffusion), water regulation (retention, availability and flood prevention) and soil conservation (prevention of erosion and degradation, carbon sequestration). These watershed services are commonly targeted by Payments for Watershed Services (PWS) schemes and they necessitate modelling at the sub-basin or basin level in order to both adequately capture the hydrogeology of the system and account for threshold effects (i.e. where a large enough area of farmland needs to be under a given set of practices for service provision to be secured).

Bennett *et al.* (2013) have provided a global review of PWS and listed 205 active programmes and 76 in development (117 and 54 of which, respectively, are in developing countries); this is a significant increase from the 127 active programmes and the 61 in development that were identified in 2008 (Stanton *et al.*, 2010). Overall, it is estimated that from 1973 to 2011 transactions costing in the order of US$66 billion have allowed 195 million ha worldwide to be under watershed services-oriented management (Bennett *et al.*, 2013). This is a growing sector of the economy, and it provides no shortage of challenges to policy makers and institutions in negotiating and implementing PWS schemes, and to researchers as providers of the science base and the models both for predicting farmer adoption and estimating the corresponding ecosystem services supply and social value of the schemes.

The purpose of this chapter is to utilize catchment-level modelling approaches to model the ecosystem services that are important to policy makers. These are applied within a developing country context, offering discussion of some of the challenges in the collection of the data that are needed for accurate model building in this context.

*E-mail: joana.ferreira@sruc.ac.uk

11.2 Review of Existing Methodologies

A growing body of literature and methodologies has attempted to integrate catchment-level models of single and multiple ecosystem service provision with econometric models. This integration can be output based, scenario based or dynamic.

11.2.1 Output-based integration

The simplest way that environmental and economic models can be integrated is by running one model and using that model's outputs as the inputs to the other model. Many PWS studies use this type of integration by first modelling the provision of watershed services, such as water supply or the reduction of soil erosion, to downstream water users (hydropower, irrigation, urban users). This is done by using a map of the current land use in the catchment – or a future land use following current trends – and a map of an alternative to be compared (e.g. conservation of native forest, terracing, agroforestry). The output of this modelling exercise (e.g. differences in water yield or silt load) is then used to calculate the costs to downstream users of the current land use, the costs that are avoided by the alternative (e.g. Kramer *et al.*, 1997) and the opportunity costs to upstream farmers of adopting the alternative land use (e.g. Emerton *et al.*, 2009).

A good example of output-based integration of environmental–economic modelling for farm systems at a watershed level is the ECOSAUT tool developed for the Andean region (Quintero *et al.*, 2006). Using SWAT (Soil and Water Assessment Tool)[1] outputs as part of its make-up, this linear optimization model was developed as a management tool that enables users to simulate various technological scenarios of land use and policy measures, and to examine their consequences in terms of farmers' net income and environmental externalities, such as the costs of sediment load in the watershed.

Lack of data of sufficient resolution, reliability or extensiveness is a key challenge of environmental–economic modelling in a developing country context, a point that was emphasized by Antle and Stoorvogel (2006) and which motivated the development of their minimum data (MD) approach, which has since been used in a number of studies (e.g. Immerzeel *et al.*, 2008; Claessens *et al.*, 2010; Valdivia *et al.*, 2012). This is a framework that explicitly focuses on the adoption decision itself, and tries to predict adoption based on expected net returns of the current practice versus the joint returns of the alternative practice and any incentives that might exist for its adoption. It is also a framework that allows for considerable sensitivity analysis, as both net returns and the provision of ecosystem services are expressed according to distribution functions in order to account for the spatial variability of both, and ecosystem service curves can be derived for adoption with different levels of incentive (see Antle *et al.*, 2010). This is particularly important not only because the adoption of a practice by farmers will be determined by the opportunity costs dictated by their own spatial circumstances (soil type, crops cultivated, access to markets, local prices, etc.), but also because the supply of watershed services will be dependent on the farm location in the catchment and the effect that a particular practice has on that location (Antle and Stoorvogel, 2006). The MD approach has been shown to be accurate at regional levels, but it is more error prone at the subregional level, where there can be issues in producing approximate values for populations that are spatially complex when more general representative ecological model or economic models are used (Antle and Stoorvogel, 2006). None the less, the approach seems to be robust enough to support policy decision making, as validated by two case studies, one in Senegal and another in Kenya (Antle *et al.*, 2010).

11.2.2 Scenario-based integration

This is an extension of the output-based integration just described, in that the outputs of complex and often multiple-service environmental models are used as inputs to generate net present values through multisectoral econometric models. The key difference is that, in this method of integration, the scenarios analysed usually go beyond simply representing the current situation and an alternative borne of practice adoption or policy measure implementations and, instead, are developed through focus groups and extensive consultation with stakeholders, across sectors and interests, with the aim of making them representative of local-to-national views and priorities.

The scenarios thus created, intersecting social, economic and environmental spheres, and usually covering mid- to long-term time frames, are then translated into GIS (geographical information system) land-use maps – either 'manually', layer by layer, or by the application of algorithms or economic models that follow scenario-established rules. These maps for baseline and alternative future scenarios are then used to generate, usually through separate environmental and economic models, the ecosystem services and economic outcomes of a particular policy design or the interaction of various policies.

Scenario-based integration of environmental and economic models has become a popular methodology for ecosystem services valuation studies in both developed and developing countries, particularly those that go beyond the basin and consider multiple catchments or even whole countries. For example, Boody *et al.* (2005) analysed two watersheds in the USA, where GIS layers of land-use change were produced following four citizen-driven scenarios. The ecological modelling used included water quality in terms of nitrogen and phosphorus contamination, lethal and sublethal effects of sediment to fish assemblages, greenhouse gas emissions from livestock and fertilizer use, and soil carbon sequestration; the socio-economic modelling excluded agricultural commodity payment schemes and included costs (variable and fixed) of present and alternative practices, net farm income (calculated using the estimated average production of the producers surveyed, adjusted to reflect soil quality and valued with a 5-year weighted average of output prices), avoided costs of sedimentation and of reduction in peak flow and flood flow volumes, contingent values of environmental benefits, and social and human capital building. The ecological models were adapted from previous literature for the area, and ADAPT (agricultural drainage and pesticide transport), a field-scale model for water table management, was used for the hydrological modelling. The study provides evidence that diversifying agriculture can provide social, economic and environmental benefits, and that this diversification can be created through incentives that fit local situations, rather than by the current conservation programmes.

A similar methodology was followed by LTS International (2013) in a multi-catchment analysis of Malawi which compared three stakeholder developed future scenarios based on different land-use priorities and the use of different agricultural practices. Avoided sedimentation costs to downstream users, conservation of soil fertility (and therefore of crop productivity) and carbon sequestration were the main ecosystem services considered. The outputs of the modelling were used for a multisectoral, multi-stakeholder cost–benefit analysis that spanned 20 years and allowed comparison of the socio-economic impacts (in terms of net present value) of the policy choices. One of the strengths of the LTS International (2013) approach is that it is capable of giving valuable insight, despite having the disadvantage (compared with the Boody *et al.* (2005) study) of having to use the limited secondary data and patchy survey data that are typically available for sub-Saharan African countries.

A later study by Claessens *et al.* (2012) is another example of the use of the type of survey, experimental and modelled data that are typically available in countries where semi-subsistence systems are predominant. Here, the Tradeoff Analysis for Multi-Dimensional Impact Assessment (TOA-MD) model was used for two areas in Kenya to simulate scenarios with and without technology adoption and estimate their associated economic, environmental and social outcomes.

11.2.3 Dynamic integration

A framework for the dynamic integration of a catchment-level on-farm model, a watershed model, a farmer-adoption behaviour model and a non-market valuation model was set out by Yang *et al.* (2007). However, the data requirements enumerated in the framework would be hard to fulfil even for a developed country, as evidenced by the lack of publication of further developments of the model after the release of the interim and final reports (Boxall *et al.*, 2008; AAFC, 2010).

In van Paddenburg *et al.* (2012), a macroeconomic system dynamics model, customized from an existing T21 model developed by the Millennium Institute, was coupled to the InVEST (Integrated Valuation of Ecosystem Services and Trade Offs) models for a multi-catchment regional assessment of Borneo. This model, described in Bassi (2012), provides a dynamic integration between the environmental and economic models, and endogenous links both within and across the economic, social and environmental sectors through a variety of feedback loops and temporal delays. Despite this model being an example from a developing country, such complex dynamic integration models, with their large data requirements, may be out of reach for many countries with more incomplete data sets.

An interesting, less data intensive, dynamic integration alternative is the methodology developed in Bekele and Nicklow (2005). In this work, the authors have effectively created a feedback between a spatial watershed-level map of agricultural land uses and practices and the outcomes of the SWAT model. An evolutionary algorithm creates random combinations of land uses and management practices, and then selects them in terms of their performance in a SWAT run. The best performing combinations are retained and 'mated' to create the next generation of combinations. To be 'best performing' combinations, outcomes must be positioned along a Pareto frontier (i.e. they should be equal in terms of welfare outcome as defined by the equation that minimizes sediment, nitrogen and phosphorus, and maximizes gross margins), representing trade-offs between the ecological and economic elements. The new generation of combinations is then selected based on a new SWAT run, for a number of predetermined iterations. The gross margins in the model are also dependent on SWAT-generated crop yield/plant growth estimates (which are, in turn, dependent on the hydrogeological constraints and the agricultural practices of a particular spatial unit) and market prices, adding another layer of feedback. Another example of the use of evolutionary algorithms and water management can be found in Arabi *et al.* (2006).

One of the best advantages of the dynamic integration method is that it allows for the evaluation of spatial interactions between practices, environmental outcomes and farmer revenues, as recommended by Antle and Stoorvogel (2006) and Nelson *et al.* (2009). As mentioned by these authors, though, there would be considerable institutional and policy challenges in implementing the probably spatially complex model solution for the most efficient supply of agricultural ecosystem services in the actual catchment(s). Nevertheless, such hypothetical ecosystem services management maps could provide valuable insights into locally relevant agricultural practices that can affect catchment-level social, economic and environmental outcomes, thereby informing policy.

11.3 Core Model

Although a number of studies have applied partially or fully integrated biophysical environment models with economic farm models at the basin or sub-basin level, the majority of these have been in developed countries, where the data required for running them can be more easily procured or produced. Despite the data limitations, some studies have emerged, as we have seen in the previous section, that link land use in developing countries to hydrogeology and to ecosystem services. A slowly growing number of PWS scheme studies are following their lead, investing in field soil erosion and water measurements and using catchment modelling, in order to strengthen the evidence base for PWS design and its delivery of environmental and social outcomes (Porras *et al.*, 2008).

Using socio-economic and agricultural data and outputs from a SWAT, we constructed a catchment model to try to investigate the levels of adoption of SLM practices by smallholder farmers at different levels of payment for watershed services and the resulting benefits to the downstream water users, taking into account climate change. In developing countries, where erosion is one of the main factors in agricultural land degradation, the adoption of SLM practices by smallholder farmers upstream can not only preserve soil productivity, but also significantly reduce silt-related costs to downstream irrigation, hydropower and urban water users. However, smallholder farmers rarely have the capital to invest in such a switch, which is why PWS could be a potentially effective instrument of land management change in such contexts. Our catchment-model, which aggregates farms at the basin level, is essentially an output-based integrated model, an overview of which can be seen in Fig. 11.1.

To assess how much adoption of SLM practices could be expected at the catchment level with different levels of payments from the watershed service buyers, a linear optimization model (linear programming or LP model) was developed. In this model, the focus was on the costs of soil erosion in terms of: (i) agricultural production lost to the farmers; and (ii) costs to downstream water users due increased siltation of infrastructures.

The use of LP optimization has been reported extensively in the literature for modelling farmers' resource management decisions, as it enables the simultaneous consideration of both current and alternative production activities, and of how a limited set of inputs can be allocated to each activity in order to deliver certain outputs (such as yield) optimally (Janssen and van Ittersum, 2007). LP models are thus versatile tools for short- and long-term exploration of different resource allocations, permitting the evaluation of the effects that changing parameters might have on the model outputs (sensitivity analysis), as well as assessments over a range of geographic and climatic circumstances (Janssen and van Ittersum, 2007). One of the major strengths of LP models is the possibility of identifying trade-offs between economic and ecological objectives, by using environmental effects, such as pollution, as an activity output.

The objective of the catchment LP model is to enable the assessment of the investment needed to trigger significant watershed services provision and whether there are net gains for the users of these services in paying incentives to farmers for the adoption of SLM practices. In the model, the net returns from upstream farmers are maximized by choosing between different land-use options, taking into account input costs (e.g. labour, seeds, capital and fixed assets costs) and the revenue accrued by the outputs (e.g. maize and legumes sold, firewood, timber, non-timber forest products). The outputs and characteristics of the farming activities will differ with slope, as higher slopes will be more prone to soil erosion, thereby: (i) resulting in higher cumulative effects of soil degradation on crop production; and (ii) demanding more

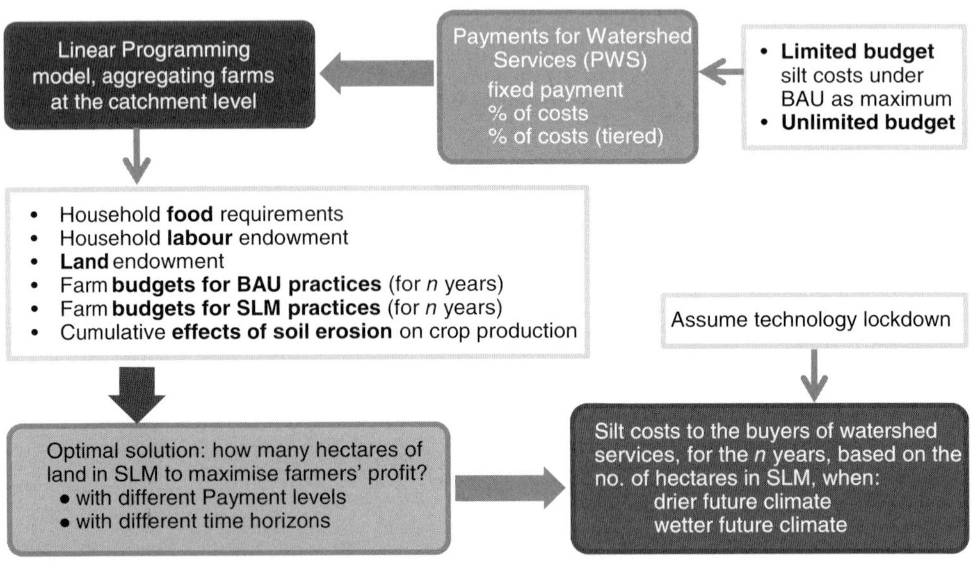

Fig. 11.1. Overview of the catchment-level farm model and how it connects to the addition of a payments for watershed services (PWS) scheme and the calculations of the costs to watershed service buyers. The model was developed for Malawi using data from LTS International (2013). Key: BAU, business as usual; SLM, sustainable land management.

capital and labour investment for effective soil conservation.

Farm budgets need to cover the n years' time period being considered. In this way, any time horizon between 1 and n years can be considered, and we can look at how optimal solutions may differ when different decision time horizons are considered. This is particularly useful for developing countries, as smallholder and subsistence farmers have high rates of time preference, and the benefits of adoption of SLM practices (and costs of soil erosion) take time to come into full effect.

We assumed smallholder farmers to be profit maximizing and that no uptake of SLM practices would take place if the opportunity costs of adoption were not covered by the payments. The optimization decision in the model is therefore:

$$\max \pi_T$$

subject to

$$\pi_t = \sum_a L_a \times \left(R_{a,t} - C_{a,t} + P_{a,t}\right)$$

where π_T is the objective function corresponding to the maximization of the aggregate profit of smallholder farmer households for the time horizon T, subject to the sum of revenue R in any given time period t from activities a minus the sum of costs C, but plus any payment P for those activities made for the watershed services provided by said activities a, multiplied by the number of hectares of land, L, used for those activities.

The optimization decision is constrained by the resources available per farm household. The model can include different farm household typologies, if there is enough information to adequately characterize them in terms of per household average available labour, land and capital endowments, as well as household nutrition requirements. However, developing countries are information scarce, and some broad assumptions will often have to be made.

Payments for watershed services are set in the model so they act primarily as input subsidies, and only act as monetary incentives after the full cost of inputs has been covered. Payments can also be set as a fixed amount, a percentage of the costs covered or a tiered percentage of the costs covered (where a premium percentage cover of costs can be given for higher slopes, for example). These payments will affect the net returns of the farmer households, therefore affecting the optimization solution.

The model is optimized at the catchment level and resources are therefore pooled. For this reason, no option was given for rent of additional land or use of credit, or hire of labour from outside the catchment.

11.3.1 Estimating silt costs to the downstream water users

Direct silt-related costs

The outputs of a SWAT model of the catchment, built using a GIS land use map for the baseline year and a soil type map of the catchment, were used in the analysis. For mid- to long-term periods, silt-cost calculations will depend on future climate assumptions. To account for the uncertainty related to climate predictions, two sets of SWAT outputs were used: one corresponding to the dry and the other to the wet extreme of a selected IPCC (Intergovernmental Panel on Climate Change) scenario. Silt costs to downstream water users were calculated using these outputs to produce linear regressions that related historic cost data for each of the different water users with the change in silt reaching downstream water users, as in the following equation:

$$\Delta c_i = b_i \times \Delta u, \quad i = \{1, 2, 3\}$$

where Δ is the percentual change from one year to the next, u is the sediment produced upstream (in tonnes), c is the cost, b is the specific coefficient for type of cost i, and i represents the type of cost (i.e. desilting, chemical treatments, repairs). With these relationships established, direct silt costs to downstream water users, s, are calculated for each year y of the time horizon of n years as:

$$s = \sum_i \sum_y c_{i,y}$$

$$c_{i,y} = c_{i,(y-1)} \times \left(b_i \times \frac{u_y - u_{(y-1)}}{u_{(y-1)}} \right), \quad i = \{1, 2, 3\}, \quad y = \{1, \ldots, n\}$$

To estimate *u*, the SWAT model was run for the baseline year and the last year of the chosen time horizon using two climate variants: the wet and dry extremes of a chosen IPCC scenario. For comparisons between different time horizons with the *n* years' time frame, yearly values for the intervening years are needed. A linear relationship was assumed between the two yearly soil yield values obtained for the baseline and the last year.[2]

So that for each hectare which had switched management, a simple calculation could be made to estimate the percentual change in sediment reaching downstream water users, a per hectare soil yield average was calculated for land in each of the slope categories under different land uses and under different land management regimes (SLM or business-as-usual, BAU), for a wetter and drier climate, for each year between the baseline and year *n*. Values for *u* were then estimated by multiplying the number of upstream hectares on different slopes and under different land uses and managements by their respective average soil yield per hectare for each year of the *n* year period.

To assess the effect of changing land management in existing farmland due to the activity of the PWS scheme, it was also assumed that land use outside the farmland included in the LP model would not change from the baseline year for this *n* year period.[3]

Indirect silt-related costs

The PWS scheme aims to reduce the direct silt-related costs to downstream users, but implementation has its own costs. These indirect silt-related costs include the payments made for watershed services by downstream users, as well as any transaction costs to downstream users of implementing the PWS scheme (which could potentially be substantial; see Porras *et al.*, 2008, for discussion). However, as the latter are hard to quantify, only the former have been included in the model.

Simulations were conducted under a constrained and an unconstrained budget for payments for watershed services. The value for the constrained budget was set at the maximum estimated silt costs to downstream water users under BAU practices. The unconstrained budget was included as a means of assessing the price effectiveness of the PWS scheme to the downstream water users beyond this maximum (constrained) budget (to assess, for example, the effects of any contributions from non-governmental organizations (NGOs) or the government that could add to the budget for the scheme).

Indirect costs in the model also assume technology lock-in, so that any land that switched to SLM within the simulated decision horizons is assumed to remain so within the remainder of the *n* years' time period without any additional payments needed.

11.4 Application

11.4.1 Case study of Dwangwa catchment in central Malawi

The methodology described was applied to the Dwangwa catchment in central Malawi, for which the necessary key data were available through the LTS International (2013) study. These data included information on the livelihoods of subsistence farmers, the private costs and benefits to farmers of engaging in different land-use practices, and the private costs to downstream water users of dealing with increasing silt loads. SWAT modelling outputs for the period of 2010–2030 under two different climate variants and three land-use scenarios were also available, making it possible to evaluate watershed service changes with changes in land use and in climate.

In the case of the Dwangwa catchment, smallholder maize and tobacco farmers are all upstream of a large downstream estate of irrigated sugarcane farming, owned by Illovo Sugar (Malawi) Ltd. The sugarcane estate would therefore constitute the private buyer of watershed services in the catchment, and the smallholder maize farmers would constitute the service sellers.

To assess the effect of climate change on silt costs and net savings to the private buyer, the time horizon of 20 years (from the 2010 baseline to 2030) used in the LTS International (2013) study was chosen, and two climate variants, corresponding to the dry and wet extremes of the IPCC 2007 A1B climate scenario (LTS International *et al.*, 2013a,b) were adopted.

Shorter time horizons within the 20 year horizon were also used in the simulation, namely: 1, 3 and 5 years after the baseline. Silt costs to the irrigation water user were calculated based on the linear regressions produced from 2007–2011 data for the Illovo operation at Dwangwa Illovo sugarcane operation (LTS International *et al.*, 2013d).

Soil yield values were used to calculate the average amount of sediment produced annually per hectare of land in different slope categories (0–2, 2–9, 9–15, 15–20 and >20%) and under different uses (built, pasture, miombo forest, maize and tobacco farming, etc.) for the 2010 baseline and the 2030 prediction for a wetter and drier climate. In the case of the 'maize' land use, values for soil yield under different land management regimes for maize farming (BAU, conservation agriculture, agroforestry and reforestation of slopes >20%) were also obtained. A linear relationship was assumed between the 2010 and 2030 values.

Simulations were run with a constrained (i.e. set at maximum silt costs under BAU to the sugarcane estate) and an unconstrained budget for payments for watershed services. After the initial run for each of the four time horizons and for both budgets (a total of 76 simulations), the values of payments and the percentages of costs and tiers were modified to improve payment efficiency. In total, 22 extra simulations were run, corresponding to 16 modified tiered payment schemes (three for 1 year and 13 for a 3 year time horizon, with only six run under a constrained budget).

11.4.2 Linear optimization model assumptions

Model activities

Smallholder farmers were given three maize land-use management options per slope category: BAU, conservation agriculture (CA) and agroforestry (AGRF). The higher the slope category, the more capital and labour investment was required for effective soil conservation, with conservation structures taking more land from maize production (CA and AGRF) (LTS International *et al.*, 2013c). An additional land use option, government-subsidized reforestation, was given for the >20% slope category, as agricultural use of such steep areas should be avoided (Bishop, 1995).

Both SLM options for maize (CA and AGRF) were assumed to stop the cumulative effects of soil erosion on maize productivity from the first year of implementation; however, they were also assumed to take 3–5 years to produce improved maize yields compared with the BAU baseline.

Farmers were given the option to use either local maize or a more productive hybrid variety, with the latter being restricted to a maximum of 50% of maize planted,[4] as smallholder farmers are very risk averse and will always plant some of the local variety, which can fare better under variable conditions such as those that are expected for the coming decades (Smale *et al.*, 1995). In the model, hybrid maize demands more labour for harvesting and shelling owing to its higher yield.

Household characteristics

Socio-economic criteria were set in the model in the form of different annual budgets and land endowments for the three wealth categories defined in the Malawi Vulnerability Assessment Committee (MVAC) Livelihood Profiles (MVAC and SADC FANR VAC, 2005) for smallholder farmers: 'poor', 'middle income' and 'better off'. In the model, farmer households need capital to buy seeds, labour and (in the case of SLM options) materials for the time period of the model runs. The budget estimated for use by the poor and middle-income households was what would be that needed to cultivate their plots of land using BAU practices and local maize varieties; the budget estimated for better-off households would cover the BAU cultivation of the more expensive, but more productive, hybrid maize. Note that the basic household budgets of the smallholder farmers (the watershed service sellers/providers) are always constrained, although any payments for services would increase those budgets. Conversely, the sugarcane estate can have either a constrained or an unconstrained budget for paying the smallholder farmers for their watershed services, but the constrained budget is always set at the maximum estimated silt costs to downstream water users under BAU practices.

Following Alwang and Siegel (1999), labour availability and household nutrition constraints were also included in the model. All three household types were assumed to constitute 4.9 people (LTS International, unpublished data) with the average gender and age distribution as described by the Malawi Household Survey of 2010–2011 (NSO, 2012). Adult-equivalent (AE) conversion factors according to age and gender were taken from Claro *et al.* (2010) and per annum consumption requirements in maize and legumes were taken from Alwang and Siegel (1999) and Muwalo *et al.* (1999; cited in Lo Monaco, 2006), respectively.

Slope of farmland

Slope is a determinant factor in the susceptibility of agricultural land to soil erosion in Malawi (Bishop, 1995). Table 11.1 shows the land area in the catchment divided into the five slope categories defined by Bishop (1995). In 2010, around 51% of the catchment area was under tobacco and maize and mixed maize–miombo systems, 27% corresponded to the Kasungu National Park and 6% was occupied by built areas, pasture, wetlands and sugarcane plantations. The remaining 15% corresponded to miombo and evergreen forests, and could potentially be areas of future expansion of farmed land.

Environmental factors affecting smallholder farmers were restricted to the effect that soil erosion has in decreasing maize yields. Maize is the main source of revenue for the smallholder farmers under the model. In the model, soil erosion has a cumulative negative effect on maize productivity, with yield decreasing with increase in slope, and affecting smallholder farmers' revenue (Table 11.2). However, because these effects are felt more strongly with time (e.g. by Year 20, maize yields are expected to decrease by 44.4% on the 0–2% sloping land and by 99.5% on land with a slope >20%), farmer revenue for Year 1 is not affected, and the maize yields for the first 3 years are only affected by a 4.0% reduction in maize yield on 0–2% slopes and a 23.4% reduction on >20% slopes, with those for the first 5 years affected by a 7.9% reduction on 0–2% slopes and a 40.3% reduction on >20% slopes.

This reduction in yield due to soil erosion is not felt in the SLM options, although both conservation agriculture (CA) and agroforestry (AGRF) have lower yields for the first 4 and 3 years, respectively, as improvements from these practices take 5 years to come into full effect (after Babalola *et al.*, 2007).

11.4.3 Results

Simulations for the 5 year and 20 year time horizons resulted in all farmers switching to SLM practices without any payments being necessary. As this indicates a lack of additionality for the private buyer, these simulations were not included in subsequent analyses. By contrast, a 3 year time horizon resulted in no uptake of SLM in the absence of a PWS scheme, and a 1 year time horizon resulted in the uptake of government-subsidized reforestation for 843 ha of land with a slope >20% (corresponding to 19% of the farmland in this slope category). Subsequent simulations for 1 year using increasing fixed payments suggest that this optimal solution results from the allocation of scarce capital and labour resources that would have been used on the >20% sloped land to the more costly, but much more profitable, hybrid maize on less sloped land (Table 11.3).

Payments for watershed services and land under SLM

As shown in Fig. 11.2, the general trend for an unconstrained budget is that higher PWS expenditure will result in more land coming under the scheme, until all land is under SLM (expenditure with PWS can continue to increase after this, with higher payments offered per hectare under SLM, but no further benefit to the private buyer will be derived from it).

This general trend is less clear for the simulations with a constrained budget. With a constrained budget, the trade-offs between getting more land into SLM and obtaining the highest silt reduction are more apparent. It is cheaper to get 1 ha of land with a lower slope to switch to SLM (low costs of soil erosion management), but it is from the higher investment in land with a high slope that the higher reductions in silt costs can be obtained. As a result, some of the simulations with higher expenditure yield lower areas

Table 11.1. Land area in the Dwangwa catchment (2010 baseline). After LTS International (2013). The data on the 2010 baseline farmed land was obtained from the LTS land-use map. The data on the 2010 farmer households' land is the land included in the optimization model.

Land	Total land (ha)	Land in each slope	Protected area (ha)	Built area, pasture, wetlands, sugarcane (ha)	Available land for farming (ha)	Land in each slope	Baseline 2010 farmed land (ha)	Land in each slope	2010 farmer households land (ha)	Land in each slope
0–2% slope	155,567	21%	70,884	10,755	73,929	15%	66,501	17%	24,285	17%
2–9% slope	456,809	60%	135,480	24,212	297,117	59%	263,075	69%	95,792	69%
9–15% slope	67,648	9%	1,057	6,832	59,759	12%	31,665	8%	11,537	8%
15–20% slope	29,551	4%	264	2,277	27,010	5%	10,423	3%	3,791	3%
>20% slope	47,645	6%	438	2,462	44,745	9%	12,233	3%	4,472	3%
Total	757,220		208,123	46,538	502,560		383,897		**139,877**	
	100%		**27%**	**6%**	**66%**		**51%**		**18%**	

Table 11.2. Average yield loss of maize per slope category following the methodology in LTS International *et al.* (2013c).

Slope category	0–2%	2–9%	9–15%	15–20%	>20%
Annual yield decrease	4.30%	8.35%	12.21%	19.18%	26.94%

Table 11.3. Uptake of sustainable land management (SLM) practices and use of hybrid maize in 1 year simulations for three fixed amount Payments for Watershed Services (PWS) schemes.

Payment for watershed services (US$/ha)	Hybrid maize cultivation		Farmland under SLM	
	Area (ha)	Slope	Area (ha)	Slope
0.00	46,345	0–2% (52%); 2–9% (48%)	843[a]	>20% (100%)
1.33	46,349	0–2% (52%); 2–9% (48%)	831	>20% (100%)
10.00	46,372	0–2% (52%); 2–9% (48%)	761	>20% (100%)
33.30	46,626	2–9% (100%)	5,456	0–2% (100%)

[a]No PWS but 843 ha afforested under government-subsidized reforestation scheme.

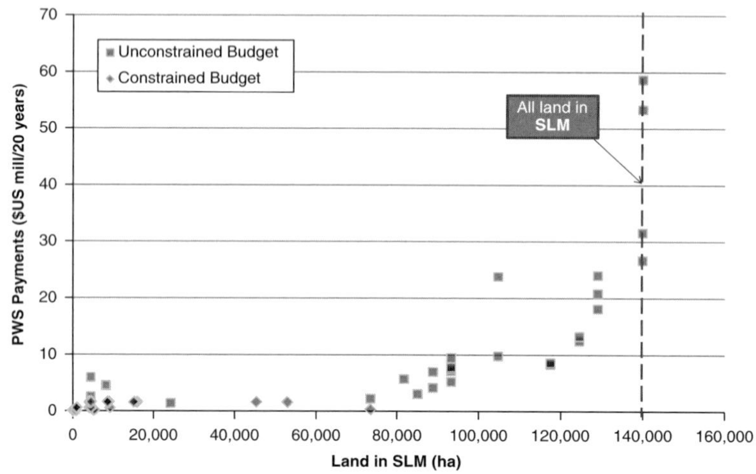

Fig. 11.2. Plot of the payments for watershed services (PWS) made by the private buyer (Illovo sugarcane estate) against the smallholder farmer land that switched to sustainable land management (SLM) as a result of payments. Data points on the graph correspond to either constrained or unconstrained PWS budgets, as simulated by the catchment-level farm model developed for the Dwangwa catchment in Malawi.

under SLM. The best performances for constrained budget simulations show the following clearly:

- If the objective is a reduction in silt costs, the best result was a reduction of 1.97% (dry) and 2.72% (wet), with 8,727 ha in SLM (6.2% of household land). However, this represented a 19.22% (dry) and a 12.43% (wet) increase in total costs for the private buyer.
- If the objective is an increase in SLM area, a maximum of 73,452 ha in SLM (52.5% of household land) was obtained, with a reduction in silt costs of 1.03% (dry) and 2.10% (wet), at the expense of an overall increase in costs of 2.89% (dry) and 0.73% (wet).

Silt costs savings and climate variation

Uncertainty over climate change will affect the reductions in silt costs that the private buyer can expect from changes in land use upstream. Figure 11.3 shows that there is a considerable difference in silt cost reductions for the same level of payments, depending on whether the climate is

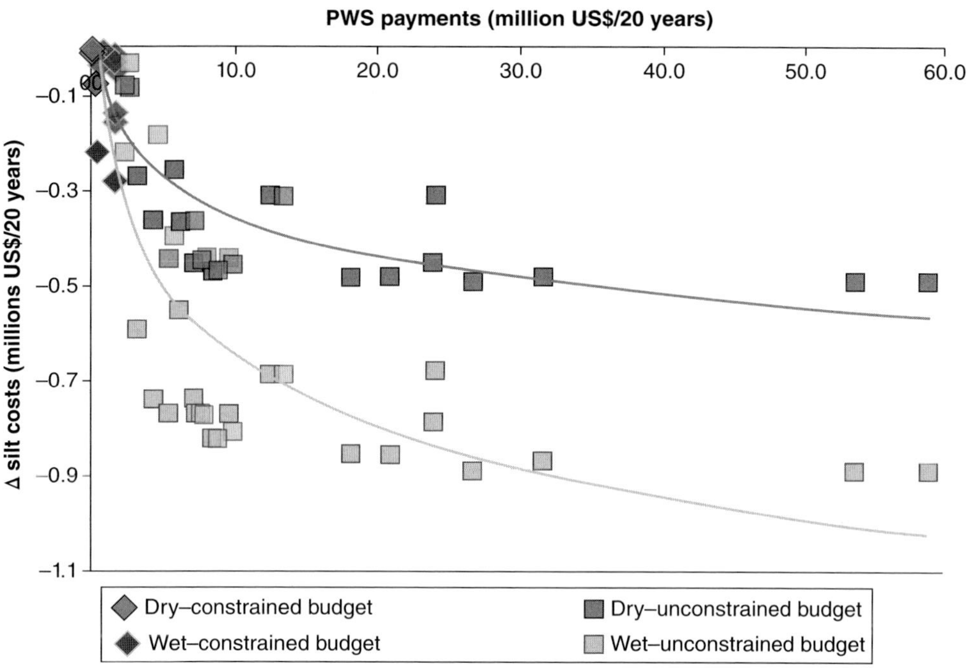

Fig. 11.3. Plot of silt costs savings (Δ or change in silt costs) for the Illovo sugarcane estate under predicted drier or wetter future climates against the payments for watershed services (PWS) made to obtain them. Data points refer to simulations made by the catchment-level farm model developed for the Dwangwa catchment in Malawi using constrained or unconstrained PWS budgets. Zero values were excluded from the logarithmic trend lines.

predicted to be drier or wetter for the 20 year period. This leads to a difference in cost values of 26.9–28.7%.

Notably, for both the wet and dry climates, the private buyer can expect higher silt costs at the beginning of the 20 year period, as even in the wetter climate there is a decrease in the annual amount of silt produced upstream (57% reduction in 2030 compared with the 2010 baseline for the dry climate and 1% reduction in the wetter climate). This means that even the annual silt costs under BAU are expected to be 61% smaller in 2030 compared with those in 2010 in the dry climate and 1% smaller in the wetter climate; this is due to a predicted 55% reduction in water yield for the dry climate and only a 4% increase for the wet climate under BAU.

Silt costs savings from engaging in PWS in perspective

Although silt cost savings can be substantial, they are comparatively small relative to background silt costs and were usually obtained in our simulations at much higher PWS expenditure. In Fig. 11.4, it can be seen that in most simulations, there are no net savings to the private buyer. The payments necessary to get upstream farmers to adopt SLM are a considerable additional cost to the private buyer that is not compensated for by the slim savings that come from the silt costs avoided.

11.4.4 Limitations

The aggregation of farm resources at a catchment level does mean that, despite farm-level constraints being observed, the optimal allocation decision is made regionally rather than at the individual farm level. This means that practices can be allocated interchangeably to any existing farmland plot, regardless of any spatial constraints in the field (such as proximity to villages, accessibility to roads, connectivity to other plots, etc.).

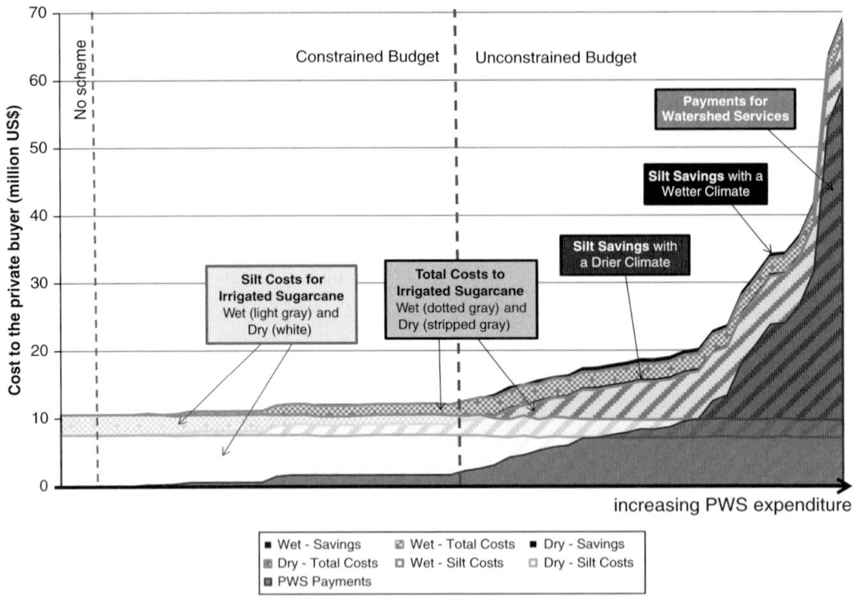

Fig. 11.4. Total costs to the private watershed services buyer (Illovo irrigated sugarcane estate) with constrained or unconstrained payments for watershed services (PWS) budgets under predicted dry and wet future climates. The total costs are the sum of silt-related costs for the sugarcane irrigation systems and the PWS made to obtain sustainable land management (SLM) adoption in upstream farmland, minus the costs avoided due to SLM adoption upstream. The first three data points ('No scheme') correspond to the counterfactual (all farmland under business-as-usual, BAU) and the simulations for 'No scheme' for the baseline year (2010), and for 1 year and 3 years. Data points are ordered in increasing amounts of expenditure on PWS. The simulations were run under the catchment-level farm model developed for the Dwangwa catchment in Malawi.

The assumption of technology lockdown used in buyer-side efficiency calculations means that there can be an underestimation of the payments needed to maintain land use in SLM after the payment contract is finished. Payment values in the model also result from a perfect knowledge of all the costs and all the benefits of the alternatives, and they were generated to create an optimal outcome for the region. However, human beings do not operate under perfect knowledge, nor are they systematic enough to consider all iterations when making judgements on opportunity costs (Landell-Mills and Porras, 2002). It is possible that lower payments can still cause farmers to adopt SLM (e.g. Quintero *et al.*, 2009), thereby increasing efficiency, but it is also possible that service sellers perceive costs to be higher than they are, which would produce the opposite effect.

The inclusion of aspects of seasonality (from the SWAT model) in the LP model used would mean that potential valuable benefits from increased water flow during the dry season could be considered under payment efficiency. This would require seasonal budgets and planting calendars for both upstream farmers and downstream users, such as irrigation farms, and the linking of these to crop growth models. As noted before, this level of detailed information might not be available in developing countries, and indeed it was not available for our case study.

The model also does not consider the transaction costs involved in negotiating a scheme with the large and diverse numbers of households throughout the catchment, nor does it consider the costs of getting the subsidized inputs to those that need them, which, considering the experience of the Malawi national fertilizer subsidy programme (Dorward and Chirwa, 2011), can be considerable.

11.5 Conclusion

11.5.1 Policy implications

Despite the evidence presented here of the potential demand for a PWS scheme in the Dwangwa catchment, there are several considerable constraints to watershed services market development in Malawi, including the essential matter of whether the sugarcane estate owned by Illovo would be a willing buyer, in addition to being an able buyer in terms of financial capital (unlike the smallholder sugarcane farmers, who also stand to benefit from watershed service provision). Nevertheless, Illovo is currently one of the participant parties in the development of a potential Green Water Credits scheme in the Shire catchment in southern Malawi, so three of the obstacles to private sector participation (lack of knowledge of market potential, the perception of PWS as a risky investment and the view that watershed services are a good to be provided solely by the government) have at least been partially overcome for this potential buyer. As for the fourth obstacle, transaction costs, there is the evidence from Malawi's national fertilizer subsidy programme (which has now been running since the mid-1970s, with mixed results) of the many challenges of an inputs subsidy scheme, which a PWS would be. These challenges include the scheme's design, farmers' outreach, logistics, cost-effectiveness of targeting, mid- to long-term sustainability, supply to local retailers and security issues related to the forgery of vouchers (Dorward and Chirwa, 2011). In spite of these problems though, the subsidy fertilizer programme has also afforded Malawi a wealth of experience that could help in the design of PWS schemes and even, possibly, provide some institutional DNA with which to create intermediary bodies that could negotiate contracts, organize the disbursement of subsidized inputs from the service buyers and allocate the materials to village farmer associations or traditional authorities.

None the less, Illovo is not the only one that stands to benefit from smallholder farmers adopting SLM practices. There are sizeable social and economic benefits to be accrued from this change in terms of improved livelihoods and increased food security that may well justify government involvement. This study highlights the fact that even though a prospective buyer may stand to benefit from a PWS scheme, these benefits may not be significant enough for a profit-maximizing company to make the investment in PWS (particularly when the predicted costs of dealing with service deterioration can be considerably smaller). Incentivizing a private sector buyer to participate may depend on it not investing in the full cost of PWS implementation, in which case, foreign donors and NGOs will most likely be necessary allies in supplementing the government's already insufficient funds to make a PWS scheme in Dwangwa a possibility.

11.5.2 General conclusions

Malawi's heavy dependence on rain-fed agriculture and grazing means that improved watershed services management has the potential to positively contribute to improved livelihoods in the country, particularly in the face of climate change and of the pressures that a growing population will exert on the very limited amount of land resources available (LTS International, 2013).

The model highlights a key constraint: the buyer-side payment inefficiency of PWS. Payment inefficiency can be minimized by refining PWS design, researching watershed dynamics at a lower scale to pinpoint where more cost-effective investments lie and, what seems to be particularly significant in this case, giving farmers the support they need to get as quickly as possible to the point where they can sustain these practices autonomously, thereby improving their livelihoods and decreasing their vulnerability to climate change. However, as much as payment inefficiency is a constraint to private buyer participation in PWS, this seems to be a comparatively minor issue in the wider context of the challenges facing PWS creation and implementation in Malawi, such as institutional support and the human capital to effectively reduce transaction costs. Despite this, the potential social, economic and environmental benefits that could accrue from a successful scheme should encourage further research into the matter and Malawi should seek to take advantage of the international momentum of PWS to secure the much needed funds and expertise to achieve it.

This is to say that although the picture painted by the results from the model might appear discouraging for future private-initiative-led PWS scheme development in Malawi, in reviewing some of the issues on the service buyer side, modelling work such as that described above will hopefully contribute to the inclusion, in future PWS scheme proposals, of a research phase that seeks to look into these aspects in more detail, so producing more solid and robust PWS designs that can effectively operate changes in watershed services protection.

Notes

[1] SWAT is a hydrological model developed by USDA to quantify the impact of diverse soil management practices in watersheds. It can model numerous physical processes, of which hydrology, plant growth, sediment and nutrient transport are the most relevant for PWS modelling. SWAT uses the modified universal soil loss equation (MUSLE) to quantify soil erosion according to elevation, soil type and climatic conditions (Bekele and Nicklow, 2005).
[2] Water yield and soil yield values are highly correlated. Water yield outputs from the LTS International (2013) SWAT model at national level (which included the values for the total 15 catchments) for the 20 year time period used in the case study show a linear progression, suggesting that this is a reasonable assumption to make about soil yield in this case.
[3] This assumption was supported, in our case study (see Section 11.4), by the outputs of the LTS International (2013) Markov-Cellular Automata (MCA) model using biophysical and socio-economic data for our chosen time horizon, which did not show considerable change in land use from the baseline during the 20 year time period that was studied.
[4] The area planted with hybrid maize was 40 to 60% of all maize planted in Dwangwa in 2008 (CountrySTAT Malawi, 2008).

References

AAFC (2010) *Watershed Evaluation of Beneficial Management Practices – WEBs. Towards Enhanced Agricultural Landscape Planning. Four-Year Review (2004/5–2007/8). Executive Summary*. WEBS Program, Agriculture and Agri-Foods Canada, Ottawa. Available at: http://www.agr.gc.ca/resources/prod/doc/pdf/WEBs_Report_2004-08Summary_Web_EN_fixed.pdf (accessed 10 March 2016).

Alwang, J. and Siegel, P. (1999) Labor shortages on small landholdings in Malawi: implications for policy reforms. *World Development* 27, 1461–1475.

Antle, J.M. and Stoorvogel, J.J. (2006) Predicting the supply of ecosystem services from agriculture. *American Journal of Agricultural Economics* 88, 1174–1180.

Antle, J.M., Diagana, B., Stoorvogel, J.J. and Valdivia, R.O. (2010) Minimum-data analysis of ecosystem service supply in semi-subsistence agricultural systems. *Australian Journal of Agricultural and Resource Economics* 54, 601–617.

Arabi, M., Govindaraju, R.S. and Hantush, M.M. (2006) Cost-effective allocation of watershed management practices using a genetic algorithm. *Water Resources Research* 42, 1–14.

Babalola, O., Oshunsanya, S.O. and Are, K. (2007) Effects of vetiver grass (*Vetiveria nigritana*) strips, vetiver grass mulch and an organomineral fertilizer on soil, water and nutrient losses and maize (*Zea mays*, L) yields. *Soil and Tillage Research* 96, 6–18.

Bassi, A.M. (2012) *Technical Background Material: Kalimantan System Dynamics Model, Heart of Borneo: Investing in Nature for a Green Economy*. Prepared for the WWF Heart of Borneo Global Initiative, Jakarta. Available at: http://www.hobgreeneconomy.org/downloads/Kalimantan_System_Dynamics_Model_Technical_Background.pdf (accessed 10 March 2016).

Bekele, E.G. and Nicklow, J.W. (2005) Multiobjective management of ecosystem services by integrative watershed modeling and evolutionary algorithms. *Water Resources Research* 41: W10406.

Bennett, G., Carroll, N. and Hamilton, K. (2013) *Charting New Waters: State of Watershed Payments 2012, Ecosystem Marketplace*. Forest Trends, Washington, DC. Available at: http://www.forest-trends.org/documents/files/doc_3308.pdf (accessed 10 March 2016).

Bishop, J. (1995) *The Economics of Soil Degradation: An Illustration of the Change in Productivity Approach to Valuation in Mali and Malawi*. LEEC [Discussion] Paper DP 95-02, Environmental Economics

Programme [formerly London Environmental Economics Centre], International Institute for Environment and Development (IIED), London.

Boody, G., Vondracek, B., Andow, D.A, Krinke, M., Westra, J., Zimmerman, J. and Welle, P. (2005) Multifunctional agriculture in the United States. *Bioscience* 55, 27–38.

Boxall, P., Weber, M., Perger, O., Cutlac, M. and Samarawickrema, A. (2008) *Results from the Farm Behaviour Component of the Integrated Economic–Hydrologic Model for the Watershed Evaluation of Beneficial Management Practices Program: Summary of Phase 1 Progress. Project Report #08-01.* University of Alberta, Edmonton, Canada.

Claessens, L., Stoorvogel, J. and Antle, J. (2010) Exploring the impacts of field interactions on an integrated assessment of terraced crop systems in the Peruvian Andes. *Journal of Land Use Science* 5, 259–275.

Claessens, L., Antle, J.M., Stoorvogel, J.J., Valdivia, R.O., Thornton, P.K. and Herrero, M. (2012) A method for evaluating climate change adaptation strategies for small-scale farmers using survey, experimental and modeled data. *Agricultural Systems* 111, 85–95.

Claro, R.M., Levy, R.B., Bandoni, D.H. and Mondini, L. (2010) Per capita versus adult-equivalent estimates of calorie availability in household budget surveys. *Cadernos de Saúde Pública* 26, 2188–2195.

CountrySTAT Malawi (2008) CountrySTAT Malawi data. Available at: http://www.countrystat.org/home.aspx?c=MWI (accessed 10 March 2016).

Dorward, A. and Chirwa, E. (2011) The Malawi Agricultural Input Subsidy Programme : 2005-6 to 2008-9. *International Journal of Agricultural Sustainability* 9, 232–247.

Emerton, L., Erdenesaikhan, N., De Veen, B., Tsogoo, D., Janchivdorj, L. *et al.* (2009) *The Economic Value of the Upper Tuul Ecosystem*. Mongolia Discussion Papers, East Asia and Pacific Sustainable Development Department, World Bank, Washington, DC. Available at: http://www-wds.worldbank.org/external/default/WDSContentServer/WDSP/IB/2011/07/12/000356161_20110712010137/Rendered/PDF/628210WP0Upper00Box0361494B0PUBLIC0.pdf (accessed 10 March 2016).

Immerzeel, W., Stoorvogel, J. and Antle, J. (2008) Can payments for ecosystem services secure the water tower of Tibet? *Agricultural Systems* 96, 52–63.

Janssen, S. and van Ittersum, M.K. (2007) Assessing farm innovations and responses to policies: a review of bio-economic farm models. *Agricultural Systems* 94, 622–636.

Kramer, R.A., Richter, D.D., Pattanayak, S. and Sharma, N.P. (1997) Ecological and economic analysis of watershed protection in eastern Madagascar. *Journal of Environmental Management* 49, 277–295.

Landell-Mills, N. and Porras, T.I. (2002) *Silver Bullet or Fools' Gold? A Global Review of Markets for Forest Environmental Services and their Impact on the Poor*. Instruments for Sustainable Private Sector Forestry Series, International Institute for Environment and Development (IIED), London.

Lo Monaco, G. (2006) Competitiveness of African pigeonpea exports in international markets. *SAT eJournal (Journal of SAT Agricultural Research)* 2(1). Available at: http://ejournal.icrisat.org/mpii/v2i1/v2i1competitevness.pdf (accessed 10 March 2016).

LTS International (2013) *Land Use Scenario Analysis Task 3 Report: Integrated Assessment of Land Use Options for Climate Change Mitigation & Adaptation*. LTS International, Penicuik, Scotland. Available at: http://www.ltsi.co.uk/malawi-land-use/Main.html (accessed 10 March 2016).

LTS International, HYDROC Consult, Bunda College of Agriculture, University of Edinburgh and Centre for Development Management (Malawi) (2013a) *Mapping Land Cover and Future Land Cover Projections: Scenario Analysis Technical Annex 1 of the Integrated Assessment of Land Use Options in Malawi, IALUO Task 3 Annexes*. LTS International, Penicuik, Scotland. Available at: http://www.ltsi.co.uk/malawi-land-use/Main.html (accessed 10 March 2016).

LTS International, HYDROC Consult, Bunda College of Agriculture, The University of Edinburgh and Centre for Development Management (Malawi) (2013b) *Watershed Modelling Methods and Results: Scenario Analysis Technical Annex 2 of the Integrated Assessment of Land Use Options in Malawi, IALUO Task 3 Annexes*. LTS International, Penicuik, Scotland. Available at: http://www.ltsi.co.uk/malawi-land-use/Main.html (accessed 10 March 2016).

LTS International, HYDROC Consult, Bunda College of Agriculture, The University of Edinburgh and Centre for Development Management (Malawi) (2013c) *Land Use Management Investments Applied in the Modelling: Scenario Analysis Technical Annex 5 of the Integrated Assessment of Land Use Options in Malawi. IALUO Task 3 Annexes*. LTS International, Penicuik, Scotland. Available at: http://www.ltsi.co.uk/malawi-land-use/Main.html (accessed 10 March 2016).

LTS International, HYDROC Consult, Bunda College of Agriculture, The University of Edinburgh and Centre for Development Management (Malawi) (2013d) *Scenario economic analysis and results: Scenario*

Analysis Technical Annex 7 of the Integrated Assessment of Land Use Options in Malawi. LTS International, Penicuik, Scotland. Available at: http://www.ltsi.co.uk/malawi-land-use/Main.html (accessed 10 March 2016).

MVAC and SADC FANR VAC (2005) *Malawi Baseline Livelihood Profiles (Version 1)*. Malawi Vulnerability Assessment Committee and Southern Africa Development Committee Food, Agriculture and Natural Resources Vulnerability Assessment Committee. Available at: http://pdf.usaid.gov/pdf_docs/Pnadg538.pdf (accessed 10 March 2016).

Nelson, E., Mendoza, G., Regetz, J., Polasky, S., Tallis, H. *et al.* (2009) Modeling multiple ecosystem services, biodiversity conservation, commodity production, and tradeoffs at landscape scales. *Frontiers in Ecology and the Environment* 7, 4–11.

NSO (2012) *Third Integrated Household Survey 2010–2011: Household Socio-Economic Characteristics Report*. National Statistical Office, Zomba, Malawi.

Porras, I., Grieg-Gran, M. and Neves, N. (2008) *All That Glitters: A Review of Payments for Watershed Services in Developing Countries*, Natural Resource Issues No. 11, International Institute for Environment and Development, London.

Quintero, M., Estrada, R.D. and García, J. (2006) *A Manual for ECOSAUT: A Model for the Economic, Social, and Environmental Evaluation of Land Use*. Centro Internacional de la Papa (CIP), Lima.

Quintero, M., Wunder, S. and Estrada, R.D. (2009) For services rendered? Modeling hydrology and livelihoods in Andean payments for environmental services schemes. *Forest Ecology and Management* 258, 1871–1880.

Smale, M., Heisey, P.W. and Leathers, H.D. (1995) Maize of the ancestors and modern varieties: the microeconomics of high-yielding variety adoption in Malawi. *Economic Development and Cultural Change* 43, 351–368.

Stanton, T., Echavarria, M., Hamilton, K. and Ott, C. (2010) *State of Watershed Payments: An Emerging Marketplace*. Ecosystem Marketplace, Washington, DC. Available at: http://www.forest-trends.org/documents/files/doc_2438.pdf (accessed 10 March 2016).

Valdivia, R.O., Antle, J.M. and Stoorvogel, J.J. (2012) Coupling the tradeoff analysis model with a market equilibrium model to analyze economic and environmental outcomes of agricultural production systems. *Agricultural Systems* 110, 17–29.

van Paddenburg, A., Bassi, A.M., Buter, E., Cosslett, C.E. and Dean, A. (2012) *Heart of Borneo: Investing in Nature for a Green Economy*. WWF Heart of Borneo Global Initiative, Jakarta. Available at: http://www.hobgreeneconomy.org/en/main-report (accessed 10 March 2016).

Yang, W., Rousseau, A. and Boxall, P. (2007) An integrated economic–hydrologic modeling framework for the watershed evaluation of beneficial management practices. *Journal of Soil and Water Conservation* 62, 423–432.

12 Modelling Food Supply Chains

Cesar Revoredo-Giha*
Land Economy, Environment and Society (LEES) Research Group, Scotland's Rural College (SRUC), Edinburgh, UK

12.1 Introduction

A supply chain is a system of organizations, people, activities, information and resources involved in moving a product or service from supplier to customer. Supply chain activities transform natural resources, raw materials and components into a finished product that is delivered to the end customer. However, within the agricultural economics literature, a food supply chain tends to be synonymous with a food system, and encompasses the processes that describe how food goes from farms to consumers, including the production processes, processing, distribution, consumption and disposal.

The modelling of food supply chains is a topic that can be covered in several disciplines, namely: economics, industrial engineering, operations research and supply chain management, to mention only a few. Each of these disciplines contribute to and analyse the chains from different angles. For instance, supply chain management and operations research have a more microeconomic focus and are interested in how to improve the performance of a set of directly linked upstream and downstream organizations (via the relationships between supplier and customer) in such aspects as the flows of products, services, finances or information. In contrast, the discipline of agricultural economics has studied supply chains from the viewpoint of their implications for public policy, in particular, food policy. Furthermore, the approach taken can be seen as macro in comparison with that of other disciplines because it has encompassed models that cover stages from the production of raw materials (i.e. farming) to the final products reaching consumers and associated with a particular agricultural market,[1] or to an agri-food sector (e.g. the soybean sector).

The modelling of supply chains in agricultural economics can be seen as one of those topics that has come with the broadening of the discipline (Ezekiel, 1937), and the need to cover not just the functioning of agricultural markets but also an understanding of food markets, which require a 'from farm to fork' approach. In fact, Myers *et al.* (2010), in their article entitled 'A Century of Research on Agricultural Markets', state that:

> demands for stronger vertical coordination in the food system as a means of satisfying increasingly diverse consumer preferences are changing the landscape facing food supply chain participants. The consolidation trend in the marketing sector seems inexorable,

*E-mail: cesar.revoredo@sruc.ac.uk

> implying that non-competitive behaviour and its effects will remain high on the research agenda. Similarly, investigating the benefits and distributional impacts of stronger vertical coordination, through contracting and other means, will also remain a fertile area for the future.
>
> (Myers *et al.*, p. 395)

The purpose of this chapter is to review supply chain models within agricultural economics, although it is important to note from the outset that there is not a prototypical supply chain model, and that the level of complexity and final characteristics of the models depend on the elements of the particular supply chain that is being modelled and, of course, the data available. Thus, the chapter will provide an overview of models – from static supply chain models (e.g. displacement models) to dynamic rational expectations models incorporating speculative inventories (dynamic stochastic supply chain models) – all of them having in common that they are partial equilibrium, multi-market models that cover several stages of a supply chain (that is, vertically related markets or activities).

The focus of the chapter is on modelling techniques and not on the analysis of different food industries, a topic that has been covered in the agricultural economics literature since the early 20th century (e.g. Warren and Pearson, 1928, who presented statistical studies on the demand and supply of agricultural commodities). In addition, the chapter will exclude models that comprise several supply chains, such as general or partial equilibrium trade models (e.g. the Global Trade Analysis Project, GTAP or the US Food and Agricultural Policy Research Institute (FAPRI) models). In addition, the discussion will concentrate on supply chain models of homogeneous commodities; examples of models of differentiated products can be found in Kaiser and Suzuki (2006).

The structure of the chapter is as follows. The next section (12.2) gives an overview of reasons for modelling supply chains in agricultural economics, and presents examples of different types of models; it considers marketing margin models, farm to retail price transmission models and market structure models aiming to measure the welfare implications of imperfect competition. Section 12.3 presents a dynamic rational expectation model for a supply chain, which can be considered as an extension of the aforementioned models. The chapter ends with final remarks.

12.2 Overview of Reasons to Model Supply Chains

The modelling of supply chains in agricultural economics has come under a number of research agendas, which have included: understanding food prices and analyses of marketing margins; assessing how shocks on any part of a food chain are transmitted from farming to retailing; and analysing the market structure of a chain to detect the presence of imperfect competition and its impact on welfare. In this section, we review these models, putting an emphasis on the modelling aspects rather than on specific results.

12.2.1 Marketing margins models

The initial research on marketing margins was primarily descriptive and intended to characterize the size of the farm–retail spread for various commodities. Many of these analyses were undertaken at the US Bureau of Agricultural Economics (BAE) in the 1930s (e.g. Waugh, 1934). Beginning around 1960, however, researchers began to undertake more detailed empirical studies with the goal of explaining the structural determinants of marketing margins. An important step was taken by Buse and Brandow (1960), who modelled marketing margins for individual commodities as a function of the average margin across all farm goods, retail price and other variables.

Gardner (1975) formalized the farmer–retail margin within a competitive model based on the Hicksian tradition, which can be found in Muth (1964). His study examined the consequences of competitive equilibrium in product and factor markets for the relationship between farm and retail food prices. The model was to generate quantifiable predictions about how various shifts in the demand for and supply of food would affect the retail–farm price ratio and the farmer's share of retail food expenditures. Gardner's model consisted of a competitive food marketing industry using two factors of production: a purchased agricultural commodity (F) and a marketing input (M) to produce food sold at retail (Q). The marketing industry's production function is given by Equation (1); the demand for the product at retail level is given by $d(\bullet)$, as shown in Equation (2), and this depends on the retail price P^r, as given by Equations (3) and (4) and the demand shifters Ω_d,[2] as given by Equations (5) and (6):

$$Q = f(F, M) \quad (1)$$

$$Q = d(p^r, \Omega_d) \quad (2)$$

$$P^F = P^r \cdot f_F \quad (3)$$

$$P^M = P^r \cdot f_M \quad (4)$$

$$P^F = g(F, \Omega_g) \quad (5)$$

$$P^M = h(M, \Omega_h) \quad (6)$$

where: f_F and f_M are the value of the marginal products of the agricultural commodity and the marketing input, respectively (both Equations (3) and (4) denote the input demands); the input supplies (expressed in an inverse form) for the agricultural commodity and the marketing input are given by $g(\bullet)$ and $h(\bullet)$, where both depend on their respective prices, P^F and P^M, and their shifters $\Omega_i (i = g,h)$.

A similar model to the one above developed by Gardner can also found in Freebairn *et al.* (1982). This model has been used for evaluating the distribution of research gains in multistage production systems, and is given by Equations (7)–(11):

$$F = \alpha_0 - \alpha_1 P^r \quad (7)$$

$$m = \beta_0 + \beta_1 F \quad (8)$$

$$P^r = P^F + m \quad (9)$$

$$F = \gamma_0 + \gamma_1 (P^F - P^i) \quad (10)$$

$$P^i = \delta_0 + \delta_1 F \quad (11)$$

where, F denotes product quantity specified at farm level and farm-level equivalents at the retail level (i.e. there is a constant rate of transformation between the farm and retail product), Equation (7) is the retail demand equation, Equation (8) relates the retail–farm price margin (m) to the farm output, Equation (9) states that the retail price is the farm price plus the margin, Equation (10) denotes the farm supply that depends on the gross margin and P^i is the cost of the non-farm input. Finally, Equation (11) is the inverse supply for the non-farm input.

Holloway (1989) extended the above models by considering a proper processing sector. His model is given by Equations (12)–(21):

$$Q^r = d(P^r, \Omega_d) \quad (12)$$

$$Q^r = \Omega_r h(Q^i, Q^d) \quad (13)$$

$$P^d = P^r \Omega_r h_d (Q^i; Q^d; \Omega_{h_d}) \quad (14)$$

$$P^d = s_d (Q^d; -\Omega_{s_d}) \quad (15)$$

$$P^i = P^r \Omega_r h_i (Q^i; Q^d; \Omega_{h_i}) \quad (16)$$

$$Q^i = \Omega_i g(Q^f, Q^p) \quad (17)$$

$$P^p = P^i \Omega_i g_p (Q^f; Q^p; \Omega_{g_p}) \quad (18)$$

$$P^p = s_p (Q^p; -\Omega_{s_p}) \quad (19)$$

$$P^f = P^i \Omega_i g_f (Q^f; Q^p; \Omega_{g_f}) \quad (20)$$

$$P^f = s_f (Q^f) \quad (21)$$

The model considers a farm sector that supplies a good Q^f to a processing industry, which combines it with quantities of processing services to produce Q^p through a linearly homogeneous technology $g(\bullet)$. The result is an intermediate product Q^i that is supplied to another downstream industry which distributes a retail product Q^r by combining the intermediate good with quantities of distribution services Q^d through a linearly homogeneous technology $h(\bullet)$. P^a, with $a = r, p, f, i$, represents the prices of the goods; $g_b(\bullet)$, with $b = p, f$, and $h_c(\bullet)$, with $c = d, i$, are the marginal products of b and c; $s_j(\bullet)$, with $j = d, f, p$, refers to the inverse supply function for good j; and $D(\bullet)$ represents the retail demand. Finally, the Ωs represent exogenous shifters.

The above model was extended to consider the effect of imperfect competition in intermediate markets (Holloway, 1991). Here, $D(\bullet)$ represents the aggregated demand faced by all the incumbent firms, where P^Q is the price of the good and Q its aggregated quantity, which is produced by n firms producing q_i.

$$Q = D\left(P^Q, \Omega_d\right) \tag{22}$$

$$Q = \sum_{i=1}^{n} q_i \tag{23}$$

$$P^Q\left(1 + \phi_i/\eta\right) = C(\bullet) \tag{24}$$

$$P^Q q_n - C(\bullet) q_n - \kappa = 0 \tag{25}$$

$$F_n = \frac{\partial C(\bullet) Q_n}{\partial P^F} \tag{26}$$

$$M_n = \frac{\partial C(\bullet) Q_n}{\partial P^M} \tag{27}$$

$$F = n \times F_n \tag{28}$$

$$M = n \times M_n \tag{29}$$

$$P^F = g\left(F, \Omega_g\right) \tag{30}$$

$$P^M = h\left(M, \Omega_h\right) \tag{31}$$

Equation (24) is derived from the first order conditions for firm *i*, where ϕ_i is the elasticity of the industry output conjecture by firms *i*, and η is the demand elasticity. Equation (25) is the condition of entry into the market of the last firm (n); $C(\bullet)$ is the firm marginal cost and κ represents fixed costs. Equations (26) and (27) are the factors F (farm) and M (marketing) demand equations. Equations (28) and (29) are the total quantity of the factors demanded (all the n firms demand the same quantity under the assumption of a symmetric equilibrium). Equations (30) and (31) are the inverse factor supply equations, where P^F and P^M are the factor prices and the Ωs are exogenous shifters.

In terms of empirical applications, while the equations from the four models that have been described in this section could be estimated econometrically, for instance equation by equation, by two stages least squares or as a simultaneous system of equations, in practice, the strategies have been to transform the system into one of elasticities and calibrate them based on available estimates.

12.2.2 Price transmission from farmers to consumers

This section reviews models – which are reduced form models – and studies the asymmetry of price transmission along the supply chain, from producers to retailers or to consumers. Some examples of the models described here can be found in Kinnucan and Forker (1987), Azzam (1999) and Vavra and Goodwin (2005).

Economists have long been concerned with the transmission of market shocks through the various stages of the supply chain, or through horizontally related markets. Much of this research has examined the role of price as a mechanism for characterizing the linkages among market levels.

The literature analysing vertical price linkages has concentrated on evaluations of the links between farm, wholesale and retail prices. These vertical price relationships have featured prominently in recent studies as commodity markets have become more highly concentrated at each level and integrated across levels. Vertical price transmission is the focus of this study also, and the literature review below briefly summarizes work in this field. The issue of asymmetric price transmission is taking on renewed prominence due to its potentially important welfare and policy implications. Meyer and von Cramon-Taubadel (2004) observe that a possible implication of asymmetric price transmission is that consumers are not benefiting from a price reduction at the producers' level – or that producers might not benefit from a price increase at the retail level. Thus, under asymmetric price transmission, the distribution of welfare effects across levels and among agents following shocks to a market will be altered relative to the case of symmetric price transmission.

Vertical price relationships are characterized by the magnitude, speed and nature of the adjustments through the supply chain to market shocks that are generated at different stages of the chain. In this context, the underlying links across agents at different levels of activity, from production to consumption, may be summarized in a single set of measures that define the speed and size of the impacts of a shock in prices at one level on the prices up or downstream. For example, if there is a positive shock that induces an upward shock to the farm price, models of price transmission would be used to analyse the size and timing of any impacts on wholesale and retail prices. Alternatively, the impacts on farm-level prices can be evaluated following a shock that started on retail or wholesale prices.

The primary focus of the models of this section has been to analyse vertical price transmission in supply chains. The aspects studied can be restated as four fundamental questions (Vavra and Goodwin, 2005):

- How big is the response at each level due to a shock of a given size at another level (magnitude)?
- Are there significant lags in adjustment (speed)?
- Do adjustments following positive and negative shocks at a certain marketing level exhibit asymmetry (nature)?
- Do adjustments differ depending on whether a shock is transmitted up or down the supply chain (direction)?

It should be noted that asymmetries can occur within any aspect of the adjustment process and they do not necessarily need to be due to market power (e.g. Weldegebriel, 2004). Price transmission might be asymmetric in its speed and magnitude, and could differ depending on whether the price shock is positive or negative and is being transmitted up or down along the chain. Figure 12.1 provides an example of these points, considering only the case of a positive asymmetry. It represents the prices of two related products, namely: an agricultural product (P_F) and a processed product (P_W). It represents a shock in the price of the agricultural product that makes it increase from P_F^0 to P_F' (and remain at that level).

Let us assume that the full price transmission from P_F to P_W happens if P_W increases from P_W^0 to P_W'' (to a point like A). If the increase is to P_W' (to a point B) and the price remains there, then there is an asymmetric price transmission in magnitude of only half of the shock. If, instead, P_W adjusts over time, we have an issue of asymmetry in the speed of the adjustment. From this point, at least two adjustment cases are possible. In this example, there is: (i) a further adjustment of P_W to the point of C, and the price remains at that level (where there is asymmetry in both magnitude and speed); and (ii) P_W increases to point D (P_W''), which

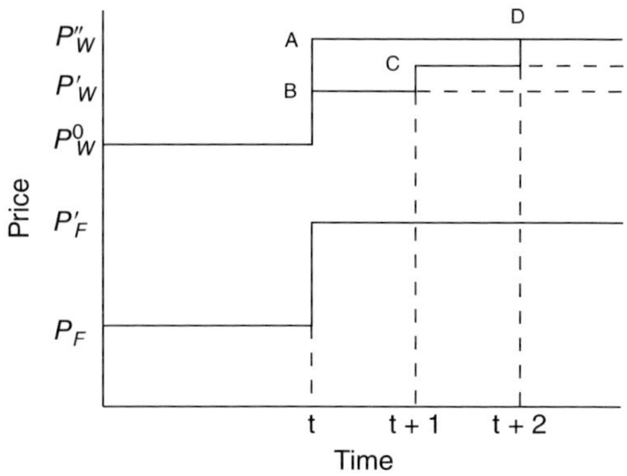

Fig. 12.1. Representation of a positive price asymmetry. The prices of two related products, an agricultural product (P_F) and a processed product (P_W) are represented, showing a shock in the price of the agricultural product at time t (or t + 1, t + 2) that makes it increase from P_F^0 to P_F' (and remain at that level), and the related, but asymmetric, possible increase(s) in the price of the processed product from P_W^0 to P_W' or P_W''. See text for further discussion.

completes the adjustment in period $t + 2$, and there is a temporary case of asymmetry in speed and in magnitude.

There are several models of price transmission along supply chains as they have evolved with the development of time series econometrics. Below, we focus on threshold autoregression (TAR) models (Goodwin and Holt, 1999; Goodwin and Harper, 2000; Goodwin and Piggott, 2001) because TAR encompasses previous contributions (e.g. the use of cointegration to deal with the presence of non-stationarity) and allows for the presence of thresholds in the transmission that may represent the existence of transaction costs.

To start with, consider the following standard linear cointegration relationship given in Equation (32):

$$P_{it} - \alpha - \beta_j P_{jt} - \beta_k P_{kt} = v_t \qquad (32)$$

where, P_{it}, P_{jt} and P_{kt} are three vertically related prices (retail, wholesale and farm level), β_j and β_k are the estimated coefficients and $v_t = \varphi v_{t-1} + u_t$ represents the residual of the relationship (i.e. a deviation from equilibrium). A cointegration relationship between the prices requires v_t to be stationary, implying $|\varphi| < 1$.

Balke and Fomby (1997) extended the case shown in Equation (32) to that where v_t follows a threshold autoregression. For the case of a threshold autoregression with three regimes, v_t is expressed as in Equation (33):

$$v_t = \varphi^{(i)} + u_t \tag{33}$$

where

$$\varphi^{(i)} = \begin{cases} \varphi^{(1)}\ if -\infty < v_{t-d} \leq c_1 \\ \varphi^{(2)}\ if\ c_1 < v_{t-d} \leq c_2 \\ \varphi^{(3)}\ if\ c_2 < v_{t-d} \leq +\infty \end{cases} \tag{34}$$

where, c_1 and c_2 represent the threshold parameters that delineate the different regimes, and v_{t-d} represents the variable relevant to the threshold behaviour (often referred as the 'forcing variable'). In most empirical applications, d is assumed to be equal to one (Goodwin and Holt, 1999), though this is a restriction that can be empirically tested within the threshold model estimation framework. The vector error correction representation of the threshold model is given by:

$$\Delta \mathbf{p}_t = \begin{cases} \sum_{i=1}^{l} \mathbf{b}_i^{(1)} \Delta \mathbf{p}_{t-i} + \gamma^{(1)} v_{t-1} + e_t^{(1)} \text{ if} -\infty < v_{t-d} \leq c_1 \\ \sum_{i=1}^{l} \mathbf{b}_i^{(2)} \Delta \mathbf{p}_{t-i} + \gamma^{(2)} v_{t-1} + e_t^{(2)} \text{ if}\ c_1 < v_{t-d} \leq c_2 \\ \sum_{i=1}^{l} \mathbf{b}_i^{(3)} \Delta \mathbf{p}_{t-i} + \gamma^{(3)} v_{t-1} + e_t^{(3)} \text{ if}\ c_2 < v_{t-d} \leq +\infty \end{cases} \tag{35}$$

where $\mathbf{P}_t$ is the vector of prices being analysed and $\mathbf{b}_i$ and γ are the vectors of the parameters to be estimated.

As pointed out by Vavra and Goodwin (2005), two methods have been proposed for estimating threshold models. Both approaches involve computationally intensive grid searches for the optimal thresholds, but they differ in their definition of the optimum. Although the methods are entirely equivalent in the case of a single variable, they may produce different threshold estimates when a group of two or more variables is considered. The first method consists of searching for the minimum value of the system regression sum of squared errors (SSE). The second method minimizes the logarithm of the determinant of the covariance matrix for the residuals. This latter approach is analogous to maximum likelihood (ML) estimation and incorporates covariance terms as well as the trace of the residual covariance matrix. In both cases, it is important to test the statistical significance of the differences in parameters across alternative regimes of price transmission. This is done using the test proposed by Hansen (1997), which uses a Chow-type test of the significance of threshold effects once these have been identified. However, the test statistic has a non-standard distribution, so simulation methods are used to approximate the asymptotic null distribution and identify appropriate critical values.

12.2.3 Market structure models

The models in this section are focused on testing the degree of competition of agents along a supply chain. Thus, the section will consider modelling vertical integration from farmers to consumers following Kaiser and Suzuki (2006).

Let us consider that there are four key players along the supply chain, farmers, processors, retailers and consumers, which have either oligopoly power (where the market has only a few sellers) or oligopsony power (where the market has only a few buyers). If a group has no market power, it is a price taker; otherwise, it is a price setter. There are many market structure possibilities along the supply chain. Here, we consider the notation due to Kaiser and Suzuki (2006), which is given in Fig. 12.2. This indicates the degree of oligopoly or oligopsony power in the system and designates price takers or setters at different points along the chain by the letters a, b, c, d, e or f.

The starting point in oligopoly models is the profit $\Pi_i(\bullet)$ maximization problem of the firm, which is given by:

$$\underset{q_i}{\text{Max}}\ \Pi_i(q_i) = P\left[q_i + \sum_{j \neq i} q_j(q_i)\right] q_i - c_i(q_i) \tag{36}$$

where, q_i is the output of firm i, $P[\bullet]$ is the inverse price function, which depends on q_i, and the total output of all the competitors firms given by: $\sum_{j \neq i} q_j(q_i)$. Note that the formulation in Equation (36) considers the reaction of the total output of the other firms to changes in the output of firm i. The cost function of firm i is given by $c_i(q_i)$.

[Farmers] => <=[Processors]=> <=[Retailers] => <=[Consumers]
(a) (b) (c) (d) (e) (f)

Fig. 12.2. Relationships along the supply chain. The symbol => indicates the degree of oligopoly power and the symbol <= indicates the degree of oligopsony power in the system. If a group has no market power, it is a price taker, otherwise it is a price setter; the groups are represented by the letters a, b, c, d, e and f. From Suzuki and Kaiser, 2006, p. 23.

In the case of the oligopsony models, assuming a one-to-one relationship between the input product and the transformed product (e.g. between raw material and the processed products or the processed product and the product at the retail stage), the profit maximization problem is given by:

$$\underset{q_i}{\text{Max}}\ \Pi_i(q_i) = P \times q_i - W\left[q_i + \sum_{j \neq i} q_j(q_i)\right] q_i \qquad (37)$$

where $W[\bullet]$ is the inverse input supply function that depends on the quantity demanded of the input. In Equation (37), P is the price of the transformed product.

In the case of a firm that has both oligopoly power and oligopsony power, the problem becomes:

$$\underset{q_i}{\text{Max}}\ \Pi_i(q_i) = P\left[q_i + \sum_{j \neq i} q_j(q_i)\right] \times q_i - W\left[q_i + \sum_{j \neq i} q_j(q_i)\right] q_i \qquad (38)$$

A variety of models can be derived considering the cases described by Equations (36)–(38) in combination for the different stakeholders of the supply chain represented by Fig. 12.2. Some representatives of these are reviewed below.

Processors' oligopoly

In these models, (c) is the price setter and (d) the price taker (see Fig. 12.2). Examples of this case can be seen in Appelbaum (1982) and Bhuyan and Lopez (1997), who produced studies estimating welfare losses generated by oligopoly for each sector of the US food-processing industry. Using the maximization in Equation (36), they obtained the first-order condition which, when expressed in terms of elasticities, gives:

$$\frac{(P - \text{MC})}{P} = \frac{\theta}{e} \qquad (39)$$

where, MC is the marginal cost, θ is the conduct parameter (which takes the value of zero in the case of perfect competition and one in the case of monopoly) and e is the absolute value of the demand elasticity.

Processors' oligopsony

In processors' oligopsony power over raw materials, (a) is the price taker, (b) the price setter, (c) the price setter and (d) the price taker. Azzam and Pagoulatos (1990) developed the basic model, which is a generalization of the monopoly/monopsony model to oligopoly/oligopsony. The first-order condition of the model is expressed as:

$$P\left(1 - \frac{\theta}{e_d}\right)\left(\frac{\partial Q}{\partial L}\right) = w\left(1 + \frac{\lambda}{e_s}\right) \qquad (40)$$

where, P is product price, w is input price, Q is product quantity, L is input quantity, e_d is price elasticity of demand in absolute terms, e_s is the inverse of elasticity of marginal cost with respect to supply, θ is conjectural elasticity (a market-power parameter) in the input market and λ is conjectural elasticity (a market-power parameter) in the product market. On the left-hand side of Equation (40), the expression $P\left(1-\frac{\theta}{e_d}\right)$ is the perceived marginal revenue from the processed product, $\left(\frac{\partial Q}{\partial L}\right)$ is the marginal product generated from the input, and so $P\left(1-\frac{\theta}{e_d}\right)\left(\frac{\partial Q}{\partial L}\right)$ is the marginal product value generated from the input.

The term $w\left(1+\frac{\lambda}{e_s}\right)$ on the right-hand side of Equation (40) is the marginal expense to buy the input.

Values for θ and λ can be obtained from (simultaneous) estimation of the demand and cost functions and Equation (40).

Farmers' cooperatives market power

In these models, (a) is the price setter and (b) the price taker. Although farmers individually can be considered to be price takers, when organized as cooperatives they can gain market power. Therefore, farmer cooperatives naturally create an imperfectly competitive market by controlling production quantities and/or the amounts sold. If farmers have two or more separate markets, such as fresh and processing or private and government, they may exercise price-discriminating marketing by reducing sales to the price-inelastic market.

When both production and selling controls are implemented, we can use a generalization of the discriminated monopoly model, which can be expressed as in Equation (41):

$$P_1\left(1-\frac{\theta_1}{e_1}\right)=P_2\left(1+\frac{\theta_2}{e_2}\right)=\text{MC} \tag{41}$$

where, P is price, MC is the marginal cost, e is price elasticity of demand, θ is conjectural elasticity and the subscripts 1 and 2 stand for the two separate markets. Values for θ_1 and θ_2 can be obtained from (simultaneous) estimation of the demand and cost functions and Equation (41).

Successive oligopoly and oligopsony models covering the whole supply chain

Sexton and Zhang (2001) (see also Kaiser and Suzuki, 2006) discussed welfare losses caused by successive oligopoly and oligopsony models covering the whole food system from farmer to consumer. They assumed that farmers and consumers are always price takers. They then presented two alternative models for the entire food system.

In the first model, they assumed that (a) is a price taker, (b) is a price setter, (c) is a price setter, (d) is a price taker, (e) is a price setter and (f) is a price taker. The successive oligopoly model used by Sexton and Zhang is as follows. The farmers' inverse supply function is given by $P^f = S(Q)$. The processors' marginal expense to raw materials plus unit processing costs is equal to the marginal revenue from sales to retailers given by Equation (42):

$$P^f\left(1+\frac{\theta^f}{e^f}\right)+c^w=P^w\left(1-\frac{\lambda^W}{\eta^W}\right) \tag{42}$$

where P^f is the price paid to farmers; e^f is the market price elasticity of supply of the farm product; θ^f and λ^W are conjectural elasticities defined as $\theta^f=\left(\frac{\partial Q^f}{\partial q}\right)\left(\frac{q}{Q^f}\right)$ and $\lambda^W=\left(\frac{\partial Q^W}{\partial q}\right)\left(\frac{q}{Q^W}\right)$; and η^W is the absolute value of the market price elasticity of derived demand for the processed product when the retailer behaves competitively. The values of θ^f and λ^W, which are between zero and one, measure the degree of market power of processors when buying from farmers and selling to processors, respectively.

The purchase price paid by supermarkets to processors (P^W), plus unit retailing costs (c^R), is equal to the marginal revenue from retail sales, $P^R\left(1-\frac{\lambda^R}{\eta^R}\right)$, and is given by Equation (43):

$$P^W+c^R=P^R\left(1-\frac{\lambda^R}{\eta^R}\right) \tag{43}$$

where P^R is the retail price, λ^R is a conjectural elasticity (between zero and one and measuring the degree of market power of retailers when selling to consumers) and η^R is the absolute value of the market price elasticity of consumers' demand. The price takers' inverse demand function of consumers is given by $P^R = D(Q)$.

In the second model, Sexton and Zhang assumed that (a) is a price taker, (b) is a price setter, (c) is a price taker, (d) is a price setter, (e) is a price setter and (f) is a price taker. In this model, the farmers' inverse supply function is given by $P^f = S(Q)$. The processors' marginal expense to raw materials plus unit processing costs is equal to the wholesale price to retailers and is given by Equation (44):

$$P^f\left(1+\frac{\theta^f}{e^f}\right)+c^w=P^w \tag{44}$$

The supermarkets' marginal expense to the product from processors plus unit retailing costs is equal to marginal revenue from retail sales shown in Equation (45):

$$P^W\left(1+\frac{\theta^W}{\eta^W}\right)+c^R=P^R\left(1-\frac{\lambda^R}{\eta^R}\right) \tag{45}$$

The consumers' inverse demand function is given by $P^R = D(Q)$.

Bilateral oligopoly between cooperatives and processors and between processors and retailers

In this model, (a), (b), (c) and (d) are all price setters. When oligopsonistic processors face oligopolistic farmer cooperatives instead of price-taking individual farmers, one has to consider a bilateral oligopoly model. Similarly, when modelling the relationship between processors and retailers, a typical assumption is that one of the two is a price taker. In the past, retailers were often assumed to be price takers when processors' oligopoly power is estimated. However, that assumption is questionable due to the growth of supermarkets. The following model follows Azzam (1996), who considers the case of bilateral oligopoly between packers and retailers in the US beef market.

On the market for the farmer product, the model assumes that farmer cooperatives have two markets (the private market and the government market) and that the government's purchase price is given; given these conditions, the necessary condition for optimal allocation between the two markets that will maximize sales is given by Equation (46):

$$P^f\left(1-\frac{\theta^f}{\eta^f}\right)=P^G \quad (46)$$

where P^f is the price paid by processors to cooperatives; P^G is the exogenously given government purchase price; η^f is the price elasticity of the processor demand that cooperatives face (in absolute value); and θ^f is a parameter representing the cooperatives' degree of horizontal competition and ranges from zero to one. Equation (46) assumes that processors have no vertical market power. Therefore, the price determined by Equation (46) means that the price cooperatives or farmers receive (P^{fU}) based on their degree of horizontal competition (θ^f) is at the maximum level. P^{fU} can be rewritten as Equation (47):

$$P^{fU}=\frac{P^G}{\left(1-\frac{\theta^f}{\eta^f}\right)} \quad (47)$$

The next step is to consider the processor level of the industry. The processors' profit-maximizing condition can be expressed as in Equation (48):

$$P^{WU}\left(1-\frac{\theta^W}{\eta^W}\right)=P^{fL}\left(1+\frac{\lambda^W}{\varepsilon^W}\right)+\mathrm{MC}^W \quad (48)$$

where P^{WU} is the maximum or upper limit price that processors can receive from retailers when retailers have no vertical market power relative to processors; P^{fL} is the minimum or lower limit price processors will pay to cooperatives when cooperatives have no vertical market power relative to processors; MC^W is the marginal processing and selling costs other than the raw material; η^W is the price elasticity of the retailer demand that processors face (in absolute value); θ^W is a parameter representing the processors' degree of horizontal competition in the processor versus retailer market (ranging from zero to one); ε^W is the price elasticity of the cooperative supply that processors face; and λ^W is a parameter representing the processors' degree of horizontal competition in the processor versus retailer market (ranging from zero to one). The minimum (lower limit) price that processors can pay to cooperatives (P^{fL}) is given by Equation (49):

$$P^{fL}=\frac{\left[P^{WL}\left(1-\frac{\theta^W}{\eta^W}\right)-\mathrm{MC}^W\right]}{\left(1+\frac{\lambda^W}{\varepsilon^W}\right)} \quad (49)$$

Note that the actual price (P^f) is determined by negotiation between cooperatives and processors, and would be somewhere between P^{fU} (the upper limit) and P^{fL} (the lower limit); this is $P^f=W^fP^{fU}+\left(1-W^f\right)P^{fL}$, and is given in full in Equation (50):

$$P^f=W^f\frac{PG}{\left(1-\frac{\theta^f}{\eta^f}\right)}+\left(1-W^f\right)\frac{\left[P^{WU}\left(1-\frac{\theta^W}{\eta^W}\right)-\mathrm{MC}^W\right]}{\left(1+\frac{\lambda^W}{\varepsilon^W}\right)} \quad (50)$$

where W^f is the parameter (ranging from zero to one) for the degree of vertical market power of cooperatives relative to processors and $(1-W^f)$ is the parameter for the degree of vertical market power of processors relative to cooperatives.

The next step is to consider the retailer stage of the market. The retailers' profit-maximizing condition can be expressed by Equation (51):

$$P^R\left(1-\frac{\theta^R}{\eta^R}\right)=P^{WD}\left(1+\frac{\lambda^R}{\varepsilon^R}\right)+\mathrm{MC}^R \quad (51)$$

where P^R is the actual retail price or the maximum (upper limit) that price retailers can get from consumers because consumers probably have no market power; P^{WD} is the lower limit price that retailers pay to processors assuming that processors have no vertical market power relative to retailers; MC^R is the marginal retailing costs; η^R is the price elasticity of consumers' demand (in absolute value); θ^R is the retailers' degree-of-horizontal-competition parameter in the retailer versus consumer market (ranging from zero to one); ε^R is the price elasticity of the processors' supply; and λ^R is the retailers' degree-of-horizontal-competition parameter in the retailer versus processor market (ranging from zero to one).

The lower limit price that retailers can pay to processors (P^{WL}) is given by Equation (52):

$$P^{WL} = \frac{\left[P^{RL}\left(1-\frac{\theta^R}{\eta^R}\right)-MC^R\right]}{\left(1+\frac{\lambda^R}{\varepsilon^R}\right)} \quad (52)$$

The maximum or upper limit price that processors can obtain from retailers (P^{WU}) is given by Equation (53):

$$P^{WU} = \frac{\left[P^{fL}\left(1+\frac{\lambda^W}{\varepsilon^W}\right)+MC^W\right]}{\left(1-\frac{\theta^W}{\eta^W}\right)} \quad (53)$$

The actual price, P^W, determined by transactions between processors and retailers, $P^W = W^W P^{WU} + (1-W^W)P^{WL}$, is somewhere between P^{WU} and P^{WL}, as in Equation (54):

$$P^W = W^W \frac{\left[P^{fL}\left(1+\frac{\lambda^W}{\varepsilon^W}\right)+MC^W\right]}{\left(1-\frac{\theta^W}{\eta^W}\right)} + (1-W^W)\frac{\left[P^R\left(1-\frac{\theta^R}{\eta^R}\right)-MC^R\right]}{\left(1+\frac{\lambda^R}{\varepsilon^R}\right)} \quad (54)$$

where W^W is the degree of vertical market power of processors against retailers (ranging from zero to one) and $(1-W^W)$ is the degree of vertical market power of retailers against processors.

Equations (50) and (54) have some common parameters. Therefore, simultaneous estimation of those equations with parameter constraints provides for all values the degree of vertical and horizontal competition among cooperatives, processors, and retailers (W^f, W^W, θ^f, θ^W, θ^R, λ^W and λ^R). However, the variables P^{WU} in Equation (50) and P^{fL} in Equation (54) are unobservable, so this approach is not a practical way to estimate these vertical and horizontal competition parameters.

Using some assumptions, Equations (50) and (54) can also be treated separately. For instance, consider the case of bilateral oligopoly between processors and retailers by adjusting Equation (54). Assuming that processors have full vertical market power against cooperatives, P^{fL} can be replaced by the actual P^f in Equation (50). When it can be assumed that processors have full vertical market power against retailers, then the term $P^{WU}\left(1-\frac{\theta^W}{\eta^W}\right)$ can be replaced by the actual P^W in Equation (49). In addition, it is possible to obtain price elasticities of supply and demand from other sources, such as Azzam (1996). Using these methods, it is possible to identify the competition parameters.

12.3 Extension: a Supply Chain Model Considering Inventories

The following application aims to extend the supply chain models to consider speculative and processors' inventories, which will render the supply chain models stochastic and dynamic. An example of this type of model can be found in Carter and Revoredo-Giha (2009) and it has its basis in the commodity storage model literature (e.g. Williams and Wright, 1991).

Let us assume that the output (Q_t, measured as the shipments of the processed good, see Ramey, 1989) of a competitive processing industry is represented by a quasi-fixed proportions production function,[3] i.e. $Q_t = \min\left\{\frac{I_t}{\lambda}, f(K_t)\right\}$, where λ is a parameter of the production function (i.e. turnover inventory), I_t is the processors' raw material inventories, K_t is a composite index of other production factors and $f(\bullet)$ is an increasing function that relates other factors with output. Under this assumption, risk neutral processors that maximize the expected profits ($E[\pi]$) at period t solve the following problem:

$$\max_{I_t, K_t} E[\pi] = E\left[P_t \min\left(\frac{I_t}{\lambda}, f(K_t)\right) - m_t I_t - w_t K_t\right] \quad (55)$$

where P_t is the processed product price (i.e. P_t^P) net of the raw material price p_{ct}, m_t is the price of raw material inventories (which has been defined slightly differently from Ramey (1989), as $m_t = (1+r)(p_{ct}+ko) - p_{ct+1}$), where r is the interest rate and ko is the storage cost), and w_t is the price of the composite factors of production. Cost minimization implies that $Q_t = I_t/\lambda$ and $Q_t = f(K_t)$. The expected cost function is then given by Equation (56):

$$E\left[C(m_t, w_t, Q_t)\right] = E[m_t]\lambda Q_t + w_t f^{-1}(Q_t) \qquad (56)$$

Now let us assume (for the purpose of obtaining the supply of storage function developed below) that $f^{-1}(Q_t) = Q_t(\ln(Q_t)-1)$. Thus, replacing the expected cost function in the profit function given in Equation (55) and maximizing with respect to the output, we obtain expressions for the output Q_t and the processors' raw material inventories I_t that depend on the expected price for the commodity:

$$Q_t = \exp\left\{\frac{P_t - \lambda\{(1+r)(p_{ct}+ko) - E[p_{ct+1}]\}}{w_t}\right\}$$
$$I_t = \lambda\exp\left\{\frac{P_t - \lambda\{(1+r)(p_{ct}+ko) - E[p_{ct+1}]\}}{w_t}\right\} \qquad (57)$$

12.3.1 Derivation of the supply of storage equation

Let us start from the storage cost function used by Miranda and Glauber (1993) and Miranda and Rui (1996). The function is written in Equation (58) as:

$$\frac{E\left[p_{c,t+1}\right]}{(1+r)} - p_{c,t} = \theta_0 + \theta_1\ln(I_t) \qquad (58)$$

Equation (58) is a reduced form for particular parameters of the processors' demand for inventories I_t for a quasi-fixed proportions production function. To show this, let us write the processors' demand for the inventories shown in Equation (57) as in Equation (59), and let us call $P_t^* = P_t/\lambda$.

$$I_t = \lambda\exp\left[\frac{\lambda}{w_t}\left[P_t^* - \left\{\begin{matrix}(1+r)(p_{c,t}+ko)\\ -E\left[p_{c,t+1}\right]\end{matrix}\right\}\right]\right] \qquad (59)$$

Let us simplify Equation (58) by introducing the parameters β_0 and β_1, where:

$$\beta_0 = \lambda$$
$$\beta_1 = \frac{\lambda}{w_t}$$
$$I_t = \beta_0\exp\{\beta_1[P_t^* - \{(1+r)(p_{c,t}+ko) - E[p_{c,t+1}]\}]\} \qquad (60)$$

Taking natural logarithms to both sides and factoring terms, we obtain Equation (61):

$$\frac{E\left[p_{c,t+1}\right]}{(1+r)} - p_{c,t} = ko - \frac{P_t^*}{(1+r)} - \frac{1}{(1+r)\beta_1}\ln(\beta_0) + \frac{1}{(1+r)\beta_1}\ln(I_t) \qquad (61)$$

Rewriting Equation (61) in terms of the parameters θ_0 and θ_1, we obtain the storage cost function, i.e. Equation (58). The values for the parameters θ_0 and θ_1 are given in Equation (62):

$$\theta_0 \equiv ko - \frac{P_t^*}{(1+r)} - \frac{1}{(1+r)\beta_1}\ln(\beta_0) \equiv ko - \theta_1\left(\frac{P_t}{w_t} + \ln(\beta_0)\right)$$
$$\theta_1 \equiv \frac{1}{(1+r)\beta_1} \qquad (62)$$

Therefore, the commodity model presented by Miranda and Glauber (1993), and Miranda and Rui (1996), represents only the inventories carried by processors, excluding speculative stockholding. The requirements for considering θ_0 and θ_1 as parameters are: (i) the relative price of the processed good (net of the price of the raw material) with respect to the price of other factors of production must be constant; (ii) the prices of other factors of production must be constant; and (iii) the period considered in the model has be relatively short to preclude technological change.

12.3.2 Econometric estimates

Table 12.1 compares estimates from four versions of the commodity storage model. All of the models have been estimated using the

Table 12.1. Parameter estimates (with standard errors) and pseudo maximum likelihood functions (PMLE).[a]

Model parameters	Commodities											
	Cocoa	Coffee	Copper	Cotton	Jute	Maize	Palm Oil	Rice	Sugar	Tea	Tin	Wheat
Model I - Deaton and Laroque (1992, 1995)												
a	0.160	0.260	0.540	0.640	0.570	0.630	0.460	0.600	0.640	0.480	0.260	0.720
	0.010	0.020	0.040	0.040	0.030	0.040	0.050	0.030	0.050	0.020	0.040	0.040
b	−0.220	−0.160	−0.330	−0.310	−0.360	−0.640	−0.430	−0.340	−0.630	−0.210	−0.170	−0.390
	0.030	0.030	0.050	0.040	0.060	0.150	0.060	0.030	0.060	0.020	0.050	0.030
g	0.120	0.140	0.070	0.170	0.100	0.060	0.060	0.150	0.180	0.120	0.150	0.130
	0.040	0.020	0.020	0.030	0.050	0.030	0.030	0.040	0.030	0.030	0.050	0.030
PMLE	125.2	111.0	73.9	29.8	44.8	32.1	22.2	26.0	−10.7	69.3	108.9	24.6
Model II – Miranda and Rui (1996)[b]												
a	0.185	0.224	0.647	0.600	0.575	0.743	0.452	0.603	0.449	0.494	0.217	0.381
	0.033	0.025	0.048	0.097	0.049	0.134	0.139	0.071	0.117	0.029	0.030	0.047
b	−0.330	−0.217	−0.654	−0.939	−0.427	−1.413	−1.155	−0.611	−0.923	−0.244	−0.289	−0.697
	0.070	0.049	0.142	0.357	0.077	1.173	0.305	0.112	0.170	0.025	0.036	0.057
k_1	−0.029	−0.018	−0.041	−0.113	−0.038	−0.073	−0.087	−0.165	–[e]	–[e]	−0.029	−0.011
	0.013	0.027	0.007	0.051	0.030	0.026	0.014	0.050	–	–	0.009	0.003
k_2	0.023	0.025	0.028	0.058	0.051	0.043	0.047	0.122	0.013	0.022	0.019	0.002
	0.010	0.022	0.003	0.027	0.023	0.026	0.007	0.031	0.010	0.003	0.005	0.001
PMLE	134.4	132.3	97.4	79.9	59.2	48.1	74.4	64.7	−2.1	78.3	160.5	33.6
Model III[c]												
a	0.142	0.255	0.552	0.681	0.634	0.690	0.595	0.768	0.545	0.510	0.397	0.757
	0.027	0.031	0.048	0.063	0.039	0.100	0.608	0.064	0.118	0.019	0.036	0.042
b	−0.234	−0.240	−0.376	−0.351	−0.348	−0.602	−0.686	−0.392	−0.758	−0.196	−0.430	−0.348
	0.036	0.052	0.066	0.051	0.053	0.107	0.245	0.077	0.116	0.023	0.116	0.032
ko	0.003	0.008	0.011	0.030	0.033	0.017	0.013	0.048	0.029	0.041	0.010	0.045
	0.002	0.003	0.004	0.008	0.012	0.008	0.036	0.022	0.016	0.010	0.003	0.007
PMLE	132.5	132.1	92.3	46.3	52.9	37.7	54.7	36.1	−4.2	75.9	144.0	26.2

Model IV[d]												
a	0.152	0.222	0.463	0.758	0.575	0.743	0.478	0.523	0.469	0.519	0.212	0.629
	0.024	0.036	0.071	0.078	0.047	0.135	0.212	0.157	0.239	0.024	0.028	0.138
b	−0.249	−0.315	−0.594	−0.977	−0.429	−1.413	−1.017	−1.048	−0.923	−0.208	−0.262	−1.274
	0.037	0.072	0.135	0.250	0.083	1.197	0.471	0.312	0.158	0.020	0.030	0.277
k_1	−0.011	−0.024	−0.153	−0.098	−0.039	−0.073	−0.076	−0.241	–[e]	–[e]	−0.028	−0.133
	0.001	0.020	0.033	0.020	0.031	0.026	0.027	0.148	–	–	0.009	0.017
k_2	0.029	0.018	0.118	0.064	0.051	0.043	0.044	0.167	0.016	0.052	0.019	0.072
	0.004	0.010	0.024	0.016	0.022	0.026	0.009	0.096	0.040	0.004	0.005	0.000
ko	0.003	0.009	0.005	0.024	0.051	0.056	0.011	0.011	0.020	0.057	0.015	0.067
	0.002	0.007	0.005	0.009	0.158	0.127	0.004	0.026	0.019	0.024	0.025	0.016
PMLE	135.8	135.4	101.4	77.3	59.2	48.1	73.9	64.7	−2.0	85.5	160.2	57.2

[a]In all of the models, the interest rate was assumed to be 5%. Standard error coefficients are given below each parameter.
[b]Miranda and Rui model estimated by pseudo maximum likelihood.
[c]Model with only arbitrageurs and substituting the shrinkage coefficient of Model I by a constant storage cost, *ko*.
[d]Model with manufacturers, arbitrageurs and constant storage cost, *ko*; includes parameters of both Models II and III.
[e]Model without intercept.

pseudo maximum likelihood estimator described by Deaton and Laroque (1995). Even if prices in the presence of speculators are not normally distributed, the estimator is consistent.

In all the cases, the harvest has been assumed to be independent and identically normally distributed with mean zero and variance one, and approximated by the same discrete distribution used by Deaton and Laroque (1992, 1995). Furthermore, in all the cases, we have assumed an interest rate equal to 5% and the consumption functions linear with Deaton and Laroque's parameters a and b.

Model I was taken directly from Deaton and Laroque (1995) and corresponds to the commodity storage model in which the cost of storage is given by the interest rate and the shrinkage coefficient (γ). Model II corresponds to the model from Miranda and Rui (1996), which was re-estimated using the original data (instead of using prices deflated by the historical means as in Miranda and Rui). The parameters of the supply of storage costs are k_0 and k_1. Model III is a variation of Model I in which the shrinkage coefficient is substituted for a fixed storage cost, *ko* (such as in Williams and Wright, 1991). Model IV has processors and speculators, and includes the parameters of Models II and III.

The econometric results, which compare the pseudo likelihood values of Models II and IV with those of Models I and III, show the importance of including a component that takes into account the phenomenon of storage under backwardation to explain commodity price dynamics. However, by comparing Models I and III, it is possible to conclude that the shrinkage coefficient imposes too high a cost compared with the fixed storage cost, and reduces the stocks carried by speculators.

The comparison between Models II and IV indicates that the enforcement of the arbitrage condition is not necessary to explain the dynamics of most of the commodity prices. The reason lies in the nature of the data, as aggregate average prices present what may be non-existent arbitrage opportunities – as has already been proved in the literature (see Working, 1961; Gilbert, 1986). This also explains the poor results of Model I, which assumes a homogeneous commodity.

The results of Model IV are mixed, despite a possible problem of identification of the five parameters involved, and it is interesting to note that it seems to perform well in the cases of cocoa, coffee, cotton and wheat. The model is expected to perform better with market level data, such as the one used in Working's empirical work.

12.4 Final Remarks: Limitations in Modelling Supply Chains

The purpose of this chapter has been to provide an overview of supply chain models available in the agricultural economics literature. The review has considered only models of homogeneous products and has excluded models considering explicitly more than one supply chain, such as in empirical partial and general equilibrium models.

We conclude by listing some of the major limitations in modelling supply chains:

- A lack of data is probably one of the most important limitations as it prevents explicit consideration of several food supply chains because of the increasing contracting that occurs at some stages of the chain (e.g. processing and retailing). In this respect, the work of Villas-Boas (2007) probably offers a way to overcome this problem.
- In comparison with models that only consider one market, supply chain models show greater complexity not only because of the aforementioned point about data availability but also because of the number of interactions among agents that need to be represented.
- The introduction of inventories in supply chain models increases the complexity of the models, bringing the so-called 'curse of the dimensionality' problem (Judd, 1998). This has implications for both model solution and the econometrics of the models.
- Finally, empirical observation of how food markets operate suggests that models, particularly on the food retailing side, need to be considerably more sophisticated than those presented in this review. In particular,

they need to account for the existence of the multi-product/multi-brand nature of food retailing and the nature of vertical contracts between manufacturers and retailers. As pointed out by Kaiser and Suzuki (2006), although data constraints may limit the researcher's ability to apply such models, the wider availability of scanner data at the retailing level now makes this type of research more plausible.

Notes

[1] Note that agricultural markets extend through the entire food system from input supply to farm production, collection, processing, packaging and transportation, and all the way to the final consumption of retail food products (Myers *et al.*, 2010).
[2] The shifters (Ω s) should be taken as vectors as they may include several variables. In the case of retail demand, one can consider, for instance, population and income.
[3] The derivation of the supply chain model considering inventories that is described here does not necessarily require the specific quasi-fixed proportions production function that is used, which was chosen because of its tractability and because it is bounded when the price of the factor of production is equal to zero.

References

Appelbaum, E. (1982) The estimation on the degree of oligopoly power. *Journal of Econometrics* 19, 287–299.

Azzam, A.M. (1996) Estimating the degree of dominance in a bilateral oligopoly. *Applied Economics Letters* 3, 209–211.

Azzam, A.M. (1999) Asymmetry in rigidity in farm–retail price transmission. *American Journal of Agricultural Economics* 81, 525–533.

Azzam, A.M. and Pagoulatos, E. (1990) Testing oligopolistic and oligopsonistic behavior: an application to the U.S. meat-packing industry. *Journal of Agricultural Economics* 41, 362–370.

Balke, N.S. and Fomby, T.B. (1997) Threshold cointegration. *International Economic Review* 38, 627–645.

Bhuyan, S. and Lopez, R.A. (1997) Oligopoly power in the food and tobacco industries. *American Journal of Agricultural Economics* 79, 1035–1043.

Buse, R.C. and Brandow, G.E. (1960) The relationship of volume, prices and costs to marketing margins for farm foods. *Journal of Farm Economics* 42, 362–370.

Carter, C. and Revoredo-Giha, C. (2009) Eastham's commodity storage model in a modern context. *Oxford Economic Papers* 61, 801–822.

Deaton, A. and Laroque, G. (1992) On the behavior of commodity prices. *Review of Economic Studies* 59, 1–23.

Deaton, A. and Laroque, G. (1995) Estimating a nonlinear rational expectations commodity price model with unobservable state variables. *Journal of Applied Econometrics* 10(Supplement S1), S9–S40.

Ezekiel, M. (1937) The broadening field of agricultural economics. *Journal of Farm Economics* 19, 96–101.

Freebairn, J.W., Davis, J.S. and Edwards, G.W. (1982) Distribution of research gains in multistage production systems. *American Journal of Agricultural Economics* 64, 39–46.

Gardner, B.L. (1975) The farm–retail price spread in a competitive food industry. *American Journal of Agricultural Economics* 57, 399–409.

Gilbert, C.L. (1986) Testing the efficient markets hypothesis on averaged data. *Applied Economics* 18, 1149–1166.

Goodwin, B.K. and Harper, D.C. (2000) Price transmission, threshold behaviour, and asymmetric adjustment in the US pork sector. *Journal of Agricultural and Applied Economics* 32, 543–553.

Goodwin, B.K. and Holt, M.T. (1999) Asymmetric adjustment and price transmission in the U.S. beef sector. *American Journal of Agricultural Economics* 79, 630–637.

Goodwin, B.K. and Piggott, N. (2001) Spatial market integration in the presence of threshold effects. *American Journal of Agricultural Economics* 83, 302–317.

Hansen, B.E. (1997) Inference in TAR model. *Studies in Nonlinear Dynamics and Econometrics* 2, 1–14.

Holloway, G. J. (1989) Distribution of research gains in multistage production systems: further results. *American Journal of Agricultural Economics* 71, 338–343.
Holloway, G.J. (1991) The farm–retail price spread in an imperfectly competitive food industry. *American Journal of Agricultural Economics* 73, 979–989.
Judd, K.L. (1998) *Numerical Methods in Economics*. The MIT Press, Cambridge, Massachusetts.
Kaiser, H.M. and Suzuki, N. (eds) (2006) *New Empirical Industrial Organization and the Food System*. Peter Lang Publishing, New York. Also available in Peter Lang Publishing Group in association with GSE Research's Sustainable Organization Library at http://gse.publisher.ingentaconnect.com/content/plpg/newemp (accessed 16 March 2016).
Kinnucan, H.W. and Forker, O.D. (1987) Asymmetry in farm—retail price transmission for major dairy products. *American Journal of Agricultural Economics* 69, 307–328.
Meyer, J. and von Cramon-Taubadel, S. (2004) Asymmetric price transmission: a survey. *Journal of Agricultural Economics* 55, 581–611.
Miranda, M.J. and Glauber, J.W. (1993) Estimation of dynamic nonlinear rational expectations models of primary commodity markets with private and government stockholding. *Review of Economics and Statistics* 75, 463–470.
Miranda, M.J. and Rui, X. (1996) *An Empirical Reassessment of the Commodity Storage Model*. Mimeo, Department of Agricultural Economics, Ohio State University, Columbus, Ohio.
Muth, R.F. (1964) The derived demand curve for a productive factor and the industry supply curve. *Oxford Economic Papers* 16, 221–234.
Myers, R.J., Sexton, R.J. and Tomek, W.G. (2010) A century of research on agricultural markets. *American Journal of Agricultural Economics* 92, 376–403.
Ramey, V.A. (1989) Inventories as factors of production and economic fluctuations. *American Economic Review* 79, 338–354.
Sexton, R.J. and Zhang, M. (2001) An assessment of the impact of food industry market power on U.S. consumers. *Agribusiness* 17, 59–79.
Suzuki, N. and Kaiser, H.M. (2006) New empirical organization theories and their application to food system analyses. In: Kaiser, H.M. and Suzuki, N. *New Empirical Industrial Organization and the Food System*. Peter Lang Publishing, New York, pp. 3–61. Also available in Peter Lang Publishing Group in association with GSE Research's Sustainable Organization Library at http://gse.publisher.ingentaconnect.com/content/plpg/newemp (accessed 16 March 2016).
Vavra, P. and Goodwin, B.K. (2005) *Analysis of Price Transmission along the Food Chain*. OECD Food, Agriculture and Fisheries Working Papers, No. 3, OECD Publishing, Paris.
Villas-Boas, S. (2007) Vertical relationships between manufacturers and retailers: inference with limited data. *Review of Economic Studies* 74, 625–652.
Warren, G.F. and Pearson, F.A. (1928) *Interrelationships of Supply and Price*. Bulletin 466, Cornell University Agricultural Experiment Station, Ithaca, New York.
Waugh, F.V. (1934) Margins in marketing. *Journal of Farm Economics* 16, 233–245.
Weldegebriel, H.T. (2004) Imperfect price transmission: is market power really to blame? *Journal of Agricultural Economics* 55, 101–114.
Williams, J.C. and Wright, B.D. (1991) *Storage and Commodity Markets*. Cambridge University Press, Cambridge, UK.
Working, H. (1961) Note on the correlation of first differences of averages in a random chain. *Econometrica* 28, 916–918.

13 Linkage of a Farm Group Model to a Partial Equilibrium Model*

Alexander Gocht,[1†] Pavel Ciaian,[2] Maria Espinosa[2] and Sergio Gomez y Paloma[2]
[1]*Thünen Institute of Farm Economics, Braunschweig, Germany;*
[2]*Joint Research Centre European Commission, Seville, Spain*

13.1 Introduction

There is a wide range of methodological and technical solutions available in the literature for modelling the impacts of agricultural policies at the farm level. The complexity of the questions and objectives that different models aim to answer, and the need to be able to validate and repeat model runs, limit the development of models towards rather specialized modelling approaches rather than the development of a super model. However, when the variables of interest and functional relationships are spread over different models, linkage of models is necessary to derive results that are relevant to policy analysis.

A prominent example of this is the modelling chain that has been established to provide impact assessments for the European Union (EU) Water Framework Directive (WFD) (Cools *et al.*, 2011; Heidecke *et al.*, 2014). It is necessary to model the flow of nitrogen from fields to the water basin, which depends on the application of nitrogen fertilizer by farmers and the hydrological (water basin) and soil properties. So a modelling chain of hydro-ecological and agricultural sector models is used to propose cost effective agri-environmental policy measures.

Another example of model linkage is the impact assessment model used for the recent reforms of the Common Agricultural Policy (CAP) of the EU. The EU changed its focus from price support to farm-specific coupled – and later decoupled – support, with the aim of reducing production and trade distortion effects and promoting environmental protection. The reorientation from the coupled support system to a farm-specific system has been the main agenda of the CAP reforms since 2003. To analyse the impact of CAP measures (including farm-specific measures), farm models at the individual farm level (e.g. Farm Boss, Gocht and Münch, 2003; DAIRYDYN, Lengers and Britz, 2012; and IFM-CAP, Louhichi *et al.*, 2015) or at the representative level as defined by representative single farms or farm groups at the regional level (e.g. FARMIS, Offermann *et al.*, 2005; AROPAj, Baranger *et al.*, 2008; FSSIM, Louhichi *et al.*, 2010 and CAPRI-FT, Gocht *et al.*, 2013), are used to understand how the farming system (agricultural supply and resource use) reacts to different policy measures. These farm models simulate changes in cropping, animal herds, investments and farm labour use, along with key drivers of economic, environmental and social indicators, but with variations

*The authors are solely responsible for the content of this paper. The views expressed are purely those of the authors and may not in any circumstances be regarded as stating an official position of the European Commission.

†E-mail: alexander.gocht@thuenen.de

among them. They are also suitable for modelling policy changes under the CAP, such as production quotas, direct payments, decoupling of payments and greening measures, as introduced by the 2013 CAP reform.

Although systematic sensitivity analyses for the input and output of farm models can be applied to derive price point elasticities, such analyses cannot be easily connected to market models for policy simulations because of the incomplete product coverage and limited regional coverage of the farm models. To overcome this problem, the Common Agricultural Policy Regionalized Impact (CAPRI) model was extended by the farm-type (FT) model (Gocht, 2010; Gocht and Britz, 2011; Gocht *et al.*, 2014) to create a farm group model (referred to as CAPRI-FT), which represents the complete farm population at the EU-27 level using 2450 mathematical programming models. The composite model covers the full range of outputs and inputs that are necessary to link it to a partial equilibrium model.

In this chapter, we describe the development of the FT model and illustrate how it is iteratively linked to the global spatial multi-commodity partial equilibrium model for agricultural products in CAPRI. In Section 13.2, we review the existing literature in which farm models are linked to partial equilibrium models. In Section 13.3, we present the approach for the sequential linkage of the FT layer and partial equilibrium model in CAPRI. Section 13.4 describes the limitations of the CAPRI approach, and Section 13.5 presents an application of the linked model (CAPRI-FT). A brief summary and conclusions closes the chapter.

13.2 Review of Existing Models

Modelling market interlinkages in farm models is relevant when the input and output prices adjust significantly in response to a simulated policy at the farm level. If market effects are not considered, farm models will usually be biased in terms of the simulated policy effects. For example, if output market adjustments in response to a policy are not taken into account, a farm model will overestimate the supply response because the negative relationship between farm output and prices is not considered. In this case, controlling for output price effects would reduce the absolute magnitude of the farm supply responses. A similar result holds for the input market interlinkage – because input prices vary inversely with input use, accounting for input price effects will reduce the adjustment of the farm input quantity to a given policy scenario. Another consideration is that the interlinkage of a farm model with a market model allows inferences from the micro level to be extended to aggregated macro responses at the regional and sectoral levels. Linking a farm model with a market model requires aggregation of the various farm-level results because aggregated results are particularly able to detect the magnitude of the combined farm effects and their magnitude relative to the overall sector.

Market models operate at regional (e.g. country, EU or global) levels and simulate aggregate impacts. Thus, a key issue when linking a farm model with a market model is the aggregation of farm-level responses to regional-level responses – the incorrect aggregation of heterogeneously behaving farms may lead to aggregation bias. The challenge is to ensure consistency between the aggregate responses and individual farm responses. Inconsistency in responses may emerge, especially if the farm behavioural parameters vary across farms and if the distribution of the farm population changes over time or as a response to a simulated policy scenario (Hosseini and Spriggs, 2009). According to Green (1964), 'Consistency means that a knowledge of the "macro-relation" ... and of the values of the aggregate independent variables would lead to the same value of the aggregate dependent variable as a knowledge of the micro relations and the values of the individual independent variables.'

Modelling farm behaviour at any level other than that of the individual farm implies problems of aggregation bias (Gómez-Limón, 2004). According to Stoker (1993), there are two approaches that are used in the empirical literature for modelling aggregate responses with respect to individual agent behaviour: (i) the representative agent approach; and (ii) joint modelling of individual and aggregate behaviours. Aggregated (e.g. by size or specialization) representative farm models – which belong to the first approach – are a common approach that is used in mathematical programming when the aim is to provide large-scale regional or sector-level analysis. An aggregated representative farm model is obtained by aggregating individual farms into a small number of farms, each of which is assumed to be endowed with the resources available to all of the farms that comprise the aggregate farm and to utilize a production technology that characterizes the 'average' technology used by the individual farms. This approach is used to avoid computational difficulty and the problem of model parameterization, but such models may suffer from aggregation bias, because the aggregate model may result in different solutions from the sum of the optimal solutions for the individual farm models. The second approach, modelling a set of farms in a combined mathematical programming model, overestimates the mobility of resources because the modelled farms are allowed to combine resources in proportions that are economically most optimal given the model constraints; however, such mobility may not be possible in real-world situations because the model implicitly assumes that resources that are owned by individual firms are traded freely within the aggregate farm (Hazell and Norton, 1986).[1]

An approach that is widely used in the literature to reduce the aggregation bias of representative farm models is to group farms into homogeneous groups with similar decision-making behaviours. This grouping requires the fulfilment of strict homogeneity requirements for aggregated individual farms, such as in their technological, climatic, soil-quality, market, policy and institutional homogeneity (Gómez-Limón, 2004). The main limitation of this approach is that its effectiveness decreases as the regional scale increases, because farm heterogeneity increases as the regional scale becomes larger. This approach is quite widely used in applied farm modelling. Studies often pool together similar farms into an aggregate of representative farms that are as homogenous as possible. The most common criteria used to group farms are size, specialization, location and production intensification (Louhichi *et al.*, 2013). Gallego-Ayala *et al.* (2013) noted that even for relatively small areas, farm homogeneity may not hold. Although farms may operate within the same agricultural system and under the same market conditions, they still may exhibit significant differences in terms of their production decisions.

Stoker (1993) suggests a second approach that considers the joint modelling of individual and aggregate behaviours. This framework permits individual and aggregate data to be modelled under one consistent format. In particular, an individual model is specified along with assumptions that permit the formulation of an aggregate model that is consistent with the individual model and also accounts for individual heterogeneity. There have been a few empirical attempts at using this approach, often in the context of modelling the agricultural sector. An application of this approach to agricultural policy (in this case the Net Income Stabilization Account programme in Saskatchewan, Canada) was presented by Hosseini and Spriggs (2009). They used estimated supply functions in a dynamic stochastic simulation of representative farms over a 10 year period, considered heterogeneity among farm types and regions, and assumed homogeneous production within a given farm type and given region.

According to Heckelei (2013), there is no need for sophisticated aggregation theory to derive aggregate product supply models because current computing power allows simulations to be conducted with many thousands of farms or representative farm models, and market equilibrium can be obtained with iterative algorithms, even in multidimensional and highly interdependent product settings (Britz, 2008). Rather, the main challenge is linking farm models with input and output markets because prices are exogenous in farm models and endogenous in market models.

The type of market interlinkage depends on the geographic coverage, type of policy analysed and time horizon, among other factors. For example, if the geographic coverage is small, the market effect of the analysed scenarios is likely to be minimal; thus, it is not necessary to link a farm model with a market model. Even if the geographic coverage is large, for certain policy analyses, the market effects are small and so they can be ignored. This is the case for modelling the impact of decoupled payments. For example, Gocht *et al.* (2013) simulated the income distribution effects of decoupled payment harmonization across farms and regions in the EU and found that the effect on the output price is rather small. However, in this exercise, the linkage with the input (land) market is more fundamental. Given that the allocation of the EU decoupled payments is linked to land, this consideration is not surprising. The payment harmonization significantly affects the rental prices of land and land use. Production and prices are affected indirectly, mainly through land adjustments and the effects of these are minimal.

There are different methods for modelling market linkage. Britz (2008) differentiated among three approaches: (i) model chains without calibration of the interlinked models; (ii) models with one-way calibration; and (iii) models with sequential calibration. The first approach shocks a model with data from another model without considering any further feedback between the interlinked models. One example of this approach is the use of the price changes from a partial equilibrium model (e.g. AGLINK or CAPRI) in a supply programming model (Bertelsmeier *et al.*, 2003). An example of the reverse linkage consists of using macro/market simulation results from a partial or general equilibrium model (e.g. Magnet or GTAP) in a farm model such as CAPRI-FT. This approach is usually implemented in a soft manner in which only some major variables are exchanged between models.

The second approach for modelling market linkage is one-way calibration, in which one model is calibrated to results generated by another model. Approaches in this category can be distinguished as either bottom-up or top-down. One example of the bottom-up approach is when the behavioural function of the market models is calibrated to economically estimated micromodels (e.g. Domínguez *et al.*, 2009). In the top-down approach, the more-disaggregated results or the response parameter are calibrated to the results and to the response parameter of the top level (Leip *et al.*, 2008). The latter approach is infrequently used for the linkage of economic models, but is often necessary for the linkage and harmonization of biophysical models with the results from economic models from the top level.

The third approach, which was proposed by Britz (2008), is based on sequential calibration. In this case, while each model uses and produces its own results, there is iterative feedback among the models. One example is the linkage of the partial equilibrium model of CAPRI and the FT layer, which is described in more detail in Section 13.3. Another example is the linkage of the GTAP model with the CAPRI model (Jansson *et al.*, 2008; Britz and Hertel, 2011; Pelikan *et al.*, 2015).

A similar approach was also applied by Deppermann *et al.* (2014), who interlinked the FARMIS model with the partial equilibrium market model ESIM. In the first step, exogenous variables, such as policies, technological progress, growth in gross domestic product (GDP), changes of labour and capital costs are harmonized in both models. In the second step, price and yield changes from ESIM are input (as exogenous shocks) to FARMIS. After a comparison and analysis of the responses of both models to the same price and yield changes, further harmonization of the model specifications is performed if the results of the two models disagree significantly. In the final step, iterations between the two models are conducted. In the FARMIS-ESIM approach, FARMIS simulates quantitative changes in crop area allocation and supply of livestock activities. These quantitative changes are then introduced into ESIM and used to create new price changes that are then used in FARMIS. This iteration between ESIM and FARMIS is repeated until convergence is reached. The key difference between the FARMIS-ESIM and CAPRI-FT approaches is that the former exchanges the relative changes in prices quantities, whereas the latter exchanges absolute changes between the farm and market models. It should be noted that sequential calibration requires iterations over model components and thus requires more computational time than stand-alone applications of single components.

13.3 The CAPRI Approach

CAPRI is a comparative static partial equilibrium model for the agricultural sector that was developed for performing policy and market impact assessments from global to regional and farm type levels. The core of CAPRI is based on the linkage of a Europe-focused supply module and a global partial equilibrium market module (Britz and Witzke, 2014).

The supply module covers a detailed representation of production activities for the EU, Norway, the Western Balkans and Turkey. It represents all agricultural production activities and related output generation and input use at the regional (NUTS2)[2] or farm type level (Gocht and Britz, 2011).[3] From a methodological perspective, the supply module consists of independent non-linear programming models. Each programming model (at the NUTS2 or farm type level) optimizes income under economic, environmental and policy constraints. The different policy instruments of Pillar I and Pillar II of the CAP for the EU are specified in detail to cover policy implementation. Prices are exogenous to the supply module and are provided by the partial equilibrium market module.

The global partial equilibrium market module is a spatial, non-stochastic, global multi-commodity model for approximately 50 primary and processed agricultural products, and it covers approximately 80 countries or country blocks. It is defined by a system of behavioural equations that represent the agricultural supply, human and feed consumption, multilateral trade relations, feed energy and land as inputs, along with the processing industry, all of which are divided into commodity and geographical units. On the demand side, the Armington approach (Armington, 1969) assumes that the products are differentiated by origin, thereby allowing simulation of bilateral trade flows and related bilateral and multilateral trade instruments, including tariff-rate quotas. This sub-module delivers the output prices used in the supply module, allows for market analyses at the global, EU and national scales, and includes a welfare analysis.

13.3.1 Farm types in the CAPRI-FT layer

The introduction of farm-specific support schemes (e.g. direct payments) as part of the CAP motivated the development of a farm type (FT) module for CAPRI – the CAPRI-FT layer. The CAPRI-FT layer is a consistent disaggregation of the regional models into FT models (Gocht and Britz, 2011). In all, 2450 FT models aim to capture the heterogeneity in farming practices within a region to reduce the aggregation bias of the CAPRI regional responses to policies and market signals. Hence, the CAPRI-FT layer is suitable for the analysis of policy instruments that are either targeted at the farm level (e.g. in the case of the historically decoupled payments) or implemented depending on farm characteristics (e.g. the modulation of direct payments) (Gocht *et al.*, 2013; Schroeder *et al.*, 2015).

Each farm type in the CAPRI-FT layer is represented by a non-linear programming model that captures all activities that are associated with the farms of a certain typology in a specific NUTS2 region. The same economic optimization principles as for the NUTS2 models are used. Specifically, each model optimizes the aggregated farm income under restrictions that are related to land balances – including a land supply curve, nutrient balances and nutrient requirements of animals, and, if applicable, to quotas and set-aside obligations. The decision variables include crop acreages, total land use, herd sizes, fertilizer application rates and feed mix. The direct payments made under the CAP are captured in detail. The allocation response depends primarily on non-linear terms in the objective function that are either econometrically estimated (Jansson and Heckelei, 2011) or derived from exogenous supply elasticities.

The main data sources for the calibration and parameterization of the models are the European Farm Structure Survey (FSS)[4] of 2007 and the Farm Accountancy Data Network (FADN)[5] for the years 2007 to 2009, along with information from the regional database in CAPRI. Each farm type is characterized by a production specialization and economic size class (ESC), as determined by the shares of the standard gross margin of the different farm production activities in the total value of the standard gross margin of the farm. CAPRI-FT considers 39 possible farm types (see Table 13.1) based on 13 production specializations and three farm economic size classes (ESCs), thus representing a compromise among the model complexity and size, robustness of the results, reporting limitations and data constraints. The three ESCs are defined in terms of economic size units (ESUs), where one ESU is equivalent to a gross margin of €1200. The smallest ESC has fewer than 16 ESUs,[6] the next ESC has between 16 and 100 ESUs and the largest has more than 100 ESUs.

To determine the production structure, we use an extraction of the FSS from Eurostat (the EU statistics database) aggregated by farm specialization and size class. Figure 13.1 shows the number of farm groups available from the FSS ordered sequentially and presented for all NUTS2 regions in the EU-27 (225 in total;

Table 13.1. Farm types (FTs) and economic size classes (ESCs), with abbreviations.

Farm type	Abbreviation
Specialist cereals, oilseed and protein crops	FT13
General field cropping + mixed cropping	FT14 + FT60
Specialist horticulture	FT20
Specialist vineyards	FT31
Specialist fruit and citrus fruit	FT32
Specialist olives	FT33
Various permanent crops combined	FT34
Specialist dairying	FT41
Specialist cattle + dairying rearing, fattening	FT42 + FT43
Sheep, goats and other grazing livestock	FT44
Specialist granivores	FT50
Mixed livestock holdings	FT70
Mixed crops–livestock	FT80
Economic size class	**Abbreviation**
<16 ESUs[a]	ESC 1
≥16 and <100 ESUs	ESC 2
≥100 ESUs	ESC 3

[a]ESU, economic size unit.

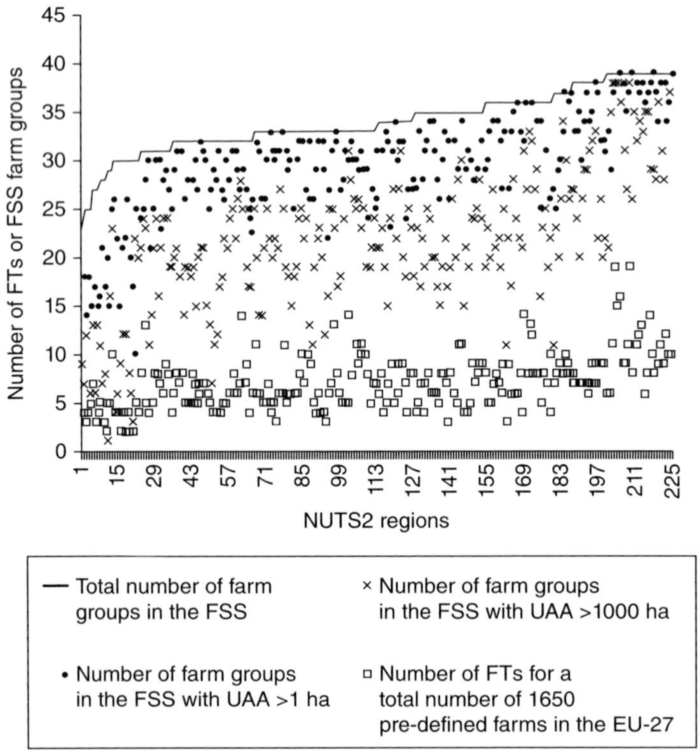

Fig. 13.1. Distribution of the Farm Structure Survey (FSS) farm groups or farm types (FTs) by NUTS2 (Nomenclature of Units for Territorial Statistics) region (there are 225), with 1650 FTs predefined for the EU-27 countries. UAA is utilized agricultural area.

Croatia is not included in the FT layer), which ranges from 23 to 39 (represented by the full stepped line at the top of the graph). The dots (which may appear as dashes when close together) represent the number of farm groups that remain when groups with no utilized agricultural areas (UAAs) are excluded. Some farm groups are excluded as a result of the data confidentiality regulations applied at Eurostat, which require the values of variables (including UAAs) to be rounded to the first digit and farm groups for which there are a low number of farm observations to be excluded. The crosses indicate the number of farm groups above the threshold of 1000 ha of UAA.

Given computational constraints and the need to keep the database and model outputs within a manageable size, it is impossible to build an FT layer that is based on all of the available farm groups in the FSS (around 8000). Our approach considers only those farm groups that are classified as important according to two equally weighted criteria: livestock units (LUs) and UAAs.[7] Rather than selecting a certain fixed number of farm types per region, we developed an optimization model that determines the number of farm types per NUTS2 region. We maximize the representation of UAAs and LUs at the EU-27 level (or minimize residual, non-selected, farm types) for a predefined total number of farm types for the EU-27 (Gocht *et al.*, 2014). For example, the squares in Fig. 13.1 represent the distribution of the selected farm types per NUTS2, with 1650 farm types predefined for the EU-27. The distribution of selected farm types ranges from two (primarily in NUTS2 regions that are large cities) to 20 per NUTS2 region.

Figure 13.2 presents data on the coverage of farm types selected with the combined UAA and LU criteria at the EU-27 level (i.e. in the total population) for different levels of the predefined total number of farm types (ranging from 1650 to 4000). The graph shows that the coverage of the selected farm types increased from 82 to 98% as the total number of farm types increases. For example, for 1800 predefined farm types, the coverage level is 85%, whereas for 2200 farm types the coverage is slightly more than 90%. The final number of farm types considered in CAPRI-FT is 2200; the remaining farm types (i.e. those that are not selected) are aggregated into a residual farm type. Thus, in total, there are 2450 farm models in CAPRI-FT (the sum of the selected and the residual farm types) and these represent the total regional agricultural production and input use in the EU-27.

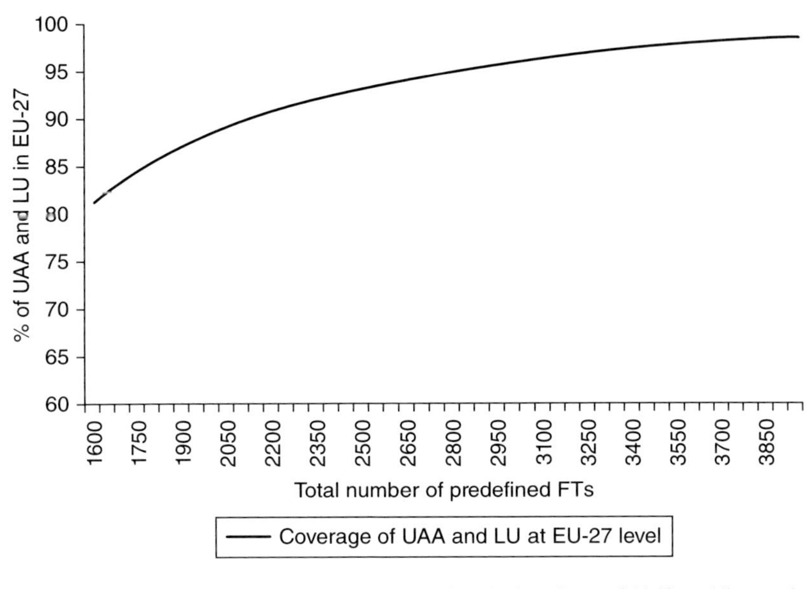

Fig. 13.2. The percentage of farms with combined utilized agricultural area (UAA) and livestock units (LU) versus the number of predefined farm types (FTs) in the EU-27 countries.

The FT layer is fully integrated with the overall CAPRI structure. The farm types that are available in the CAPRI-FT model are fully consistent with regional statistics on production quantities, activity levels and input use. The top-down consistency of the FT layer ensures a data set that is harmonized across regional scales and farm types. The top-down data consistency integrates the FT models in the overall system, thereby ensuring their interoperability with the global partial equilibrium market model. Given that all regional levels, from the Member States top level down to the NUTS2 regions and further down to farm types, are consistent, the activity levels (production and land use, for example) sum up to the regional aggregate value. Consequently, the model linkage with the partial equilibrium market model is straightforward and does not require the application of a procedure for scaling the farm type results to the regional and Member State levels.

13.3.2 Output market linkage

As mentioned above, the CAPRI model comprises two modules: the global multi-commodity partial equilibrium market module and the FT supply module. The market module determines the equilibrium prices and supply quantities by solving the interaction of supply and demand functions, while the supply module solves for output quantities by taking the output prices as given. Linking the market and supply modules requires exchanging quantities and prices between the two modules in a sequential iterative process, as shown in Fig. 13.3 (Britz, 2008).

Figure 13.4 shows the sequential model iteration between the supply and market modules. The implicit marginal cost curve (i.e. the supply curve) of the FT supply models is indicated by MC. To simplify the explanation, MC is assumed to remain constant in the iteration process (Britz, 2008). MC reflects how the FT supply responds to different price levels. Note that because the parameters of the marginal cost curve are unknown, other iterations are not necessary and the optimal solution can be determined by solving the point of intersection between the demand and MC functions.

In the first iteration, the supply model is solved at price p_0 and yields a supply of Q_0. In the next step, a hypothetical linear supply function, S_0, that passes through the point (p_0, Q_0) is assumed. This function is used to solve for the price p_1 in the market module, which is given by the intersection between the supply S_0 and the demand D. In a second iteration, a new simulation

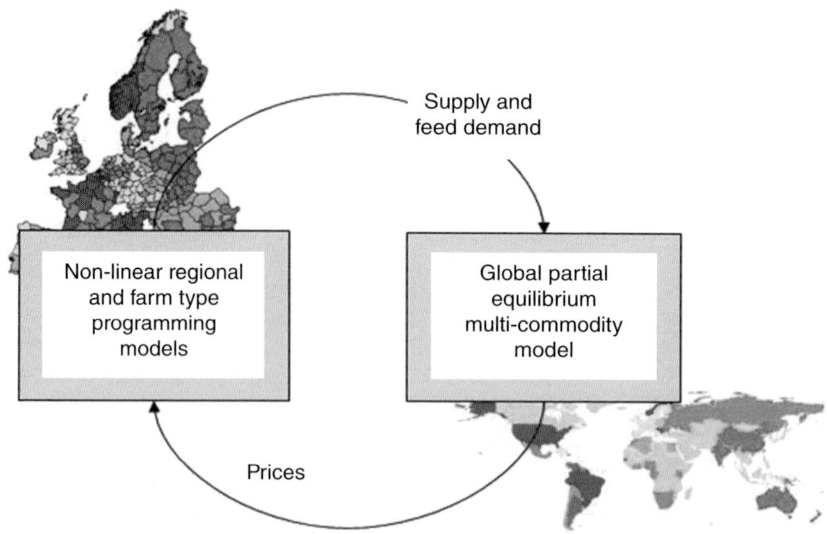

Fig. 13.3. The structure of the CAPRI model. Modified from the version at http://www.capri-model.org/dokuwiki/doku.php?id=capri:concept:market (Gocht and Witzke, 2014).

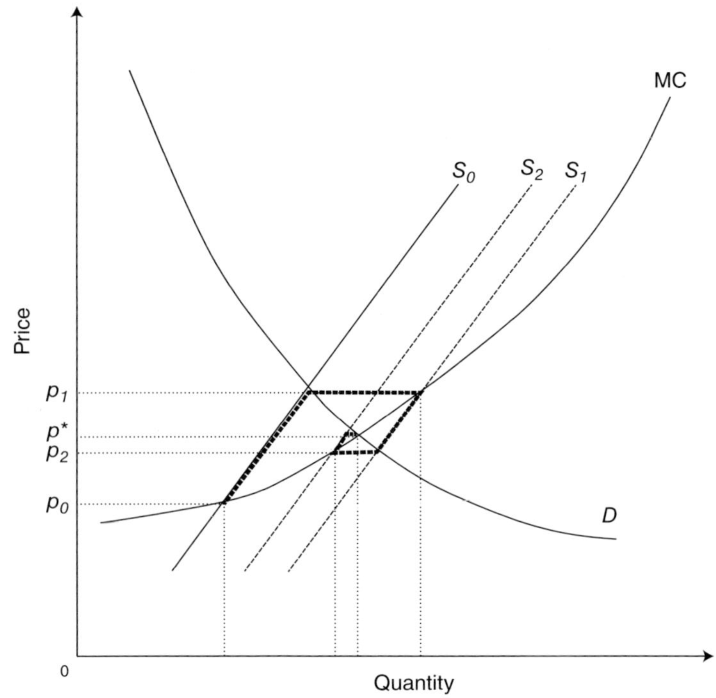

Fig. 13.4. Sequential iteration between the market and supply modules of the CAPRI model. To simplify the exposition, the figure shows the iteration for one commodity only. In reality, CAPRI is a multi-commodity model; hence, prices and quantities are determined for multiple products. Key: *D*, demand curve; MC, marginal cost curve; p_0, p_1, p_2, p^*, prices; S_0, S_1, S_2, supply curves. See text for further explanation.

with the supply module yields the new supply quantity Q_1, which is specified by the intersection of p_1 with the MC curve. The supply curve is then shifted to S_1 such that it crosses the demand curve at the point (p_1, Q_1). A new solution of the market model returns the price p_2, and the supply module solution delivers supply quantity Q_2. This process is repeated until the solution converges to the market equilibrium point (p^*, S^*) (Britz, 2008).

13.3.3 Land market linkage

A key methodological question when simulating the impact of CAP payments is how to capture the interaction between the economic returns to land from farming and the land supply to agriculture. In a mathematical programming farm model such as CAPRI-FT, with a fixed land endowment considered, any change in CAP payments or market returns will result in adjustment of the land shadow prices, leaving the total land use unchanged.

CAPRI has recently been extended to include land supply and transformation functions; the latter describe substitution between arable and permanent grassland. These behavioural functions are driven by economic returns from agricultural land uses (including subsidies) and are parameterized based on the results of Golub *et al.* (2006) and van Meijl *et al.* (2006), and adapted to the regional resolution of CAPRI based on geographic information system (GIS) analysis and on simulation experiments performed using the Dyna-CLUE model (Verburg *et al.*, 2010). Each farm type in CAPRI has its own land supply and therefore its own shadow price. CAPRI assumes a full transmission of changes in the shadow prices of land to rental rates, which is a reasonable assumption in our comparative static framework with a medium-term modelling horizon (an adjustment period of approximately one decade). This modelling of land markets results in full rental price adjustment to economic returns within a farm type, but not between farm types. Thus, this model yields segmentation of land markets along CAPRI farm types. No exchange of land between different farm types is considered in CAPRI-FT, whereas full land mobility implicitly occurs within farms that belong to a given farm type.

13.4 Limitations

The main advantage of CAPRI-FT compared with other FT models (e.g. FARMIS, Offermann *et al.*, 2005; the AROPAj system, Baranger *et al.*, 2008) lies in the individual programming models. CAPRI-FT covers all major farm types in the EU, its representation is consistent with the aggregated data at the regional level (Gocht and Britz, 2011) and it is iteratively linked to a global market model for agricultural products (Britz, 2008). Furthermore, CAPRI-FT considers endogenous land supplies, thus allowing accounting for the land market effects of simulated scenarios.

One of the main drawbacks of the CAPRI-FT approach is the use of structurally identical, stylized and relatively simple template models, in which differences between farm types and regions are expressed solely by parameters. This approach may not completely capture the full diversity of farming systems in Europe. In particular, the evaluation of policy measures that have an impact on farm management decisions, such as manure handling or feeding practices, requires the development of models that include these factors as decision variables. The relatively simple representation of agricultural technology in CAPRI compared with approaches that are parameterized on the basis of biophysical models limits the applicability of CAPRI to situations in which agricultural technology plays a significant role.

13.5 Application

CAPRI is extensively used for policy support, especially in the area of CAP and trade scenario analyses. Several applications have recently been conducted in which CAPRI was used to assess the impacts of CAP reform and trade liberalization on EU farms. For example, Gocht *et al.* (2013) simulated the impact of the harmonization of decoupled payments across farms, as proposed by the 2013 CAP reform. They analysed the extent to which the incomes of different farm types are affected by the reform. Compared with other partial equilibrium models, CAPRI-FT is very suitable for this type of analysis at the EU scale because its main strength is in its ability to capture farm heterogeneity, including variation in decoupled payments across farm types, regions and Member States. Gocht *et al.* (2011) applied the CAPRI-FT model to assess the implications of a macroeconomic environment by focusing on the impact of recovery from the economic crises on the EU farming sector. Simultaneously, they investigated the effects of trade liberalization considered under the current World Trade Organization (WTO) negotiation round (the Doha Round).

Here, we illustrate a key result for the latter scenario (i.e. trade liberalization). The scenario is chosen to highlight the specific advantages of the FT layer with respect to its linkage with the partial equilibrium market module. Specifically, the scenario aims to quantify the impact of the Doha Round multilateral negotiations based on the revised draft modalities presented to the WTO Agriculture Committee in December 2008 (WTO, 2008). In this Doha Round negotiation, several measures were proposed: a general tariff reduction based on a tiered formula, a reduction of tariff rate quotas (TRQs) for quota tariffs; and the possibility of excluding certain products, called sensitive products, at the cost of the extension of TRQs for sensitive import products.[8]

The tariff reductions that are analysed here are based on tiered formulas with four bands, implying that tariffs that are high will be cut more aggressively than tariffs that are low and therefore distort trade less (Table 13.2). For example, if the initial tariff, i.e. the percentage of the value of imported goods, is greater than 75%, developed countries will be obliged to reduce the tariff by 70%. The same initial tariff is reduced by 38% for developing countries because of the different thresholds applied for developed and developing countries.[9]

For quota tariffs, a reduction of 50% (a minimum of 10% for developed countries) is proposed, and all tariffs that are already less than 5% should be set to zero. For developing countries, the only rule is that quota tariffs should be reduced by 15%.

It should be noted that CAPRI differentiates between specific and *ad valorem* tariffs. Specific tariffs represent fixed amounts of €/t that are charged on imports; *ad valorem* tariffs are charged as a percentage of the value of imports. The tariff cuts are applied to *ad valorem* equivalents (AVEs), and *ad valorem* tariff equivalents must be calculated for specific tariffs. This calculation is accomplished by dividing the specific tariffs by the average import price tariffs for the respective product.

The Doha proposal also allows each country to define certain products as sensitive. If a product is defined as sensitive, a smaller tariff cut is used at the cost of an extension of the respective TRQ. The proposal includes the options to reduce the tariff cut in the formula by 33, 50 or 66%, with corresponding TRQ extensions of 3, 3.5 or 4% of domestic consumption of that good in the importing country. We have chosen the lowest tariff decrease for sensitive products (66%) combined with the highest tariff rate quota extension of 4%. For developing countries, the extension is only two thirds of this 4%. We define the following as sensitive products for the EU-27: cheese, butter, beef, pork and poultry meat, sugar and rice. The current scenario also assumes that all export subsidies are removed.

Table 13.3 reports the percentage changes relative to the baseline for TRQs and MFN (most favoured nation) tariffs (both *ad valorem* and specific) for each product group over all EU imports. The change for each product group is the result of the different tariff regimes for the countries that import into the EU. For example, the global TRQs for meat and dairy products and secondary products increase because of the expansion mechanics for TRQs. Additionally, the TRQs for beef, pork and poultry meat increase up to 4% of domestic consumption, which includes human consumption, feed use and processing. This result leads to an increase of 873% for meat and 273% for dairy

Table 13.2. Tariff reduction formula in the World Trade Organization Doha Round trade liberalization agreement.

Developed countries		Developing countries	
Initial tariff level	% reduction	Initial tariff level[a]	% reduction
0 ≤ 20	50	0 ≤ 30	33.33
0 ≤ 50	57	30 ≤ 80	38.00
50 ≤ 75	64	80 ≤ 130	42.67
>75	70	>130	46.67

[a]Tariffs are translated to *ad valorem* equivalents (AVEs) to determine the reduction percentage that applies.

Table 13.3. Changes in tariff rate quotas (TRQs) and most favoured nation (MFN) tariffs aggregated for all EU 27 imports for the World Trade Organization (WTO) Doha Round trade liberalization scenario.

	Global TRQs		*Ad valorem* MFN		Specific MFN	
Commodity	WTO scenario (1000 t)	% to baseline	WTO scenario (%)	% to baseline	WTO scenario (€/t)	% to baseline
Cereals	3276	–	2.5	−18	31	−66
Other arable crops	–	–	9.0	−50	0	–
Vegetables/permanent	–	–	3.0	−65	–	−59
Coffee, teas, cocoa	–	–	0.0	–	0	–
Other crops	–	–	10.3	−48	0	–
Meat	2071	873	3.9	−42	977	−19
Animal products	158	–	1.1	−64	249	−64
Dairy products	677	273	6.1	−29	892	−36
Oils	–	–	–	−54	43	−47
Secondary products	1027	738	–	–	234	−22

products (cheese and butter are sensitive products in this aggregate product group). Among secondary products, sugar is the cause of a 738% increase in the global TRQs. The corresponding tariff reductions range between 50 and 70% for products that are not defined as sensitive. For meat, sugar and dairy products, the decrease is less pronounced.

13.5.1 Results

We focus on the income and welfare effects of the trade liberalization scenario in EU-27 countries as predicted by the CAPRI-FT model. The aggregate welfare effects are presented in Table 13.4. The table reports the absolute values for the baseline and the liberalization scenario (WTO, https://www.wto.org/english/tratop_e/dda_e/dda_e.htm, accessed 14 July 2016), along with the absolute differences between the two scenarios. According to the results, the WTO liberalization leads to total EU welfare losses of 1 billion €. The primary gains are experienced by the consumers (13.5 billion €) because they benefit from lower price levels, and the taxpayers, who benefit from cost reductions (0.046 billion €). Farmers lose 10.6 billion € as a result of stiffer external competition and the induced lower output prices. A direct effect of tariff reduction is a reduction of 3.5 billion € in tariff revenues.

The reduction of agricultural income in the EU-27 countries is caused primarily by the total production reduction (16.5 billion €) which is partially offset by the input use decrease (−5.9 billion €). Note that agricultural income is calculated as total output value minus value of inputs plus subsidies. Subsidies decrease only slightly, by 74 million €, because of the small change in land use. Animal production is more affected than crop production, and they decrease by 9 billion € and 7.5 billion €, respectively. The key driver of production reduction in the EU-27 countries is the drop in agricultural prices. Almost all agricultural product prices come under pressure when tariffs are reduced and trade competition is improved for non-EU trade partners. The highest price reductions are observed for sheep and goat meat (15%), other fruits (15%), butter (6%), olive oil (7.2%), citrus fruits (5.7%)

Table 13.4. EU-27 country welfare position changes in million € showing baseline figures and figures based on the World Trade Organization (WTO) Doha Round trade liberalization scenario as predicted by CAPRI-FT.

Welfare or agricultural parameter, some denoted by a letter in col. 2, totals in bold		Derivation of totals based on letters in col. 2	Million € Baseline	Million € WTO	Absolute change to baseline
WELFARE					
Consumer welfare	A	–	20,443,866	20,457,360	13,494
Agricultural income	B	–	185,761	175,121	−10,639
Profit of dairies	C	–	25,333	24,902	−431
Tariff revenues	D	–	7,959	4,436	−3,523
Tax payers cost	E	–	51,605	51,558	−46
Total welfare	–	**A + B + C + D − E**	**20,611,314**	**20,610,261**	**−1,053**
AGRICULTURE					
Agricultural income	–	**G − K + F**	**187,591**	**176,926**	**−10,665**
Total value of subsidies	F	–	53,733	53,659	−74
Gross value added	–	**G − K**	**133,858**	**123,267**	**−10,591**
Total output value	**G**	**H + I**	**488,762**	**472,175**	**−16,587**
Total input value	**K**	**L + M**	**354,904**	**348,908**	**−5,997**
Value added of crop activities	–	**H − L**	**134,730**	**127,663**	**−7,067**
Crop output value	H	–	230,472	222,921	−7,551
Crop specific input value	L	–	95,742	95,258	−484
Value added of animal activities	–	**I − M**	**83,821**	**79,901**	**−3,919**
Animal output value	I	–	258,291	249,254	−9,036
Animal specific input value	M	–	174,470	169,353	−5,117
Other inputs	N	–	84,693	84,297	−396

and beef (5%). The prices of the remaining commodities decrease by less than 5%.

One of the key strengths of the CAPRI-FT model is that it can provide results for different farm types. Table 13.5 reports income effects disaggregated by farm type for the EU-27 countries. Absolute income losses are the largest for farm types specializing in olive cultivation (1.5 billion €), followed by general field and mixed cropping (1.3 billion €) and dairying (1 billion €). When considering farm size, medium-sized farms are most affected, with losses of 3.1 billion €. In relative terms, the olives farm type loses the most income, at 17%, followed by mixed permanent crops and pig and poultry fattening (granivores), with losses of 11% for each.

The impacts reported in Table 13.5 might hide sizeable effects within the farm types. To gain further insight, as well as reporting the income change over all farm types, Fig. 13.5 reports the income distributional effects by farm type over all 2450 farm types (normalized to percentage terms, where 1 = 100%) available in CAPRI-FT. The figure shows the absolute income changes relative to the baseline (in million €) for the EU-27 countries. The main curve, for all farms (the bold black curve) is obtained by sorting, in ascending order, all of the farm types according to the size of the income change and cumulating the amount until all farm types (100%) are reported. Similar logic is applied for each farm type, except that the curves are placed successively on the top horizontal (y-) axis until all are reported. The figure also reports the per hectare income loss (given in parentheses in the legend).

The figure reveals that for the majority of farm types, the negative slope dominates, which means that, for the most part, income losses are realized across all farms in the EU-27. This is also reflected by the aggregate curve which cumulates the income changes over all farm types available in CAPRI-FT (the bold curve) (i.e. most farm types are located on the downward-sloping segment of the curve). The total aggregated income of all farm types (which is represented by the end point of the bold curve) drops by approximately 6%, or 10.6 billion € (the same value as given in Table 13.5). A particularly sizable income loss is observed for the farm types specializing in olives, in general field and mixed cropping and in dairying. Note that the total income effect of each farm type is

Table 13.5. Aggregated income change in the EU-27 countries by farm type based on the World Trade Organization (WTO) Doha Round trade liberalization scenario as predicted by CAPRI-FT.

	Million €			
Farm type/economic size class	Baseline	WTO	Absolute change to baseline	% to baseline
FARM TYPE				
Cereals, oilseed and protein crops	20,108	19,388	–720	–4
General field and mixed cropping	29,604	28,365	–1,239	–4
Dairying	27,200	26,135	–1,071	–4
Cattle + dairying rearing, fattening	7,459	6,871	–587	–8
Sheep, goats and other grazing livestock	7,145	6,586	–558	–8
Granivores	4,822	4,270	–552	–11
Mixed livestock holdings	4,750	4,410	–340	–7
Mixed crops-livestock	13,721	12,928	–793	–6
Vineyards	7,249	7,150	–99	–1
Fruit and citrus fruit	5,015	4,499	–516	–10
Olives	8 798	7,261	–1,537	–17
Permanent crops mixed	2,361	2,098	–263	–11
Horticulture	1,422	1,379	–44	–3
Residual farms	43,589	41,339	–2,249.7	–5.2
Total farms	183,249	172,679	–10,570	–5.8
ECONOMIC SIZE CLASS				
<16 ESU (ESC 1)	30,613	28,366	–2,247	–7
≥16 and <100 ESU (ESC 2)	51,152	48,033	–3,119	–6
≥100 ESU (ESC 3)	57,896	54,942	–2,953	–5

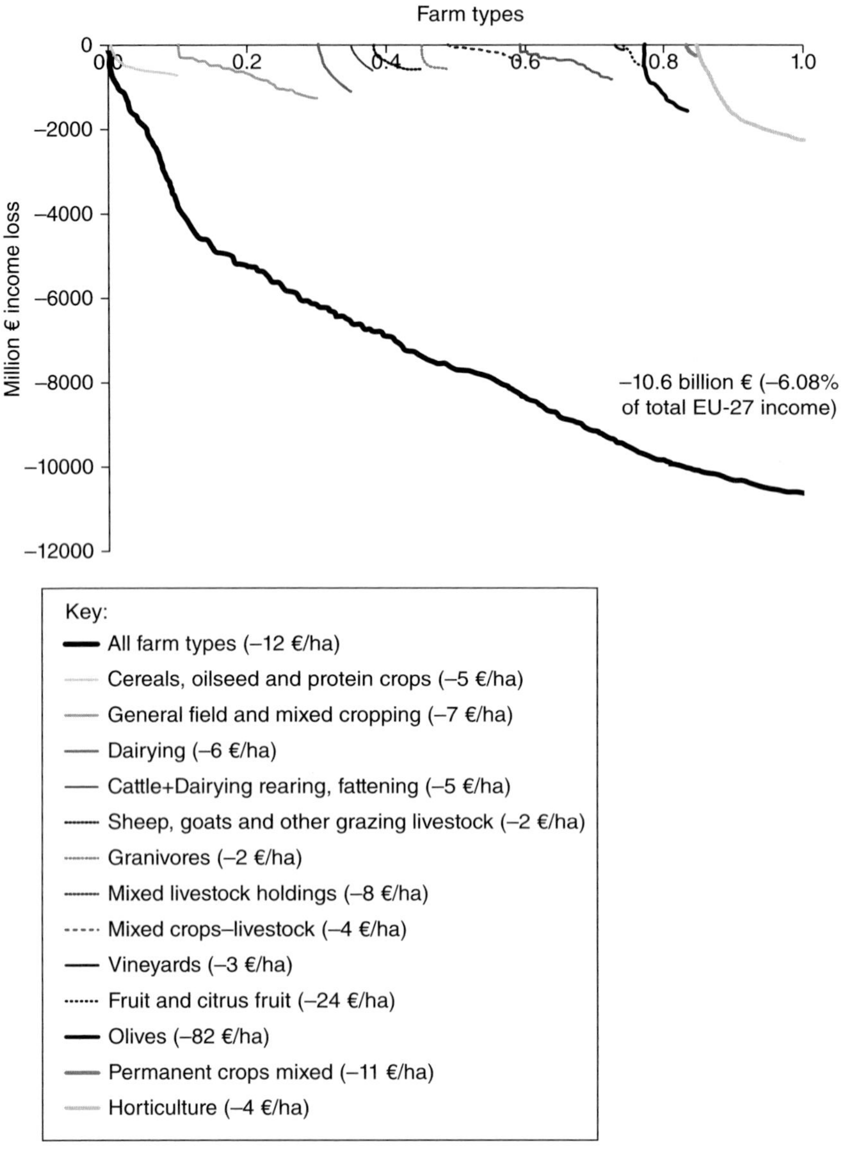

Fig. 13.5. The distribution of income change by farm type in EU-27.

indicated by the endpoint of its respective curve and is identical to the total income loss listed in Table 13.5. The largest income loss is observed for farm types specializing in olives and fruits (82 and 24 €/ha, respectively), while the average income loss over all farms is 12 €/ha.

13.6 Summary and Conclusions

After a general introduction to farm models, we have illustrated an application of the CAPRI-FT model to a policy scenario that cannot typically be simulated with a stand-alone farm model. To model the impact of trade liberalization, the model must be capable of capturing not only price feedbacks but also trade policies, which is not usually possible with most other farm models. The linkage between the farm type layer model and the partial equilibrium model developed in CAPRI allows the capture of these types of policy scenarios and provides results not only at the aggregate regional level but also at the disaggregated farm type level.

Notes

[1] This may result in unrealistic extreme specialization in production (especially in linear programming models) in the aggregated model, and the aggregate production may significantly differ from the sum of individual farm production (Önal *et al.*, 2009). McCarl (1982) developed the 'historical crop mix approach' in mathematical programming models to reduce overspecialization. This approach assumes that the expected production decisions fall within the historical ranges of crop mixes. In acknowledging the restrictiveness of the historical crop mix approach, several studies have extended historical mixes by generating crop mixes, either by using expert opinion in utilizing an auxiliary farm level model that allows increased planting flexibility (e.g. Adams *et al.,* 1985), or by generating 'synthetic' crop mixes that are estimated based on observed land allocation responses to a farm's own prices and cross prices (Chen and Önal, 2012). Howitt (1995) introduced positive mathematical programming (PMP) as an alternative method to treat extreme product specialization in mathematical programming supply response models. The PMP approach avoids extreme product specialization regardless of the aggregation level.

[2] Some 280 NUTS2 regions are represented in the CAPRI supply model. NUTS is an acronym for Nomenclature of Units for Territorial Statistics, and refers to the territorial subdivisions of the member states of the EU.

[3] The programming models used in CAPRI are a type of hybrid approach. They combine a Leontief technology for variable costs that incorporates low- and high-yield variants for the different production activities with a non-linear cost function that captures the effects of labour and capital on farmers' decisions. The non-linear cost function allows for perfect calibration of the models and a smooth simulation response that is rooted in observed behaviour.

[4] The FSS provides harmonized data on the structure of agricultural holdings in terms of land use, livestock numbers, farm labour force, machinery and equipment, and participation in rural development programmes. The complete agricultural census is updated every 10 years (with intermediate sample surveys).

[5] The FADN provides accounting data for a sample of commercial agricultural holdings. The survey is conducted annually.

[6] The minimum size of a farm considered in a farm type is derived from the Member-State-specific regulations for recording the FSS; it also depends on the application of the data protection rules by Eurostat.

[7] For a detailed description of the methodology, see Gocht and Britz, 2011.

[8] All countries have agreed, in the WTO context, not to increase the rate of duty beyond a certain tariff level (bound tariff); these tariffs are also called most favoured nation (MFN) tariffs. MFN tariffs exist for different products and are not differentiated by importer (i.e. countries exporting into an importing country). The tariffs that are applied are those that are actually applied by the country group. There are exceptions to the rule that all importers face the same import MFN tariff. The first exception is that the least developed countries under the everything-but-arms agreement face lower import tariffs. The second exception is the European Commission and African, Caribbean and Pacific (EC-ACP) partnership programme for ACP countries. The third exception is the free-trade areas (bilateral agreements) with no tariffs. The last exception is the application of TRQs. A TRQ is a defined quota of an imported good that is available either bilaterally to a specific trading partner or globally to all trading partners, to which a so-called lower 'in quota tariff' is applied. For imports beyond the quota, MFN tariffs are applied. Another exception to the rule of equal tariffs is the instrument of variable import levies, which reduces tariffs if the import price is greater than a specified minimum border price in the EU.

[9] The developed countries are the EU-15, EU-10, Norway, Bulgaria, Romania, the Western Balkans, the rest of Europe, Russia, Belarus, Ukraine, the USA, Canada, Japan, Australia and New Zealand. All other countries are developing countries.

References

Adams, R.M., Hamilton, S.A. and McCarl, B.A. (1985) An assessment of the economic effects of ozone on U.S. Agriculture. *Journal of the Air Pollution Control Association* 33, 938–943.

Armington, P.S. (1969) A theory of demand for products distinguished by place of production. *IMF Staff Papers* 16, 159–178.

Baranger, E., Clodic, M., Galko, E., Jayet, P. and Zakharov, P. (2008) Improvement of the AROPAj model covering a large range of agricultural activities at wide (UE) and high resolution (mapping of farm types) scales. *Paper presented at the 107th EAAE Seminar 'Modeling of Agricultural and Rural Development Policies', Sevilla (Spain), January 29th–February 1st, 2008.* Available at: http://ageconsearch.umn.edu/bitstream/6664/2/cp08ba02.pdf (accessed 16 March 2016).

Bertelsmeier, M., Kleinhanss, W. and Offermann, F. (2003) Aufbau und. Anwendung des FAL-modellverbunds für die politikberatung. *Agrarwirtschaft/German Journal of Agricultural Economics* 52, 175–184.

Britz, W. (2008) Automated model linkages: the example of CAPRI. *Agrarwirtschaft/German Journal of Agricultural Economics* 57, 363–367.

Britz, W. and Hertel, T.W. (2011) Impacts of EU biofuels directives on global markets and EU environmental quality: an integrated PE, global CGE analysis. *Agriculture, Ecosystems and Environment* 142, 102–109.

Britz, W. and Witzke, P. (2014) CAPRI model documentation 2014. Available at: http://www.capri-model.org/docs/capri_documentation.pdf (accessed 16 March 2016).

Chen, X. and Önal, H. (2012) Modeling agricultural supply response using mathematical programming and crop mixes. *American Journal of Agricultural Economics* 94, 674–686.

Cools, J., Broekx, S., Vandenberghe, V., Sels, H., Meynaerts, E. *et al.* (2011) Coupling a hydrological water quality model and an economic optimization model to set up a cost-effective emission reduction scenario for nitrogen. *Environmental Modelling and Software* 26, 44–51.

Deppermann, A., Grethe, H. and Offermann, F. (2014) Distributional effects of CAP liberalisation on western German farm incomes: an ex-ante analysis. *European Review of Agricultural Economics* 41, 605–626.

Domínguez, I.P., Bezlepkina, I., Heckelei, T., Romstad, E., Lansink, A.O. and Kanellopoulos, A. (2009) Capturing market impacts of farm level policies: a statistical extrapolation approach using biophysical characteristics and farm resources. *Environmental Science and Policy* 12, 588–600.

Gallego-Ayala, J., Gómez-Limón, J.A. and Arriaza, M. (2013) Towards a sustainable irrigated agriculture: analysis of water pricing instruments. *Paper presented at the XVI Applied Economics Meeting, Granada, 6–7 June 2013*. Available at: http://encuentros.alde.es/anteriores/xiiieea/trabajos/pdf/020.pdf (accessed 16 March 2016).

Gocht, A. (2010) *Update of a Quantitative Tool for Farm Systems Level Analysis of Agricultural Policies (EU-FARMS) – JRC Scientific and Technical Reports*. Publications Office of the European Union. Available at: http://ftp.jrc.es/EURdoc/JRC53518.pdf (accessed 14 July 2016).

Gocht, A. and Britz, W. (2011) EU-wide farm type supply models in CAPRI – how to consistently disaggregate sector models into farm type models. *Journal of Policy Modeling* 33, 146–167.

Gocht, A. and Münch, T. (2003) Application of strategic management software for large scale farms in eastern Germany to assess MTR impact. In: Balmann, A. and Lissitsa, A. (eds) *Large Farm Management. Studies on the Agricultural and Food Sector in Central and Eastern Europe*, Volume 20. AgriMedia, Halle, Germany, pp. 180–194.

Gocht, A. and Witzke, H.-P. (2014) *CAPRI Modelling System – Common Agricultural Policy Regionalised Impact Modelling System: Market model.* Available at: http://www.capri-model.org/dokuwiki/doku.php?id=capri:concept:market (accessed 17 March 2016).

Gocht, A., Britz, W., Adenäuer, M., Ciaian, P. and Gomez y Paloma, S. (2011) *Farm Level Policy Scenario Analysis*. Report EUR 24787 EN, JRC Scientific and Technical Reports, European Commission Joint Research Centre. Publications Office of the European Union, Luxembourg.

Gocht, A. Britz, W., Ciaian, P. and Gomez y Paloma, S. (2013) Farm type effects of an EU-wide direct payment harmonisation. *Journal of Agricultural Economics* 64, 1–32.

Gocht, A., Witzke, H.P., Espinosa-Goded, M. and Gomez y Paloma, S. (2014) *CAPRI-FARM Methodological Improvements: Selection Routine, Input Allocation and GHG Reporting*. Report EUR 24787 EN, JRC Scientific and Technical Reports, European Commission Joint Research Centre. Publications Office of the European Union, Luxembourg.

Golub, A., Hertel, T.W., Lee, H.-L. and Ramankutty, N. (2006) *Modeling Land Supply and Demand in the Long Run.* GTAP Resource No. 2111, Global Trade Analysis Project, Center for Global Trade Analysis, West Lafayette, Indiana.

Gómez-Limón, J.A. (2004) Water pricing: analysis of differential impacts on heterogeneous farmers. *Water Resources Research* 40, 1–12.

Green, H.A.J. (1964) *Aggregation in Economic Analysis*. Princeton University Press, Princeton, New Jersey.

Hazell, P.B.R. and Norton, R.D. (1986) *Mathematical Programming for Economic Analysis in Agriculture*. Macmillan, New York.

Heckelei, T. (2013) General methodological issues on farm level modelling. In: Ciaian, P., Espinosa, M., Gomez y Paloma, S., Heckelei, T., Langrell, S., Louhichi, K., Sckokai, P., Thomas, A. and Vard, T. (eds) *Farm Level Modelling of CAP: A Methodological Overview*. Report EUR 25873 EN, JRC Scientific and Policy Reports, European Commission Joint Research Centre. Publications Office of the European Union, Luxembourg, pp. 29–34.

Heidecke, C., Wagner, A., Kreins, P., Venohr, M. and Wendland, F. (2014) Options for meeting WFD targets beyond 2015 in a highly polluted river basin in Germany. *Poster paper prepared for presentation at the EAAE 2014 Congress 'Agri-Food and Rural Innovations for Healthier Societies', August 26 to 29, 2014, Ljubljana, Slovenia.* Available at: http://ageconsearch.umn.edu/bitstream/182931/2/EAAE_heidecke.pdf (accessed 16 March 2016).

Hosseini, S.S. and Spriggs, J. (2009) A methodological aggregation-consistent individual level supply response model. *Journal of Agricultural Science and Technology* 11, 101–114.

Howitt, R. (1995) Positive mathematical programming. *American Journal of Agricultural Economics* 77, 329–342.

Jansson, T. and Heckelei, T. (2011) Estimating a primal model of regional crop supply in the European Union. *Journal of Agricultural Economics* 62, 137–152.

Jansson, T., Kuiper, M., Banse, M., Heckelei, T. and Adenäuer, M. (2008) Getting the best of both worlds? Linking CAPRI and GTAP for an economywide assessment of agriculture. *Paper prepared for the Eleventh Annual Conference on Global Trade Analysis (GTAP) 'Future of Global Economy' Helsinki, June 12–14, Finland.* Available at: https://www.gtap.agecon.purdue.edu/resources/download/3756.pdf (accessed 16 March 2016).

Leip, A., Marchi, G., Koeble, R., Kempen, M., Britz, W. and Li, C. (2008) Linking an economic model for European agriculture with a mechanistic model to estimate nitrogen and carbon losses from arable soils in Europe. *Biogeosciences* 5, 73–94.

Lengers, B. and Britz, W. (2012) The choice of emission indicators in environmental policy design: an analysis of GHG abatement in different dairy farms based on a bio-economic model approach. *Review of Agricultural and Environmental Studies* 93, 117–144.

Louhichi, K., Kanellopoulos, A., Janssen, S., Flichman, G., Fonseca, M.B. *et al.* (2010) FSSIM, a bioeconomic farm model for simulating the response of EU farming systems to agricultural and environmental policies. *Agricultural Systems* 103, 585–597.

Louhichi, K., Espinosa, M., Ciaian, P. and Gomez y Paloma, S. (2013) Farm-level models for EU policy analysis: review of recent literature and comparison of most relevant models. In: Ciaian, P., Espinosa, M., Gomez y Paloma, S., Heckelei, T., Langrell, S., Louhichi, K., Sckokai, P., Thomas, A. and Vard, T. (eds) *Farm Level Modelling of CAP: A Methodological Overview.* Report EUR 25873 EN, JRC Scientific and Policy Reports, European Commission Joint Research Centre. Publications Office of the European Union, Luxembourg, pp. 17–27.

Louhichi, K., Ciaian, P., Espinosa, M., Colen, L., Perini, A. and Gomez y Paloma, S. (2015) *An EU-Wide Individual Farm Based Model for Common Agricultural Policy Analysis (IFM-CAP). First Application to Crop Diversification Policy.* Report EUR 26910 EN, JRC Scientific and Policy Reports, European Commission Joint Research Centre. Publications Office of the European Union, Luxembourg.

McCarl, B.A. (1982) Cropping activities in agricultural sector models: a methodological proposal. *American Journal of Agricultural Economics* 64, 768–772.

Offermann, F., Kleinhanss, W., Hüttel, S. and Küpker, B. (2005) Assessing the 2003 CAP reform impacts on German agriculture using the farm group model FARMIS. In: Arfini, F. (ed.) *Modelling Agricultural Policies: State of the Art and New Challenges.* Monte Universita Parma Editore, Parma, Italy, pp. 546–564.

Önal, H., Chen, X., Khanna, M., and Huang, H. (2009) Mathematical programming modeling of agricultural supply response. *Paper presented at the Agricultural and Applied Economics Association's 2009 AAEA and ACCI Joint Annual Meeting, Milwaukee, WI, July 26–28, 2009.* Available at: http://ageconsearch.umn.edu/bitstream/49469/2/AAEA%202009%20Onal%20et%20al.pdf (accessed 16 March 2016).

Pelikan, J., Britz, W. and Hertel, T.W. (2015) Green light for green agricultural policies? An analysis at regional and global scales. *Journal of Agricultural Economics* 66, 1–19.

Schroeder, L.A., Gocht, A. and Britz, W. (2015) The impact of Pillar II funding: validation from a modelling and evaluation perspective. *Journal of Agricultural Economics* 66, 415–441.

Stoker, T.M. (1993) Empirical approaches to the problem of aggregation over individuals. *Journal of Economic Literature* 31, 1827–1874.

van Meijl, H., van Rheenen, T., Tabeau, A. and Eickhout, B. (2006) The impact of different policy environments on agricultural land use in Europe. *Agriculture, Ecosystems and Environment* 114, 21–38.

Verburg, P., van Berkel, D.B., van Doorn, A.M., van Eupen, M. and van den Heiligenberg, H.A. (2010) Trajectories of land use change in Europe: a model-based exploration of rural futures. *Landscape Ecology* 24, 217–232.

WTO (2008) *Revised draft modalities for agriculture.* World Trade Organization. Available at: www.wto.org/english/tratop_e/agric_e/agchairtxt_feb08_e.doc (accessed 16 March 2016).

14 Conclusions: The State-of-the-art of Farm Modelling and Promising Directions

Thomas Heckelei*
Institute for Food and Resource Economics, University of Bonn, Bonn, Germany

14.1 A Look Across the State-of-the-art: Why All This Development?

The chapters in this book demonstrate the impressive advances that have been made in the mathematical modelling of the farm in recent years. Interestingly, however, the single farm itself is rarely in focus. Almost all of the empirical applications that have been presented aim to assess policy that is relevant for a larger scale than the individual farm. This is demonstrated in the application of the methodologies to a larger sample of farms (e.g. Rault and Hennessy, Chapter 5; or Gillespie *et al.*, Chapter 6), by having a modelling structure that links farms in some way (e.g. Barnes *et al.*, Chapter 10; Revoredo-Giha, Chapter 12), or by providing a linkage to joint policy scenarios or higher scale models (Ferreira *et al.*, Chapter 11; Gocht *et al.*, Chapter 13). Not even one of the methodologies presented aims to provide information for better farm management, even though the origin of the techniques presented and discussed may lie in farm management tools such as mathematical programming (Arfini *et al.*, Chapter 2; Shrestha and Vosough Ahmadi, Chapter 3).

So, clearly, the book is policy oriented. The question then arises: why farm-level modelling? Why not rather continue with the aggregate, representative economic modelling that has been the approach of choice during many years of impact assessment of agricultural policy, both in Europe in assessing the Common Agricultural Policy (CAP) (see M'barek *et al.*, 2012) and in other OECD (Organisation for Economic Co-operation and Development) countries (van Tongeren *et al.*, 2001); why not continue with an approach that seemed more targeted at the geographical scale of the policies?

Blanco started this discussion in Chapter 1 by examining the widening role of farm-level modelling in policy impact assessment and looking at the types of models that are needed for this assessment. In the next section, 'Fostering Farm-level Modelling', I would like to examine four possible reasons for the increased attention that is now being paid to farm-level policy analysis:

- The policy instruments relevant for the agricultural sector are nowadays more targeted at the farm directly and are differentiated by farm characteristics.
- Political attention has turned more towards the distribution of indicators, and often to the tails of these distributions rather than being satisfied with averages, as before.
- There is growing awareness that the bio-physical and economic processes generating

*E-mail: thomas.heckelei@ilr.uni-bonn.de

challenges for policies and the impacts of those policies do not linearly aggregate to a larger scale.

- The simultaneous development of methodologies, databases and computational power allows farm-level analysis for larger scales that was not possible two decades ago.

14.2 Fostering Farm-level Modelling

14.2.1 Changes in (agricultural) policy instruments

For the first three decades of the CAP, the main agricultural policy objective was income support (although it was not the only one proclaimed). The policy instruments pursuing this target were market-level policies embedded in common market organizations. They raised market and farm-gate prices through intervention and border protection, later accompanied by export subsidies when self-sufficiency rates moved above 100%. Quantitative policy assessment with models did not play a role in the European Union (EU) until perhaps the late 1980s (Henrichsmeyer and Wolf, 1992). The models developed were based on sectoral accounting, and slowly developed into the suite of partial and general equilibrium models that are available today for pursuing the impact assessment of market and trade policies for the agricultural sector (M'barek *et al.*, 2012). A similar analytical capacity in an environment of comparable policy goals developed in the USA, but the USA was generally ahead of Europe in the active use of policy simulation models (Robinson *et al.*, 1990; FAPRI, 2015). It is not necessarily a natural choice to assess income support policies with only aggregate or average indicators (see further below), but at least the main policy instruments that were used provided good arguments for a market-level analysis.

The MacSharry reform of the CAP in 1992 marked the fundamental change of European agricultural policy away from market and trade policies to the implementation of instruments at the farm level. Even though the rules for the calculation of direct payments and set-aside percentages were general, the actual values were conditional on farm characteristics and choices, and thereby farm specific and, in part, even endogenous. During the same period, the increasing awareness of agri-environmental problems led to a manifestation of a new set of policy instruments within the newly created second or 'rural development' pillar of the CAP. As environmental problems associated with agricultural production strongly depend on farm, local and regional characteristics, it was a natural development that the share (number and budget wise) of policy instruments directly targeted at individual farm management started to slowly rise (EU Commission, 2013, p. 4). Climate change concerns with respect to mitigation and adaptation have become the dominant societal and policy concerns in the environmental context in recent years, and related measures were introduced into the second pillar of the CAP as well. Similar general policy developments occurred in the USA, although farm bills (legislation) had already comprised farm-specific elements (for example, base acreages in market loan programmes) for some time (Effland, 2000).

Increasing food safety concerns over the last two decades gave rise to discussion or implementation of regulations (e.g. standards) that integrated the whole food supply chain, and thus included farms as the origin of most raw material for food production. Food safety risks and optimal strategies to prevent and deal with food safety crises depend on the interplay of all actors in the chain. Some of these strategies may be supported under the competitiveness axis of the rural development measures for quality improvements (ENRD, 2015). Finally, animal welfare issues have surfaced more and more strongly in recent years and have led to consideration of animal welfare standards in private certification schemes, but also to the possibility of applying animal welfare payments as part of Pillar 2 in the most recent CAP reform (ENRD, 2015).

Moreover, the first pillar of the CAP has undergone further reform steps with increased farm-level implications. Cross-compliance (CC) measures and their probabilistic control scheme were introduced with the Agenda 2000 reform package. The latest reform of the CAP (EU Commission, 2013) has brought further farm-level dimensions into the design of Pillar 1 instruments. The 'Greening of the CAP' made 30% of the direct payments to the farmer conditional upon compliance with rules on crop diversity, the setting aside of an Ecological Focus Area

(EFA) and grassland preservation on the farm. Exceptions to this (e.g. small farms) also depend on specific farm characteristics. Consequently, the overall impact of the policy depends on the distribution of farm characteristics across Europe, as the level of change necessary for compliance depends on the state of the farm before policy implementation and this is impossible to assess based on averages across farms.

The development of some of the modelling approaches discussed in this book were supported by this change in policy concerns and a need for the assessment of related options and implemented instruments. Positive mathematical programming (PMP) type models thrived in the late 1990s and after 2000 because they allowed the combination of a more robust and empirically based behavioural specification with the technology-rich formulation of mathematical programming models. This rendered the technique specifically suitable for assessing policy measures that are linked to biophysical (i.e. production and environmental) indicators at the farm level (Heckelei and Britz, 2005; Heckelei *et al.*, 2012). The contribution by Arfini *et al.* (Chapter 2) shows an advanced development in this area in which the PMP type model is specified using an econometric specification of a total cost function. Other techniques relevant for farm-level agri-environmental analysis are integrated farm type models in the CAPRI system (Gocht *et al.*, Chapter 13) as well as the modelling of greenhouse gas (GHG) emissions (MacLeod and Eory, Chapter 7). Modelling approaches to animal health and wider biosecurity risks are presented by Rault and Hennessy (Chapter 5). The development of the microsimulation literature reported by Donoghue (Chapter 9) was most probably little affected by changing agricultural policy environments, but it certainly offers relevant contributions to the tools for analysis.

14.2.2 Changes in relevance and understanding of policy impact indicators

In Section 14.2.1, changes in policy issues and related instruments were discussed as reasons for the increased interest in farm modelling techniques. Here, we would like to add that the interest in the type of information needed, i.e. the indicators for policy assessment, may have changed as well. We also start with the income support policies mentioned above: we might wonder why the objective of income support in the CAP did not directly turn political attention to the farm household level. Where, if not at the farm household level, can one assess the need for income support or the likely impact of policy instruments on the deviation of income from a target (Hill, 2000)? Originally, there were probably several reasons for leaving such analysis to market-level models. First, representative farm household data were (and are still) not available – whereas income generated by farm activities is less of a data issue, other income sources are not part of typical farm databases. Neither are farm households sufficiently represented in general household surveys, nor do they contain sufficient information on agricultural activities. Secondly, there was little political interest in confirming that a large part of the agricultural 'income' support budget did not go to households in need. Finally, the aggregate market-level analysis was the natural approach for the depiction of the dominating policy instruments at the time, and a connection to farm household level models was too challenging in the infancy of quantitative agricultural policy assessment.

However, public interest in the EU has changed towards a deeper understanding of who benefits from public support by the CAP. Moreover, any changes in the system of direct payments will visibly change the amount of money received by farms from the government, thus rendering single farm and distributional impacts more interesting than under a price-support scheme. Such changes in the direct (in)equity impacts of payments not only refer to differences between poor and non-poor farmers in a country, but also to differences across EU Member States, as in recent CAP reform discussions on redistributional effects. Representative farm-level modelling quite naturally allows the calculation of such distributional indicators or drawing of the appropriate graphs. This was done, for example, by Keeney (2000) and Gocht *et al.* (2013) using the farm type layer in CAPRI presented by Gocht *et al.* (Chapter 13) in this book.

As environmental concerns related to agricultural activities have grown over recent decades, environmental indicators as model outputs have also gained importance. Large-scale models such as CAPRI have been developed (Heckelei and Britz, 2000; Britz and Witzke, 2014), and have added agri-environmental indicators such as regional nutrient balances, GHG emissions, or even (crop) biodiversity indicators. At the same time, single-farm modelling approaches have gone into detailed representation of technological and management processes to better capture environmental externalities (Janssen and van Ittersum, 2007). At first, both of these approaches had significant limitations for the evaluation of policy challenges and impacts, as they did not deliver relevant indicator distributions across space. The large-scale approach relied on indicator averages across larger regions, hiding environmental 'hotspots' behind an averaging-out effect. For example, local and subregional nitrate pollution problems may be masked by moderate regional nitrate balance surpluses. The modelling of single (or a few) farms did not give the necessary representativeness for political decision makers to design instruments at a larger administrative scale. By now, the two sides have moved strongly towards each other, as we may infer from Gocht *et al.* (Chapter 13) and from the higher spatial resolutions of regional modelling results (Leip *et al.*, 2008) on the one hand, and from the increasing number of farms represented in farm-level modelling exercises as visible, for example in O'Donoghue (Chapter 9) or Louhichi *et al.* (2015), on the other hand.

Here, we have argued that a change of interest by the public and by policy makers towards a certain type of information – specifically, the distribution of indicators across farms or space – might have triggered changes in modelling approaches. In Section 14.2.3, we look at the growing awareness (of modellers and policy makers) in the inability of aggregate models to depict biophysical and economic processes such that certain indicators are not calculated with the desired degree of accuracy, even at the aggregate level.

14.2.3 Key biophysical or economics processes do not aggregate linearly

The need to calculate meaningful environmental indicators is another reason for the development of farm-level, sometimes even spatially explicit, models. Farm-level modelling may be necessary to provide the desired degree of accuracy of indicators for policy challenges and impacts even at the aggregate level. Nitrate emissions may serve (again) as an example here: Farm-level conditions such as soil properties and weather, as well as the quantity, type and timing of nitrogen fertilization, determine the occurrence and extent of nitrate leaching from farming activities through non-linear processes. Therefore, the application of aggregate models using average values for parameters or driving forces will lead to potentially significant errors of calculated indicators. This is why McLeod and Eory (Chapter 7) consider certain Intergovernmental Panel on Climate Change (IPPC) approaches to estimating regional GHG emissions from agricultural activities as a method of 'last resort' and suggest, rather, the farm-level modelling of GHG emissions and subsequent aggregation of these. Similar arguments hold for other biophysical processes generating other environmental externalities. This general insight also leads to farm-level modelling activities that offer a substantial representativeness at a larger scale, such as the AROPAj model (De Cara and Jayet, 2011) or the new individual farm model activities by Louhichi *et al.* (2015) that have potentially high policy relevance.

The principle discussed here does not relate only to biophysical processes, but also to economic processes. We know from economic aggregation theory in supply that only very restrictive assumptions on technological heterogeneity in a firm's population allows the transfer of individual behavioural models to the aggregate without errors (Önal and McCarl, 1989). Analysts that believe in such restrictive assumptions can hardly be the reason for the strong presence of micro-based aggregate models of policy impact analysis in the (agricultural) economics literature. It seems rather to be the experience (or hope) that parameterizing aggregate models based on aggregate empirical observations may capture some of the inherent aggregation error and that this structure is stable. This hope breaks down, however, if the structure of the underlying population changes or if relevant interactions between individual farms occur over the range of analysis. When this happens, a representation of farm (technological) heterogeneity and interaction is required to capture appropriately the impacts of policies and other economic conditions on aggregated production activities and related indicators. This gives another reason for farm-level modelling, even if results at a larger, policy-relevant scale are targeted.

Heterogeneity in technology and efficiency is the focus of Gillespie *et al.* (Chapter 6). They give an overview and example from the extensive literature on farm-efficiency analysis that has emerged in the last two decades. Such an analysis is crucial for understanding how productivity at the aggregate level arises. An implied possibility to decompose productivity changes into changes in technical and allocative efficiency, as well as technological changes, is required for understanding and modelling the impact of policies targeting productivity. In combination with spatial or network analysis, this may provide the informational basis for analysing the adoption of new technologies.

Land rental and purchasing behaviour, and the resulting prices, are another example of the relevance of farm interactions when moving from individual to territorial modelling: the supply of land to each individual farm does depend on the behaviour of other farms. We know from econometric analyses that these mainly local interactions co-determine the conditions for long-term farm development and decide to what extent landowners or operators benefit from direct payments (Kilian *et al.*, 2012), and how direct payments influence farm survival (Storm *et al.*, 2015).

The adoption of new technologies and land–market interactions are good examples of dynamic economic processes that do not simply aggregate linearly. Farm interactions in technology adoption and land markets may lead to so-called emerging properties that go beyond the effects captured by simply summing over individual decisions. Social networks and spatial interactions render the outcomes of technology adoption or land transactions as dynamically depending on other farm conditions and behaviours. Such complex scaling effects may in principle be treated with agent-based modelling (ABM), as discussed by Barnes *et al.* (Chapter 10), even though such models focus more on the connection to a biophysical model and a realistic way of representing decision making. Generally, agent-based models allow the explicit modelling of individual farms and the structural developments arising from their interactions and, at the same time, allow for scaling up to territorial coverage that is policy relevant. The pioneering work by Balmann (1997) and Happe *et al.* (2006), and subsequent developers and users of the agricultural policy simulator, AgriPoliS, originally put this approach on to the agricultural economics map. The current main limitations are data availability and empirical validation.

14.2.4 Simultaneous development of databases, computing power and techniques

All of the issues mentioned above that focus on an increased demand for results from policy-relevant farm-level modelling would lead to little if the supply side of modelling approaches had not done its part. The techniques presented in this book are a reflection of the considerable advances made in the last two decades. They are possible through a simultaneous interplay of changes in data availability, computing power and methodological development.

Individual farm data is nowadays accessible to a wider range of scientists, not only through the accumulation of case-specific surveys, but also in the context of databases regularly updated by public statistical offices. For example, the individual sample farm data of the Farm Accountancy Data Network (FADN) database of the European Commission was originally only accessible to researchers from institutions who are responsible for administering the national segment of the data network. Today though, researchers involved in projects under the research framework programmes of the EU may ask and are usually granted access to individual farm data if they can make the usefulness of the data plausible in the context of the project's objectives, and if they commit to confidentiality rules. Moreover, some national statistical offices offer detailed individual samples or even census data information on farms, including their location. Such databases are highly valuable for improving empirical understanding of farm-level decisions and developments, and also for the representative populations used in simulation models. Another very important development for agri-environmental farm-level modelling is the enormous push in constructing and improving *ex post* maps of biophysical indicators (soil properties, time series of weather variables, irrigation, etc.) and projected weather and yield distributions for the future in the context of the analysis of climate change impacts. Here, economists can benefit from the work of colleagues in other disciplines.

Another development that has eased the application of farm-level modelling is the progress in the hardware and software that is available for computation. Advances in computational speed now allow many thousands of moderately complex linear or quadratic programming models to be solved in a few seconds, so there is no longer any relevant limit to linking them to comparative static market-level models (Gocht *et al.*, Chapter 13). Dynamic mixed-integer farm models and ABMs that include farm interaction will generally require longer solution times. However, it is more likely that a lack of data or the complexity of developing and maintaining larger modelling systems in an era of short-term academic incentives and limited institutional continuity constitute the relevant challenges. Some help certainly comes from recent developments on the software side that exploit the features of algebraic modelling environments for debugging and the analysis of results by graphical user interfaces (Britz, 2014).

Progress in data availability and information technology (IT) would still not help if the methodological and theoretical developments of our discipline did not allow the use of these new opportunities. Every single chapter in this book witnesses these achievements and makes farm-level modelling an exercise with increasing relevance for the larger scale modelling of policy impacts.

14.3 Subjective View on Current Limitations and Promising Future Directions

Having so far described and praised the recent achievements of farm modelling presented in the chapters of this book and beyond, I would like to turn in the final paragraphs of this concluding chapter to what I consider to be important limitations to, and relevant future research directions in, farm-level modelling. The choices that I have made in discussing some things and leaving out others are clearly subjective, but I have tried to be guided by relevance for an improved economic analysis of societal concerns and policies related to the farming sector.

1. There is still some way to go to consistently connect individual farm-level modelling to market-level impacts and feedbacks on a larger scale. Ding *et al.* (Chapter 8) and Gocht *et al.* (Chapter 13) offer two approaches to scaling up to a territorial coverage at the level relevant for current agricultural policy design. They do this by employing a 'representative' or 'farm type' approach, while sacrificing some of the existing individual heterogeneity of farms. Furthermore, they stick to a comparatively static setting with a short to medium term horizon. Louhichi *et al.* (2015), in contrast, currently develop an individual farm-level model for the EU based on the FADN database. They have progressed significantly in specifying the comparatively static farm models with exogenous output and variable input prices. However, further investments in linking with market-level models for price feedback are needed (and ongoing) and are crucial for rendering such large-scale, farm-level modelling efforts meaningful.

2. The first section of the book on individual and separate farm-level models illustrates the strong advances that have been made in modelling farm-level heterogeneity with respect to adaptation, risk behaviour, biosecurity management, efficiency and GHG emissions. The second section of the book describes significant innovations in upscaling farm-level behaviour, although the upscaling is generally restricted to less complex issues and involves considerable simplification. Such simplification either ignores relevant local and dynamic farm interactions or restricts attention to unrealistically homogenous input and output markets (Ding *et al.*, Chapter 8; O'Donoghue, Chapter 9; Gocht *et al.*, Chapter 13), and relies on rather abstract modelling techniques with considerable distance from real-world heterogeneity and structural complexity (Revoredo-Giha, Chapter 12). Alternatively, the analysis is restricted to specific issues of interest (Barnes *et al.*, Chapter 10; Ferreira *et al.*, Chapter 11) to make the modelling of dynamic and/or spatial interactions feasible. Farm interactions are highly relevant to the joint response of the farming sector to policies at local and regional level, especially in the long run, as the examples in Section 14.2 have illustrated. Therefore, significant effort should be invested to improve modelling in this respect. It is highly desirable to move towards the same level of issue-related relevance in upscaling as already exists at the individual farm level. Such progress would move farm-level modelling to the next level of policy relevance. In my view, the only general approach currently available that promises significant advances in this direction are ABMs, i.e. the employment of simulation models representing the behaviour of individual behavioural agents and mirroring the actual recursive dynamic nature of real-world interactions.

3. One of the reasons why ABMs raise a lot of scepticism in the (agricultural) economics community – which is dominated by applied econometricians – is the lack of satisfactory validation approaches (see also Barnes *et al.*, Chapter 10), and future acceptability will depend on this. Progress here may come from several simultaneous developments. It may start by looking at other disciplines, such as behavioural finance, which are further ahead in empirically specifying stylized ABMs (e.g. Westerhoff, 2012; Grosche and Heckelei, 2014). Improved validation may also come from a more systematic theoretical and empirical specification of individual behaviour in such models. Finally, we need to understand ABMs more as computational laboratories and perhaps start to investigate the consequences of alternative specifications in massive simulation exercises. This may open the possibility of comparing the joint distributions of relevant variables with those observed in reality.

4. The improvement in data availability that has occurred has already been seen as an important driver for advances in farm-level modelling. However, there are still databases out there that are highly relevant for improving policy impact analysis but have not been made available to researchers because of administrative or confidentiality reasons. For example, the EU-wide databases related to the administration and control of direct payments (IACS, Integrated Administration and Control System; or, in German, InVeKoS, Integriertes Verwaltungs- und Kontrollsystem) would be an invaluable source of information as it includes spatially explicit and even field-level data that would be relevant to many analytical and policy issues. Here, possible advances are not in the hands of researchers (apart from them explaining the importance of the data) but rather depend on people in the right position being willing to formulate access conditions in compliance with the regulations.

References

Balmann, A. (1997) Farm-based modelling of regional structural change: a cellular automata approach. *European Review of Agricultural Economics* 24, 85–108.

Britz, W. (2014) A new graphical user interface generator for economic models and its comparison to existing approaches. *Agrarwirtschaft/German Journal of Agricultural Economics* 63, 271–285.

Britz, W. and Witzke, P. (2014) CAPRI model documentation 2014. Available at: http://www.capri-model.org/docs/capri_documentation.pdf (accessed 29 November 2015).

De Cara, S. and Jayet, P.A. (2011) Marginal abatement costs of greenhouse gas emissions from European agriculture, cost effectiveness, and the EU non-ETS burden sharing agreement. *Ecological Economics* 70, 1680–1690.

Effland, A.B.W. (2000) U.S. farm policy: the first 200 years. *Agricultural Outlook* 269, 21–25.

ENRD (2015) *Axes & Measures [for Rural Development*]. European Network for Rural Development, Brussels. Archived data available at: http://enrd.ec.europa.eu/enrd-static/policy-in-action/rural-development-policy-overview/axes-and-measures/en/axes-and-measures_en.html (accessed 29 November 2015).

EU Commission (2013) *Overview of CAP Reform 2014–2020*. Agricultural Policy Perspectives Brief No. 5*/ December 2013, Unit for Agricultural Policy Analysis and Perspectives, Directorate-General for Agriculture and Rural Development, Brussels. Available at: http://ec.europa.eu/agriculture/policy-perspectives/policy-briefs/05_en.pdf (accessed 18 March 2016).

FAPRI (2015) FAPRI models. Food and Agricultural Policy Research Institute, Ames, Iowa. Available at: http://www.fapri.iastate.edu/models/ (accessed 29 November 2015).

Gocht, A., Britz, W., Ciaian, P. and Gomez y Paloma, S. (2013) Farm type effects of an EU-wide direct payment harmonisation. *Journal of Agricultural Economics* 64, 1–32.

Grosche, S. and Heckelei, T. (2014) *Price Dynamics and Financialization Effects in Corn Futures Markets with Heterogeneous Traders*. Food and Resource Economics, Discussion Paper 2014:5, Institute for Food and Resource Economics, University of Bonn, Bonn, Germany.

Happe, K., Kellermann, K. and Balmann, A. (2006) Agent-based analysis of agricultural policies: an illustration of the agricultural policy simulator AgriPoliS, its adaptation and behavior. *Ecology and Society* 11(1): 49.

Heckelei, T. and Britz, W. (2000) Concept and explorative application of an EU-wide regional agricultural sector model (CAPRI-Projekt). In: Heckelei, T., Witzke, H.P. and Henrichsmeyer, W. (eds) *Proceedings of the 65th EAAE Seminar on Agricultural Sector Modelling and Policy Information Systems, Bonn (Germany), March 29–31, 2000*. Vauk Verlag, Kiel, Germany, pp. 281–290.

Heckelei, T. and Britz, W. (2005) Models based on positive mathematical programming: state of the art and further extensions. In: Arfini, F. (ed.) *Modelling Agricultural Policies: State of the Art and New Challenges*. Monte Università Parma Editore, Parma, Italy, pp. 48–73.

Heckelei, T., Britz, W. and Zhang, Y. (2012) Positive mathematical programming approaches – recent developments in literature and applied modelling. *Bio-based and Applied Economics* 1, 109–124.

Henrichsmeyer, W. and Wolf, W. (1992) *SPEL System, Overview of the SPEL System*. Statistical Office of the European Communities, Brussels and Luxembourg.

Hill, B. (2000) *Farm Incomes, Wealth, and Agricultural Policy*, 3rd edn. Ashgate, Aldershot, UK.

Janssen, S. and van Ittersum, M.K. (2007) Assessing farm innovations and responses to policies: a review of bio-economic farm models. *Agricultural Systems* 94, 622–636.

Keeney, M. (2000) The distributional impact of direct payments on Irish farm incomes. *Journal of Agricultural Economics* 51, 252–265.

Kilian, S., Antón, J., Salhofer, K. and Röder, N. (2012) Impacts of 2003 CAP reform on land rental prices and capitalization. *Land Use Policy* 29, 789–797.

Leip, A., Marchi, G., Köble, R., Kempen, M., Britz, W. and Li, C.C. (2008) Linking an economic model for European agriculture with a mechanistic model to estimate nitrogen losses from cropland soil in Europe. *Biogeosciences* 5, 73–94.

Louhichi, K., Ciaian, P., Espinosa, M., Colen, L., Perni, A. and Gomez y Paloma, S. (2015) The impact of crop diversification measure: EU-wide evidence based on IFM-CAP model. *Paper presented at the International Association of Agricultural Economists 2015 Conference, August 9–14, 2015, Milan, Italy*. Available at: http://ageconsearch.umn.edu//handle/211542 (accessed 18 March 2016).

M'barek, R., Britz, W., Burrell, A. and Delince, J. (2012) *An Integrated Modelling Platform for Agro-economic Commodity and Policy Analysis (iMAP) – A Look Back and the Way Forward*. Report EUR 25267 EN, JRC Scientific and Policy Reports, European Commission Joint Research Centre. Publications Office of the European Union, Luxembourg.

Önal, H. and McCarl, B. (1989) Aggregation of heterogeneous firms in mathematical programming models. *European Review of Agricultural Economics* 16, 499–513.

Robinson, S., Kilkenny, M. and Hanson, K. (1990) *The USDA/ERS Computable General Equilibrium (CGE) Model of the United States*. ERS Staff Report No. AGES-9409, Agricultural and Rural Economy Division, Economic Research Service, US Department of Agriculture, Washington, DC.

Storm, H., Mittenzwei, K. and Heckelei, T. (2015) Direct payments, spatial competition, and farm survival in Norway. *American Journal of Agricultural Economics* 97, 1192–1205.

van Tongeren, F., van Meijl, H. and Surry, Y. (2001) Global models applied to agricultural and trade policies: a review and assessment. *Agricultural Economics* 26, 149–172.

Westerhoff, F. (2012) Interactions between the real economy and the stock market: a simple agent-based approach. *Discrete Dynamics in Nature and Society* 2012, Article ID 504840.

Index